D1201987

THE
INTERNATIONAL SERIES
OF
MONOGRAPHS ON PHYSICS

SEMICONDUCTOR CONTACTS

An approach to ideas and models

by

HEINZ K. HENISCH

The Pennsylvania State University
University Park, Pennsylvania

CLARENDON PRESS · OXFORD
1984

Oxford University Press, Walton Street, Oxford OX2 6DP
London New York Toronto
Delhi Bombay Calcutta Madras Karachi
Kuala Lumpur Singapore Hong Kong Tokyo
Nairobi Dar es Salaam Cape Town
Melbourne Auckland

and associated companies in
Beirut Berlin Ibadan Mexico City Nicosia

Published in the United States
by Oxford University Press, New York

OXFORD *is a trade mark of Oxford University Press*

British Library Cataloguing in Publication Data
Henisch, Heinz K.
Semiconductor contacts.—(International series of monographs on physics)
1. Semiconductors
I. Title   II. Series
537.6'22   Q611
ISBN 0-19-852016-6

Library of Congress Cataloguing in Publication Data
Henisch, Heinz K.
Semiconductor contacts.
(The International series of monographs on physics;
Bibliography: p.
Includes index.
1. Electric contacts.   2. Semiconductors.
I. Title.   II. Series: International series of monographs on physics (Oxford, Oxfordshire;
TK7872.C68H46 1984   621.3815'2   83–25100
ISBN 0-19-852016-6

Printed in United Kingdom by
The Universities Press (Belfast) Ltd

PREFACE

Rectifying semiconductor physics was published in 1957, after the flush of excitement that accompanied the advent of transistor physics, but before the development of modern experimental methods, and before the impact of computers on this kind of work. The passing years dealt with the book more kindly than I had any reason to expect, but by now the volume has long been out of print and even longer out of date, and the time has come for its replacement.

The electrical properties of contacts, and specifically of contacts to semiconductors, have been under investigation for almost 150 years. Indeed, the discovery of semiconductors itself has been ascribed to the casual detection of their role in governing the resistance of metal–metal interfaces. Despite this long history, and the substantial advances made in recent years, many aspects of contact behaviour remain unclear, and many others are widely misunderstood. The purpose of this book is to discuss the present status of the field in terms of concepts and models. Apart from their theoretical interest, contact properties are of great technological significance, and their understanding is therefore a matter of substantial importance.

The purpose of a contact theory is, of course, to aid design and to promote the interpretation and understanding of experimental results. However, a theory which encompasses every conceivable complication and side-effect would be too unwieldy to be of any use. It is better to treat the simplest case in the first instance, and then to isolate 'significant situations' in which one factor or another exerts the most important perturbing influence, without giving equal weight to every possible refinement. Even then, computer-derived numerical solutions are often the only resort. Because computers are now firmly with us, analytical elaborations which were justified in earlier terms are not always justified now. On those grounds, many have in fact been omitted, and it is hoped that the resulting simplification will make this intermediate-level text accessible to a wider audience. As it is, our capacity to make models has far outstripped our capacity to verify them by experiment.

The new text is intended to bridge the gap between the high level abstractions of the modern theory of solids and the phenomenological treatments that have wide currency in the realm of device physics. It does not set out to be a comprehensive review of the semiconductor contact literature and, specifically, it is *not* a summary of the vast amount of information amassed by experimenters in many contexts and in many parts of the world; this is manifestly not a data book, nor indeed does it set out to offer arbitration for every experimental conflict on record. For all that, a book of this kind is bound to draw heavily upon the work of colleagues, and I hope I have succeeded in presenting a cogent account of their insights. Friends whose findings have been unwittingly misrepresented (or, indeed, unwittingly omitted) have my apologies. Many dilemmas of selection and approach had to be faced, and in an attempt to resolve them I have concentrated on matters which I regard as essentially defensible, rather than on 'conventional wisdom' as such. If this procedure leaves some gaps, it should also be useful in drawing attention to new opportunities.

I am deeply indebted to friends and co-workers, without whose interest and support over the years the new book would not have been written: Cornel Popescu, Jean-Claude Manifacier, Yves Moreau, Roberto Callarotti, Pierre Schmidt, Sir Nevill Mott, and Stanford Ovshinsky. Grateful thanks are also due to S. Rahimi, E. H. Rhoderick, R. Grigorovici, A. Many, N. Croitoru, R. Zallen, W. Pickin, Chun Chiang, Yih Shun Gou, and the late B. R. Gossick for much helpful discussion and correspondence, and to Bonny Farmer for keeping the ship afloat with skill and charm. Lastly, a tribute to my wife Bridget, for providing the inspiration while helping me to keep semiconductors in their proper place.

Pennsylvania
March 1984

H.K.H.

CONTENTS

List of symbols xiii

1 Introductory survey 1
1.1 Basic observations 1
1.1.1 Contact voltage; contact resistance 1
1.1.2 Contact rectification 6
1.1.3 Geometrical aspects; radial systems 11
1.1.4 Geometrical aspects; non-radial systems 14

1.2 Current-controlled non-equilibrium processes 18
1.2.1 Current–composition ratios 18
1.2.2 Electrical consequences of non-equilibrium carrier concentrations 20
1.2.3 Optical consequences of non-equilibrium carrier concentrations; injection luminescence 23
1.2.4 Space–charge-controlled conduction 24

1.3 Origin of interface barriers 25
1.3.1 Barriers due to work-function differences 26
1.3.2 Barriers due to surface states 29
1.3.3 Surface and interface state distributions 34
1.3.4 Barriers due to third phases 38

2 Single carrier systems 42
2.1 Static barriers 43
2.1.1 The classic Schottky barrier 43
2.1.2 The free-carrier contribution; accumulation layers 50
2.1.3 Problems of space-charge discontinuity 54
2.1.4 The Mott barrier 57
2.1.5 High field boundary effects; image forces 58
2.1.6 The wave mechanical tunnel effect 61
2.1.7 Barrier height, bulk structure, and interface relationships 64
2.1.8 Deep donors; incomplete ionization 69

2.2 Current-carrying barriers; analysis of basic models 73
2.2.1 Barrier contours in the presence of current flow 74
2.2.2 Voltage–current relationships and concentration profiles; general 77
2.2.3 Voltage–current relationships; Mott barrier 82
2.2.4 Voltage–current relationships; Schottky barrier 84

2.2.5 Barriers in the high forward direction 86
2.2.6 Thermionic emission; diode theory 89
2.2.7 'Hot' carrier effects 94
2.2.8 High-field enhancement of reverse currents 97
2.2.9 Back-to-back barrier systems; field distribution and voltage–current characteristics 100

2.3 Current-carrying Schottky barriers; analysis of complex and hybrid models 105
2.3.1 General considerations; mechanisms of current flow 105
2.3.2 Emission-controlled boundary conditions; diffusion-diode theory 108
2.3.3 The tunnelling contribution; parametric description of rectifying characteristics 113
2.3.4 Analytic and numerical approaches to the tunnelling problem 117
2.3.5 Current-dependent barrier heights 119
2.3.6 Laterally non-homogeneous barriers 122
2.3.7 Barriers with non-uniform doping profiles 124
2.3.8 Voltage–current relationships in theory and experiment 129

2.4 Current-carrying barriers; numerical solutions 134
2.4.1 Semi-infinite Schottky systems 134
2.4.2 Finite Schottky systems 138
2.4.3 Thick composite barriers 143

3 Two-carrier systems 146
3.1 Inversion layers 146
3.1.1 Inversion layer profiles; minority-carrier injection 146
3.1.2 Inversion layer profiles 151
3.1.3 Static barriers on intrinsic material 153
3.1.4 The injection ratio 156

3.2 Injection into bulk material; lifetime semiconductors 161
3.2.1 General considerations; modulation of conductivity 161
3.2.2 Transport relationships in a two-carrier system without traps 166
3.2.3 Analytic treatment; linearized theory; trap-free case 168

3.2.4 Analytic treatment; linearized theory; case with traps 171

3.3 Injection into bulk material; relaxation semiconductors 173
3.3.1 Relaxation and lifetime regimes 173
3.3.2 Solutions for material with and without traps 176

3.4 Minority-carrier exclusion and extraction; bulk material 180
3.4.1 Carrier exclusion 180
3.4.2 Carrier extraction 182

3.5 Two-carrier contact relationships; lifetime semiconductors 186
3.5.1 Analytic approximations; hemispherical, semi-infinite systems 186
3.5.2 Metal–semiconductor–metal planar systems 192
3.5.3 Contacts on intrinsic material 194
3.5.4 Contacts on extrinsic material with an inversion layer 197
3.5.5 Metal–insulator–semiconductor systems 199

3.6 Contacts with highly disturbed and amorphous semiconductors 206
3.6.1 Bulk relationships; screening lengths 206
3.6.2 Contact barriers in the presence of deep traps 209
3.6.3 Contacts with organic materials 213

4 Reactive and frequency-dependent contact properties 215
4.1 Capacitance of trap-free contact systems 215
4.1.1 Equivalent circuits 215
4.1.2 Barrier capacitance with complete donor ionization 218
4.1.3 Barrier capacitance with incomplete donor ionization 223
4.1.4 Barrier capacitance and surface state occupancy 224
4.1.5 Capacitance of back-to-back barrier systems 226
4.1.6 Effect of inversion and interface layers on contact capacitance 227

4.2 High-frequency and microwave effects 229
4.2.1 Relaxation of donors 229
4.2.2 Asymmetry of contact characteristics 231

4.3 Capacitance of systems with traps 233
4.3.1 Capacitance–voltage relationships 233

4.3.2 Frequency dependence of the barrier capacitance 238

5 Metal–semiconductor contacts under illumination 241
5.1 Opto-electronic principles; application to simple barriers 241
5.1.1 Qualitative survey of photo-effects; terminology 242
5.1.2 Illustrative models; analytic and quasi-analytic solutions 247
5.1.3 Illustrative models; power output and fill factor 254
5.1.4 The collection velocity 259
5.1.5 Computed photoresponse characteristics 260
5.1.6 Response of potential probes to non-equilibrium situations 262
5.1.7 Illumination of an injection region; interpretation of decay times 267

5.2 Photoresponse of composite barriers 269
5.2.1 Function of interface dielectrics; general considerations 269
5.2.2 Analytic models and observations 272
5.2.3 Computations on solar cell models 276

6 Special topics and problems 281
6.1 Lateral conduction; field effect systems 281
6.1.1 Modulation of lateral (interface) conductance 281
6.1.2 Evaluation of interface charge and trap densities 286
6.2 Electrothermal effects 289
6.2.1 Self-heating; thermal turnover; temperature gradients 289
6.2.2 Thermal inertia and self-oscillations 291
6.3 Contacts with semi-insulators 293
6.3.1 Semi-infinite systems; classical Mott and Gurney case 293
6.3.2 Thick metal–‘insulator’–metal systems 295

6.4 Semiconductor–semiconductor contacts 301
6.4.1 Single contacts between semiconductors of the same type 301
6.4.2 Heterocontacts; heterojunctions 308
6.4.3 Intergranular contacts in microcrystalline systems 312
6.4.4 Intergranular contacts under illumination 318

6.5 Low-resistance contacts 319
6.5.1 General principles; carrier degeneracy 319
6.5.2 Thin barriers 323

6.6 Electrical breakdown 326
6.6.1 Avalanche and Zener breakdown; general principles 326
6.6.2 Avalanche-supported filament formation; light emission 333

6.7 Mechanical effects; deformation 336
6.7.1 Voltage–current characteristics under pressure; contacts on Si, GaAs, Ge 336
6.7.2 Voltage–current characteristics under pressure; contacts with potassium tantalate 338

References 340

Index 369

SYMBOLS

Symbol	Meaning	First mentioned in
A	constant, approximately $= \tau_D/\tau_0$	3.3.1
A_n, A_p	parameters defined by eqn (3.2.17)	3.2.2
A^*	quasi-classical Richardson's constant	2.2.6
A^{**}	quantum-mechanical Richardson's constant	2.3.4
A_T	parameter defined as $(2\pi\lambda_a\beta/h)(2m_n\epsilon)^{1/2}$, with β as a small correction term (of the order of unity)	6.5.2
$\mathscr{A}$	total contact area	2.3.6
a	exponent of 'Varistor' characteristic: $I\alpha V_T^a$	6.4.3
$a(\nu)$, $a(\lambda)$	optical absorption constant at frequency ν or wavelength λ	1.2.3
a_1, a_2	ionization and trapping rate constants	2.3.7
	(also) serial expansion constants	4.2.1
B	frequently used subscript, denoting 'barrier related'	1.1.2
B	'mean-free-path' constant used in eqns (2.3.13) and (2.3.14)	2.3.2
B_c	pre-exponential factor defined by eqn (3.1.17a)	3.1.3
b	mobility ratio μ_n/μ_p	3.2.2
b_{ls}	a constant, proportional to l_s	6.6.1
b_z	'parameter of convenience', defined by eqn (1.1.4)	1.1.4
C	integration constant with dimensions of energy	2.1.1
C_B	dynamic barrier capacitance per unit area	4.1.1
C_{BV}	voltage-dependent component of C_B	4.2.1
$C_{B\omega}$	frequency-dependent component of C_B	4.2.1
C_{cs}	circuit shunt capacitance per unit area	4.1.1
C_{IS}	capacitance element associated with interface states	4.1.1
C_M	measured capacitance per unit area	4.1.1
C_{ox}	polarization capacitance of oxide layer	4.1.1
C_s	system capacitance	4.1.3
D_a	ambipolar diffusion constant; defined by eqn (3.2)	3.2.3
D_{Fn}	effective diffusion constant of electrons at high field saturation	2.2.7
D_n, D_p	diffusion constants of electrons and holes	2.1.1
E	electron energy in general	1.3.1
E_a	energy corresponding to acceptor levels	2.1.8

		First mentioned in
E_c	energy designating the bottom of the conduction band of the semiconductor	1.1.3
E_d	energy corresponding to donor levels	2.1.8
E_{cm}	mobility edge of the conduction band in amorphous material	3.6.1
ΔE_d	defined as $(E_c - E_d)$	2.1.7
E_F	energy of the Fermi level (taken here as zero in the metal)	1.3.1
E_g	(thermal) band-gap energy	1.3.1
E_i	impact ionization energy	6.6.1
ΔE_s	barrier lowering due to Schottky effect	2.1.5
E_T	energy corresponding to a trapping level	3.6.2
E_v	energy designating the top of the valence band of the semiconductor	1.3.2
E_{vm}	mobility edge of the valence band in amorphous material	3.6.1
$\boldsymbol{E}$	normalized (to kT) electron energy	2.4.1
e	charge of an electron	1.1.2
FF	fill factor, defined in eqn (5.1.14)	5.1.3
F	field in general	2.1.1
F_{av}	average field	6.4.3
F_{BD}	breakdown field	2.2.9
F_i	field in the insulating layer	6.1.2
$\boldsymbol{F}$	normalized field	2.4.1
$\boldsymbol{F}_\infty$	normalized field at infinity	2.4.1
$\mathcal{F}$	force in general	2.1.5
$\mathcal{F}_{1/2}(\xi)$	generalized Fermi–Dirac integral for the parameter ξ; see eqns (6.5.2) and (6.5.3)	6.5.1
$\mathcal{F}_m$	Fermi–Dirac distribution function in the metal (non-integrated)	3.5.5
$\mathcal{F}_{sc}$	Fermi–Dirac distribution function in the semiconductor (non-integrated)	3.5.5
G	conductance in general	1.1.4
G_s	lateral specimen conductance	6.1.1
G_{sm}	minimum value of G_s; $\Delta G_s = G_s - G_{sm}$	6.1.1
$G(z)$	correction factor, defined by eqn (2.2.30)	2.2.5
g_0	pre-exponential constant, defined by eqn (5.1.9)	5.1.2
g_L	optical carrier generation rate	5.1.2
H_{TC}	defined as $h_{TC}J_c/T_a$	6.2.1
h	Planck's constant	2.1.5

		First mentioned in
$\hbar$	$h/2\pi$	2.2.8
h_{TC}	thermal coupling constant	6.2.1
I	defined as $-i$; 'variable of convenience', introduced to permit the voltage V_B applied to a barrier to be *positive* in the reverse direction of current flow	3.5.1
I_{Lo}	short-circuit value of I (positive)	5.1.2
I_n, I_p	defined as $-i_n$ and $-i_p$, respectively	3.5.1
I_s	saturation value of I; (positive)	5.1.2
i	total current $= i_n + i_p$, conventionally defined	3.5.1
J	defined as $-j$; 'variable of convenience', introduced to permit the voltage V_B applied to a barrier to be *positive* in the reverse direction of current flow	2.2.2
J_0	pre-exponential term; current density (constant)	2.3.8
J_c	current-density constant, related to J_s	6.2.1
J_m	current density at maximum power	5.1.3
J_s	saturation value of J; (positive)	2.2.4
$\boldsymbol{J}$	normalized J	2.4.2
j	current density (in general)	1.1.1
j_+, j_-	current components of thermionic electron flux	2.2.6
j_D	diffusion current density	2.2.1
j_F	field current density	2.2.1
j_n, j_p	current-densities of electrons and holes	1.2.1
$\boldsymbol{j}$	normalized current density	2.4.2
K	equilibrium constant	6.3.2
K_{th}	thermal conductivity	6.2.1
k	Boltzmann's Constant	2.1.1
L	specimen thickness	1.1.4
L_c, L_w	length and width of a rectangular contact	1.1.4
L_{Da}	ambipolar diffusion length	3.3.2
L_n, L_p	(lifetime) diffusion lengths for electrons and holes	2.1.1
L_s	thickness of thin film resistor	1.1.4
L_{sc}	characteristic screening length	3.3.2
l	free path length	6.6.1
l_s	mean free path (scattering length)	2.3.2
l_x, l_y, l_z	direction cosines	2.2.6
M_0	trap density	3.3.2
$\boldsymbol{M}$	normalized carrier concentration in traps	3.2.4
$\boldsymbol{M}_0$	trap density, normalized to n_e	3.2.4

		First mentioned in
$\boldsymbol{M}_{\mathrm{e}}$	normalized *equilibrium* carrier concentration in traps	3.2.4
$\mathcal{M}$	electron multiplication factor, defined by $j_{\mathrm{n}}(\lambda_{\mathrm{B}}) = \mathcal{M} j_{\mathrm{n}}(0)$	6.6.1
m	standard electron mass	2.2.6
m_{n}, m_{p}	effective masses of electrons and holes	1.2.3
m_{np}	effective mass of either electrons or holes	6.1.1
m_{x}, m_{y}, m_{z}	components of the effective mass tensor	2.2.6
N_{a}	concentration of acceptor centres	1.3.2
N_{c}	temporary integration constant	3.5.1
N_{d}	concentration of donor centres	1.3.2
N_{dM}	concentration of ionized donors within a Mott barrier	2.1.4
$\boldsymbol{N}$	normalized electron concentration n/n_{e}	2.4.1
$\boldsymbol{N}_{\mathrm{e}}$	$= n_{\mathrm{e}}/n_{\mathrm{e}} = 1$	3.2.3
$\boldsymbol{N}_{\mathrm{T}}$	normalized (to n_{e}) electron concentration *in traps*	6.3.2
$\mathcal{N}_{\mathrm{c}}$	effective density of states in the conduction band	2.2.2
n, $n(x)$	free electron concentration as a function of x	1.2.2
$n_{0(x)}$	x-dependent equilibrium value of n; $n_{0(\infty)} = n_{\mathrm{e}}$	4.2.1
n_1	parametric electron concentration in the Shockley–Read model	3.2.4
n_{d}	concentration of electrons at donor sites in bulk	2.1.8
n_{dB}	concentration of electrons at donor sites within the barrier	4.2.1
n_{e}	equilibrium value of n	2.2.2
n_{i}	intrinsic carrier concentration	3.1.4
n_{m}	electron concentration in the metal above barrier height	2.3.2
n_{s}	boundary concentration of electrons at the surface	3.5.1
n_{sc}	electron concentration in the semiconductor above barrier height	2.3.2
$\tilde{n}$	amplitude of a small periodic change in n	4.2.1
$\tilde{n}_{\mathrm{dB}}$	amplitude of a small periodic change in n_{dB}	4.2.1
P_{c}	temporary integration constant	3.5.1
$\boldsymbol{P}$	normalized (to n_{e}) hole concentration	3.2.3
$\boldsymbol{P}_{\mathrm{e}}$	$= p_{\mathrm{e}}/n_{\mathrm{e}}$	3.2.3
p, $p(x)$	free hole concentration as a function of x	3.1.1
p_1	parametric hole-concentration in the Shockley–Read model	3.2.4
p_{a}	concentration of holes at donor sites in bulk	2.1.7
p_{e}	equilibrium value of holes	1.2.1
p_{s}	boundary concentration of holes at the surface	3.5.1

		First mentioned in
Q	charge per unit contact area	4.1.2
Q_{ss}	charge (per unit area) in interface states	2.3.5
Q_T	total charge (per unit area) within a Schottky layer	6.1.1
Q_t	charge density of carriers in traps	3.2.4
$\boldsymbol{Q}_t$	normalized (to en_e) charge density in traps	3.2.4
R	resistance in general	1.1.4
R_B	barrier resistance per unit contact area	4.1.1
R_b	bulk resistance per unit contact area	2.2.1
R_{IS}	recombination resistance associated with interface states	4.1.1
R_s	spreading resistance	1.1.1
R_{sh}	effective shunt resistance of photovoltaic cell	5.1.3
R_T	total resistance	1.1.4
R_{TS}	tunnelling resistance associated with oxide layer	4.1.1
r	radial distance	1.1.3
r_0	radius of contact curvature	1.1.3
r_B	radius of curvature corresponding to edge of the barrier	3.5.1
S	parameter $(\mathrm{d}\phi_{ns}/\mathrm{d}\chi_m)$	2.1.7
S_c	'parameter of convenience', defined by eqn (3.1.9)	3.1.3
S_{cv}	collection velocity	5.1.4
$\boldsymbol{S}$	defined as $\mathrm{d}\boldsymbol{N}/\mathrm{d}\boldsymbol{X}$	2.4.1
T	absolute temperature (K)	2.1.1
T_0	a 'non-ideality pseudo-temperature'	2.3.3
T_E	wave-mechanical transparency (transmission coefficient) of a barrier at energy E	2.1.6
$T_{\Delta E}$	wave-mechanical transparency of a barrier at an interval ΔE from its top	2.1.6
T_a	ambient temperature	6.2.1
T_c	contact temperature	6.2.1
t	time	4.2.1
u	'variable of convenience'	3.1.3
	(also) real component of distance in the complex w-plane	1.1.4
$V, V(x)$	potential as a function of distance	1.1.1
V_a	applied forward voltage, referred to the energy minimum	2.4.1

		First mentioned in
V_B	barrier voltage on a planar system (positive in reverse direction)	1.1.2
V_{B1}, V_{B2}	barrier voltages across components of a two-barrier system	2.2.9
V_{BD}	breakdown voltage	2.2.9
V_b	voltage across bulk material outside barrier region	2.2.1
V_c, V'_c	'contact voltages' (see Figs 1.1 and 1.3)	1.1.1
V_D	diffusion potential (difference) across quiescent barrier in the semiconductor	1.1.2
V_{Dem}	Dember e.m.f.	5.1.1
V_{Di}	diffusion potential (difference) across 'insulating' interface layer	1.3.4
V_E	external voltage applied to metal–insulator–semiconductor system for conductance modulation	6.1.2
V'_E, $V'_E(0)$	calculated modulation voltages	6.1.2
V_{FB}	'flat band' voltage	2.2.9
V_I	'interface voltage' (see Fig. 1.3)	1.1.1
V_f	floating potential	5.1.1
V_i	portion of V_E across the insulator layer	6.1.2
V_{L0}	open circuit photo-voltage	5.1.1
V_{RT}	'reach-through' voltage	2.2.9
V_T	total voltage across (contact + bulk) system	2.2.1
$\mathbf{V}_a$	defined as $V_a/(kT/e)$	2.4.1
$\mathbf{V}_T$	defined as $V_T/(kT/e)$	2.4.1
v	imaginary component of distance in the complex w-plane	1.1.4
v_{dn}	drift velocity of electrons	2.2.7
v_{sdn}	saturation value of v_{dn} at high fields	2.2.7
$\bar{v}$	mean electron thermal velocity, given by $(kT/2\pi m_n)^{\frac{1}{2}}$	2.2.6
$\bar{v}_m$	value of $\bar{v}$ in the metal	2.3.2
$\bar{v}_{sc}$	value of $\bar{v}$ in the semiconductor	2.3.2
v_T	thermal velocity	6.6.1
W_n, W_p	field integrals (hemispherical systems) in darkness	3.5.1
W'_n, W'_p	same under illumination	5.1.2
w	distance in complex plane $u + iv$	1.1.4
$\mathbf{X}$	normalized distance (e.g. to λ_0 or λ_D)	2.4.1
x	distance from interface in general	1.1.1
x_0	characteristic decay length of electron concentration $= \lambda_D$ at zero current	2.1.2
x_f	distance of fitting point which marks the onset of space-charge exponential decay	2.1.2

		First mentioned in
x_g	width of metal–semiconductor gap before contact is established	1.3.1
x_j	value of x of electron energy minimum	3.5.3
x_M	value of x for which the field has a maximum	6.6.1
x_m	value of x for which electron energy has a maximum	2.1.5
x_s	critical scattering distance	6.3.1
x_t	tunnelling distance	2.1.6
x_T	distance over which the barrier energy drops by kT	2.3.2
y	distance in the y-direction	1.1.4
y_c	defined as $\{e(V_D+V_B)/kT\}^{1/2}$	2.2.5
Z	constant, inversely proportional to m_n and τ	1.2.3
z	distance in the complex $x+iy$ plane	1.1.4
	(also) integration limit $=\{e(V_D+V_B)/kT\}^{1/2}$	2.2.5
	(also) empirical constant governing observed reverse characteristic	2.3.8
z_1, z_2	'constants of convenience', eqn (1.2.2)	1.2.3
	and eqn (2.3.6)	2.3.2
α	dimensionless proportionality factor; eqn (3.5.12)	3.5.1
α_i	ionization probability	6.6.1
α_s	constant of proportionality, inversely proportional to the density of interface states per unit energy	2.3.5
β	dimensionless parameter used in Wilson's theory of tunnelling through barriers; $1-\beta=1/\eta$	2.3.3
β_c	proportionality constant, with the dimensions of distance	2.2.8
γ_0	bulk current composition ratio in general	1.1.1
γ_1, γ_2	current composition ratios prevailing at contacts 1 and 2, ordinarily on n-type material	1.2.1
γ_n	current composition ratio prevailing in the (equilibrium) bulk of n-type material	1.2.1
γ_p	current composition ratio prevailing in the (equilibrium) bulk of p-type material	1.2.1
ϵ	low-frequency permittivity of semiconductor	2.1.1
ϵ_0	permittivity of free space	2.2.8
ϵ_∞	high-frequency permittivity of semiconductor	2.1.5
ϵ_i	permittivity of insulating (e.g. oxide) layer	1.3.4

		First mentioned in
η	dimensionless non-ideality factor	2.2.5
κ	dielectric constant	2.1.1
λ	wavelength	1.2.3
λ_0	thickness of the quiescent (planar) Schottky barrier ($V_B=0$)	1.3.1
λ_a	thickness of an artificial (e.g. oxide) barrier layer	1.3.4
λ_B	barrier thickness at a planar contact with voltage V_B applied	2.2.1
λ_c	characteristic decay length of experimental field relationship	5.1.2
λ_D	Debye length in semiconductor defined by the bulk concentrations of a single-carrier species (λ_{Dn} for electrons; λ_{Dp} for holes, when differentiated)	2.1.1
λ_{DM}	Debye length in metal	1.3.1
λ'_D	Debye length defined by the boundary concentration $n(0)$ of a single-carrier species (here, electrons)	2.4.2
λ_{Da}	ambipolar Debye length, determined by n_e and p_e for equilibrium in bulk	3.1.3
λ_{Di}	Debye length in intrinsic material	3.5.3
λ_I	width of the inversion layer	3.1.1
λ_M	thickness of a Mott barrier	2.1.4
λ_T	barrier parameter defined in Fig. 3.37(a)	3.6.2
λ_m	barrier thickness referred to the energy minimum; forward direction only	2.4.2
λ_{s0}	quiescent barrier thickness at a hemispherical contact; zero voltage applied	2.1.1
λ_{sB}	barrier thickness at a hemispherical contact with voltage V_B applied	2.2.1
μ_{Fn}	saturation value of μ_n in a high field	2.2.7
μ_n, μ_p	electron-and-hole-mobilities	1.2.1
μ_{ns}, μ_{ps}	values for μ_n and for μ_p close to the surface	6.1.1
ν	frequency (generally)	5.1.2
ν_S	reciprocal of effective surface lifetime	3.5.0
ν_T	trap escape frequency	4.3.2
ξ	defined as ϕ_n/kT	3.5.1 6.5.1
ρ	charge density	2.1
σ	conductivity in general	3.2.1

		First mentioned in
σ_b	bulk conductivity	1.1.3
σ_R	'Randschicht' conductivity $\mu_n en(0)$	2.2.3
σ_{ext}	conductivity under total minority-carrier extraction	3.4.2
τ	mean collision time	6.1.1
τ_0	diffusion length lifetime in the bulk	1.2.2
τ_D	dielectric relaxation time in the bulk $=\epsilon/\sigma_b$	1.2.2
τ_f	filament lifetime	3.5.0
τ_n	lifetime of electrons	2.1.1
τ_p	lifetime of holes	3.1.4
τ_T	trapping time	4.3.2
ϕ_n	energy interval between conduction band and Fermi level in equilibrium (bulk term)	1.3.1
ϕ_e	height of a complex barrier, as 'seen' by electrons coming from the conduction band of the layer	3.5.5
ϕ_p	energy interval between valence band and Fermi level in equilibrium (bulk term)	1.3.1
ϕ_h	height of a complex barrier, as 'seen' by holes from Fermi level of metal (Fig. 3.33)	3.5.5
ϕ'_n, ϕ'_p	non-equilibrium values of ϕ_n and ϕ_p	3.5.3
ϕ_m	thermionic work-function of the metal	1.3.1
ϕ_{ns}, ϕ_{ps}	values corresponding to ϕ_n and ϕ_p at the metal semiconductor interface	1.3.1
ϕ_{mi}	energy interval between the conduction band of an insulating material and the Fermi level of the contacting metal	1.3.4
ϕ_{ns0}	value of ϕ_{ns} (if not constant) at zero current	2.3.5
χ	electron affinity	1.3.1
χ_m	electronegativity of the metal	2.1.7
ω	angular frequency	4.1.1

1

INTRODUCTORY SURVEY

1.1 Basic observations

1.1.1 Contact voltage; contact resistance

SINCE, in the nature of things, every contact involves at least two materials (three, when interface layers are present) the separation of bulk and contact properties always poses a problem. The classic experiment is shown in Fig. 1.1. Potential differences $V(x)$ are measured at various points on a specimen between the two end contacts, either by a movable probe (often mounted on a micromanipulator) or else by at least two fixed probes in known positions. Since no macroscopic probe can ever come infinitely close to an end contact, some degree of extrapolation is always involved. However, this extrapolation *is* simple only when we are dealing with a material that is structurally, compositionally, and thermally homogeneous, and when 'non-equilibrium processes' of the kind discussed in Section 1.2 and Chapter 3 are either absent or negligible. Such conditions cannot be taken for granted; they must be the subject of careful tests if reliable results are to be obtained. For the moment, we shall assume that the extrapolation *is* straightforward. When performed, it reveals apparent discontinuities at the two metal–semiconductor interfaces, but it will be clear enough that potential discontinuities as such are conceptually impossible. The potential must be univalued and must vary continuously. We are actually confronted by two very thin regions within which the potential gradients are extremely high. In some contexts (e.g. Section 2.1.5) these gradients are significant, but for many practical purposes the potential step can be regarded as abrupt.

By repeating measurements of this kind and the appropriate extrapolations for different current densities j in both directions of current flow, the current–voltage relationships associated with each interface can in principle be established. In general terms, they will look like the curve shown in Fig. 1.2. Conduction is easy in one direction of current flow, difficult in another; the contacts are said to 'rectify'. It will be shown below that these contact properties can be ascribed to the existence of potential barriers at the interfaces, and that the direction of rectification depends on the nature of the conducting materials.

Figure 1.1(b) represents an idealized situation. In many cases (indeed, in *all* cases, if the matter were pursued with ultimate precision) the potential profile has some curvature near the contacts, even in a homogeneous material. That curvature is produced by the current-dependent non-equilibrium processes already mentioned (but not yet

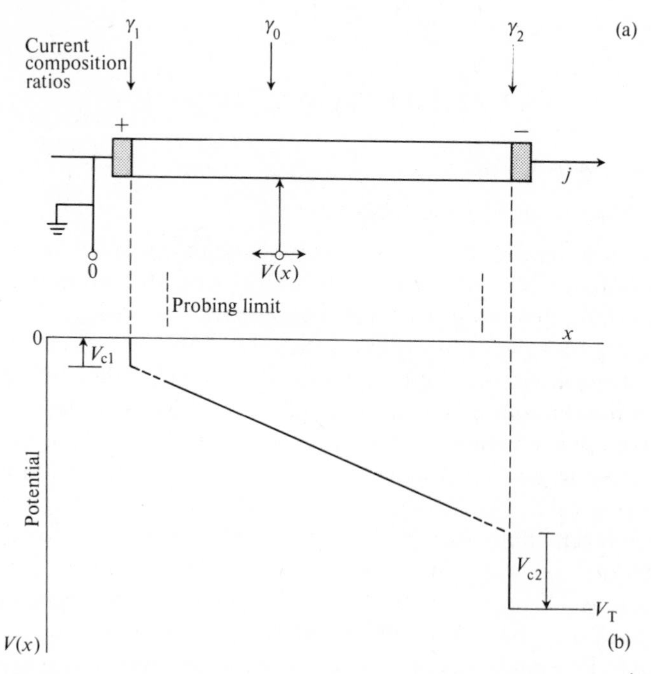

FIG. 1.1. Classical method for measuring bulk conductivity and contact properties by means of a movable potential probe. Linearity of the potential contour signifies bulk homogeneity. γ_0, γ_1, and γ_2 are current composition ratios explained in Section 1.2.1. $V(x)$ = local potential; j = current density; V_{c1} and V_{c2} are (idealized) 'contact voltages'; V_T = total voltage applied. (A comprehensive list of symbols will be found in the preliminary pages of the book.)

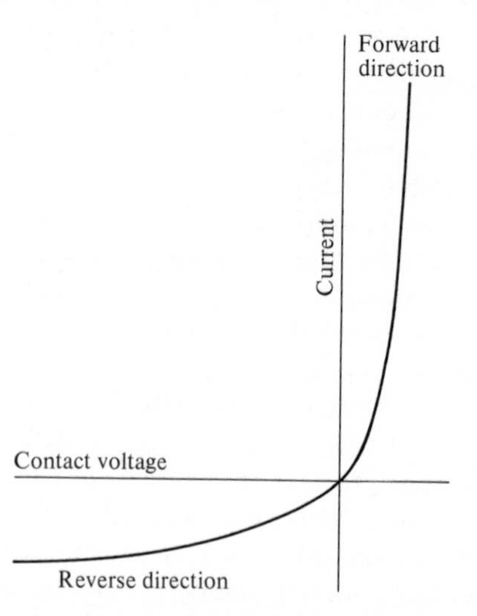

FIG. 1.2. Schematic voltage–current characteristic of a metal–semiconductor contact.

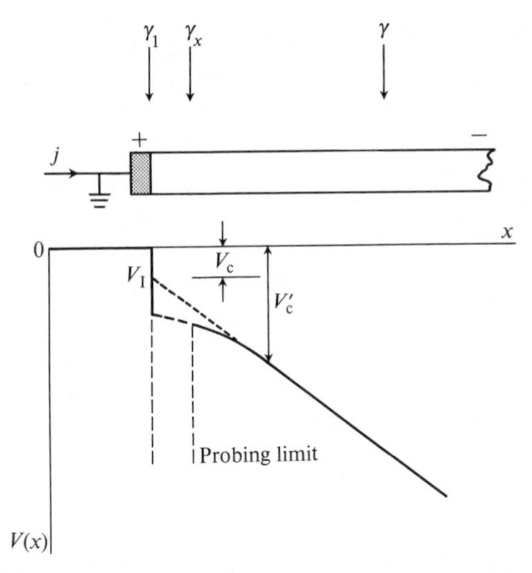

FIG. 1.3. Potential contour near a metal–semiconductor contact, in the presence of a current-controlled non-equilibrium effect. The curvature, as shown here, implies a local conductivity increase. See Section 1.2 and Chapter 3. V_c and V_c are notional contact voltages. V_I could be called an interface voltage. (The terminology of these quantities and relationships has never been properly standardized.)

described) above. These disturbances extend some way into the material, and may or may not be large enough to be traced out by the potential probe (Fig. 1.3). If they are not large enough to be traced out, the potential probe measurement will miss them altogether. The linear extrapolation will then appear to be legitimate, but will yield a contact voltage V_c that is of no real significance. Through the use of very thin probes and very sensitive micromanipulators, the problem can be minimized, but it cannot be ultimately eliminated. Inferences drawn about regions which cannot be probed are of necessity indirect and, to that extent, less reliable.

When the curved region *is* large enough to be traced by the probe, it may be possible to designate (albeit rather arbitrarily) some location at which contact effects are deemed to end, and bulk effects take over. That would yield a different contact voltage V_c', less securely defined, but still somewhat useful for comparative purposes. The full interpolation of the curved region, if and when experimentally feasible, would lead to an interface voltage V_I, also shown in Fig. 1.3. What we mean by 'the voltage–current characteristic of a contact' thus depends rather crucially on what we mean by the voltage, and in the analysis of reported results, this must be carefully borne in mind.

So must another caution: a potential probe yields the potential of the locality with which it is in contact *only* if the semiconductor is in electronic equilibrium. If it is not in equilibrium (e.g. because it is illuminated, or for some other reason, then a so-called *floating potential* (difference) is generated at the probe contact, which interferes with the potential measurement (see Section 5.1.6). Sophisticated experimenters take steps to minimize this effect, through suitable probe contact design and/or surface treatments. The floating potential is defined as an open-circuit voltage. When that voltage arises from illumination, we refer to it as a *photovoltage*, and to the principle of its generation as the *photovoltaic effect* (see Chapter 5). However, its existence is by no means limited to situations in which light is incident; its inherent cause is electronic disequilibrium, no matter how produced. In the present context, such disequilibrium is often produced without illumination by the passage of current through an interface, as described in Section 1.2 and Chapter 3.

Once such problems are resolved, one expects the term 'voltage–current characteristic of the contact' to refer to properties which are independent of specimen length. On the other hand, theoretical models, for reasons of simplicity, often postulate semi-infinite systems. When they do so, they concentrate attention on the regions near the contact over which the potential varies non-linearly with x. In many other cases it is assumed that the voltage drop in the bulk (which, in obedience to Ohm's Law, varies linearly with current) is negligible over the small bulk regions involved in systems under measurement. All these notions have to be handled with caution; they may be justified in specific situations, but are not the outcome of some general rule. We shall see in the following sections and chapters what bearing these assumptions have on the interpretation of experimental results. One aspect is totally clear: though contact effects may be confined to regions which are thin compared with specimen dimensions, they cannot in their fundamental nature be limited to interfaces as such. The extent to which contacts make their presence felt in the associated bulk regions will be the subject of detailed discussion in the following chapters.

The frequently used term *contact resistance* evidently denotes the ratio of some voltage, V_c, V_c', V_I, to the current density, and is therefore subject to the same uncertainties of definition. Indeed, the important distinction between V_c and V_I has rarely been established in practice, and used for interpretational purposes. In the circumstances, the occasional confusion found in the literature is not at all surprising. Even the nomenclature of the subject has not yet been standardized. To sharpen the concepts and make their use more rigorous is one of the cardinal purposes of this book.

The experiments described in Figs 1.1 and 1.3 are rightly described as 'classics', and the truth is that they suffer from the degree of neglect that is often associated with that term. All too often, their outcome is taken for granted, or else is inferred from optimistic short-cut procedures. There *are* other ways of measuring the bulk conductivity, but when it comes to contact properties, there is no satisfactory alternative. In some situations (e.g. in connection with transverse conduction in thin films) geometry makes the use of potential probes impossible, and less direct methods must then be used. However, the outcome is always a wider margin for conjecture, and a correspondingly greater uncertainty about the nature of the prevailing transport relationships.

In systems which prohibit the use of probe methods, the characteristics of single contacts cannot be reliably ascertained. Figure 1.4 shows such a case, in which the bulk thickness has been kept small. Its voltage–current relationship is given in Fig. 1.4(c). In consists of three components (Fig. 1.4(b)), with voltages additive as a result of simple series connection of the two contact regions (necessarily biased in opposite directions) and the bulk layer. The result is a symmetrical composite characteristic which is always curved but can appear as virtually linear in many circumstances, and is thus open to gross misinterpretation. It could, for instance, suggest

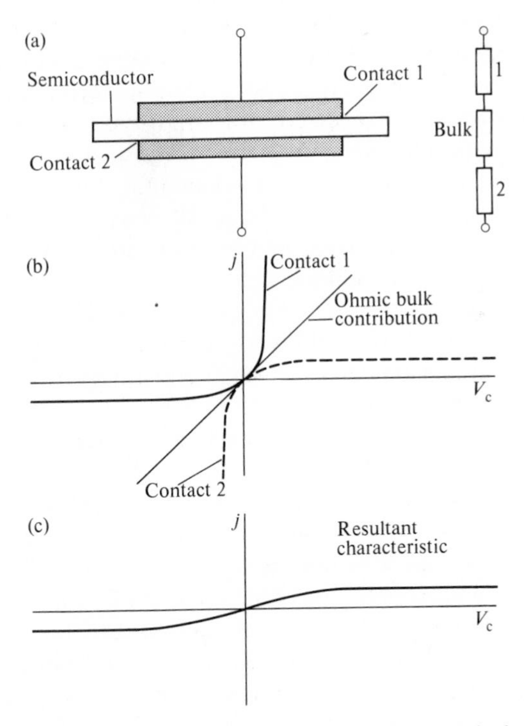

FIG. 1.4. Voltage–current characteristics of a symmetrical two-terminal system: (a) schematic structure; (b) characteristics of the individual circuit components; (c) their sum; voltages added for each current.

to the casual observer that contact properties are in this case altogether unimportant, and nothing would be more misleading. It follows at once (and not only for present reasons) that linearity as such is not a reliable criterion for the absence of contact effects. (See Section 3.3, and also George and Bekefi (1969)† for a practical example involving contacts on InSb.)

Unless otherwise indicated, the assumption is always that the two contacts on a system (e.g. as in Fig. 1.1(a)) are far apart and therefore non-interacting. On macroscopic specimens, this may be true, but complications are bound to arise when contacts are made to thin films and currents are passed in the transverse direction (see Section 2.2.9). The two contact regions then influence each other.

1.1.2 *Contact rectification*

A typical V_c–j relationship is shown in Fig. 1.5, displayed on two different scales. The terms 'forward' and 'reverse' have traditionally been used to describe the directions of easy and difficult current flow. 'Blocking' is sometimes used in place of 'reverse', but has more complicated connotations. The absolute values on the axes are not intended to have any particular significance; our main concern here is with shape.

For metal contacts on n-type and p-type material, the sign of the rectification is ordinarily reversed; for an n-type contact the reverse direction corresponds to metal negative, for a p-type contact to metal positive. The grounds for this are fundamental, but the way in which polarity is assigned to the contact voltage V_c (and, as we shall see below, to the barrier voltage V_B) is governed by conventions. These matters will be further discussed in Section 2.2.

As long as the asymmetry of the V_c–j relationship is pronounced, the distinction between forward and reverse currents is obvious enough, but there are many circumstances in which the distinction becomes blurred. The dotted line in Fig. 1.5(a) suggests such a case, which is particularly likely to arise when we deal with contacts of very low resistance (Section 6.5).

The origin of high contact resistances was a substantial mystery to researchers from the very beginnings of semiconductor physics during the second half of the last century, a mystery which remained without convincing explanation right into the 1930s. For many years, it was considered that some foreign layer of high resistivity at the interface would somehow have to be responsible, but independent means of detecting (let alone controlling) such layers were not yet well-developed.

† These designations refer to the references at the end of the book. Suggestions for further reading will be found at the end of many sections.

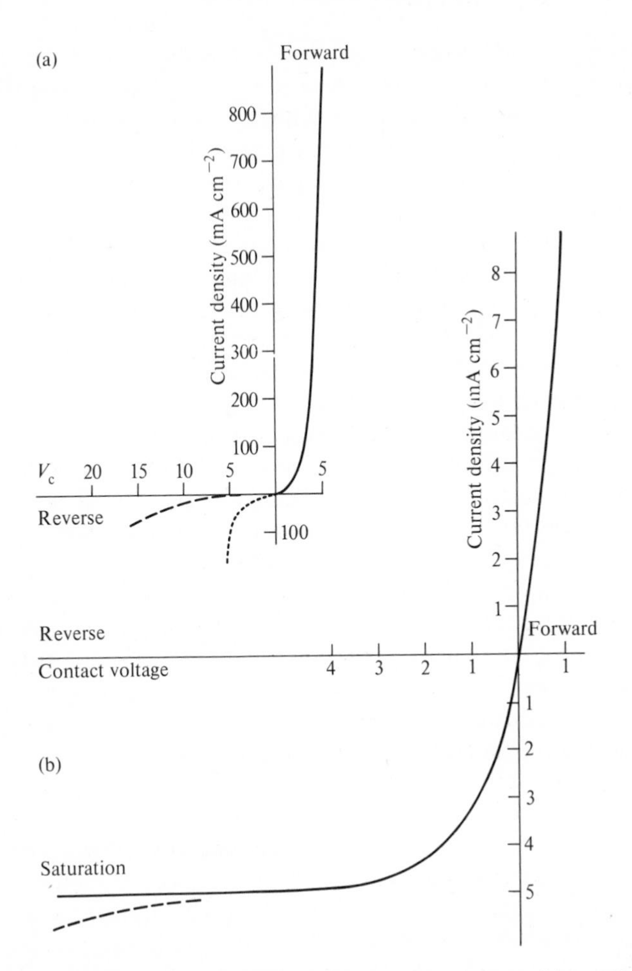

FIG. 1.5. Typical voltage–current characteristics of a single contact. (a) Full line: reverse-current saturation (ideal case). Broken line: reverse-current enhancement by one of the mechanisms described in Section 2.2.8. "Soft characteristic. Dotted line: case in which the distinction between the forward and reverse directions of current flow is blurred. (b) Region near origin enlarged.

Polarities are here ignored. With the sign conventions used in the analysis, reverse voltages and currents are actually positive (forward voltages and currents negative) for contacts on n-type material.

At any rate, the corresponding investigations yielded inconsistent and often conflicting results. This is how matters remained until Davidov, and Schottky and his colleagues independently demonstrated in 1938–9 (on the basis of work begun many years earlier) that a high contact resistance could also arise *without* a third phase. It could come from the presence of space charges close to the interface. Such charges would give rise to a *potential barrier*, and the regions in which they make their effects felt became known as *barrier layers*. It was soon recognized that the equilibrium contact conditions at *every* contact between different materials involve a charge exchange, and that every interface must therefore be associated with a double-layer. The part situated in the metal is of little consequence, because the conductivity of the metal is so high. The part situated in the semiconductor is all important, because it modulates the local conductivity. Moreover, it is associated with carrier concentration gradients which (inevitably) give rise to diffusion currents, and thus to departures from Ohm's Law. *All* interfaces have this feature in common; whether it has practical consequences depends entirely on the magnitude and distribution of the space charges.

With these contentions, which were readily substantiated, the entire subject of contact behaviour entered into a new phase. Foreign interface layers were not actually shown to be irrelevant, but it was a cardinal insight to realize that they were not necessary to account for high contact resistances or for their non-linear behaviour. The argument is not reversible; non-linear conduction effects as such are not a totally safe criterion for the existence of space-charge layers. Thus, Gossick (1969a) has shown that asymmetrical and non-linear conduction must be expected in varying degree from *any* lattice discontinuity, on wave-mechanical grounds alone. However, space-charge effects tend in practice to be enormously more effective in generating non-linear resistive behaviour, which is also why so much of the present treatment is essentially classical, apart from the use of band models.

In any event, space charges are always present at interfaces between materials of different conductivity. This may be seen from the disarmingly simple argument represented in Fig. 1.6. Two materials of different conductive properties (for whatever reason) share a common current. The bulk fields must therefore be different, and Poisson's relationship decrees that a space charge must reside where the field changes. One can at times

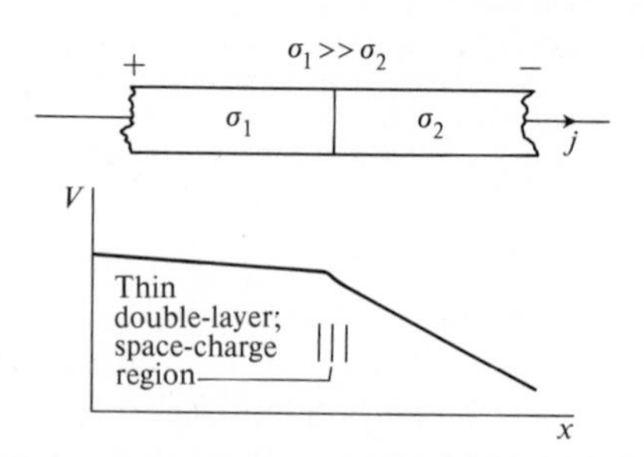

FIG. 1.6. Potential contour at any current-carrying contact between two materials of different carrier concentration (conductivity). Note existence of space-charge double-layer at interface, as shown by the S-shaped potential contour.

argue about the exact accommodation and thickness of that charge layer, but it is clear that carrier concentration gradients must be associated with it in principle, and hence diffusion currents. In that (ultra-refined) sense, there can be no such thing as an 'ohmic' contact (meaning 'obedient to Ohm's Law'), not even between metals. However, the resistive consequences arising from such an interface between metals are in practice very small. Such systems do not involve potential barriers, and interfaces that do involve them have far more interesting properties. Barrier behaviour is therefore our principal topic.

When voltages are applied to a barrier system, the barrier profile becomes distorted, in general asymmetrically, and rectifying characteristics are the direct outcome of this asymmetric distortion. This can be shown with ease, if not actually with precision (!), by reference to the simple representation in Fig. 1.7. Its detailed features and limitations will be discussed in later chapters; for the moment it is sufficient to note that something like a parabolic energy barrier does in fact exist at the metal-semiconductor interface, and that its shape can be altered by the application of external voltages, roughly as shown. The arrows above the barrier indicate the corresponding electron flows. Accordingly, a relatively large current flows in Fig 1.7(a) because the barrier as 'seen' from the semiconductor side is low. A relatively small current is implied by Fig. 1.7(c), because the barrier as 'seen' from the semiconductor side is high. In this representation (but not always in fact), the barrier encountered by electrons coming from the metal remains unchanged. Again, the polarities would be reversed if we were dealing with a high barrier on p-type material. Barriers may in practice have profiles which differ from that shown, but the same principles apply. The parabolic shape here assumed

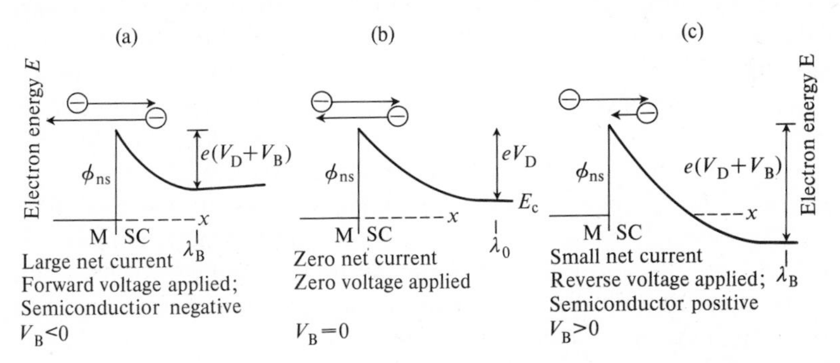

FIG. 1.7. Plots of electron energy versus distance; asymmetric distortion of a potential barrier by an applied voltage (here V_B); principle of contact rectification.
The quiescent barrier (b) is characterized by a built-in diffusion potential V_D (see also Section 1.3.2). Metal (M) on the left; semiconductor (SC) on the right. Contact barrier height ϕ_{ns} as 'seen' from the metal side remains constant. Horizontal arrows symbolize electron fluxes. E_c denotes the bottom of the conduction band.

represents an ideal case, as discussed in Section 2.1, and serves as a useful starting point for the analysis of more complex structures.

Figure 1.5(b) shows a reverse characteristic (full line) which saturates, and this is again an ideal case. If everything were simple, all reverse currents would saturate in this way, but additional factors often perturb the simplest barrier models and give rise to additional currents (mostly unwanted on technological grounds). See Sections 2.1.5 and 2.2.8.

If pronounced rectification is to be observed in a two-terminal system, one of the contacts must be suppressed. This can be done in two ways: by altering barrier structure or by changing the geometry in such a way that one of the contact resistances is low compared with the other in both directions of current flow. Both methods have been widely used in the past; the point-contact rectifier represents such a solution (Fig. 1.8). The contribution of bulk resistance must also be eliminated as far as possible, and this is done by making the bulk region thin. In a highly rectifying system of this kind the base contact is often mistakenly described as 'ohmic'. In fact it may be as non-linear and rectifying as the point contact itself, but its larger area ensures that it carries the total current for very small contact voltages, and these may very well confine its operation to the voltage region $\pm kT/e$ from the origin (where k = Boltzmann's constant, T = absolute temperature, and e = numerical value of the electronic charge). Over such an interval, all contact resistances can be shown to be linear. In any event, when a series resistance is very small, the question of whether it is linear or non-linear is unimportant in most practical contexts.

There remains the general didactic point, already alluded to, that 'low resistance' and 'ohmic' are by no means identical concepts. Problems of nomenclature often provoke more controversy than issues of substance, for the most human reason in the world: everyone likes his own terms. There are, nevertheless, objective reasons for adopting one term rather than another. For a description of contact characteristics clarity demands that a distinction be made between terms that refer to relative magnitude (e.g. 'low resistance') and terms that refer to a form of behavior (e.g. 'ohmic resistance', meaning independent of voltage). In due course (Section 1.2) we shall come across yet another designation, namely 'non-injecting', and that term, though often confused with the other two, refers to something entirely different, namely the *mechanism* of charge transport through the interface.

In what follows, the transport mechanism will be discussed in considerable detail, and this can be done in several ways, of which each is associated with a set of models. Charge transport characterized by a very small mean-free path is the basis of all *drift-diffusion models*, and because these are conceptually simplest, they will be given principal emphasis. Charge transport characterized by mean-free paths which are long compared with the barrier structures is the basis of *thermionic diode and ballistic models* (see Section 2.2.6). There are also *tunnelling models* which emphasize the wave-mechanical tunnel penetration of barriers, i.e. charge transport through barriers, rather than over them (see Sections 2.1.5, 2.2.8, 2.3.3 and Section 6.5). These are the interpretational frameworks most frequently discussed. However, it should be noted that another and very different approach is possible, which has been called the *flux method.*

The flux method is based on highly general considerations which apply to the transmission, absorption, and reflection of *anything* by a layer or series of layers. 'Anything', in this context, could be radiation or a stream of electrons, and the application of these considerations to semiconductor barrier problems was first suggested by McKelvey *et al.* (1961). For a recent overview, see Thomchick and Buoncristiani (1980, 1981). Such a general approach leads to a family of equations whose members are parametrically differentiated from each other. Each such equation represents an expression of basic conservation laws to which, in fact, all other models must conform. Within that framework, every layer of material is fully described by an absorption constant (measuring recombination) and a reflectivity, but because these descriptors are not directly accessible to experimental tests, they must ultimately be brought into a relationship with more familiar parameters, like barrier height, barrier thickness, donor density, electron mobility, etc. This is not an easy problem, which is one of the reasons for the fact that flux methods are rarely used. They are, however, profound, and whereas their total impact remains unclear, they should not be prematurely dismissed. Indeed, their more intensive exploration would probably be rewarding.

Superimposed on the voltage–current characteristic there is in practice a varying amount of *electrical noise.* In many contexts, this feature is utterly negligible, but when contact devices are used in RF or microwave circuitry, the noise characteristics become important factors. The problems involved are not covered in this book; the reader is referred to specialist treatises, e.g. van der Ziel (1970), Bosman and Zijlstra (1982), van Vliet *et al.* (1975a,b), and to the many papers cited by these authors.

1.1.3 Geometrical aspects; radial systems

In the idealized textbook case, the contact interface is an infinite plane, a situation which is not likely to confront the experimenter in practice! Even so, the assumption of zero edge effects (and thus area-independent 'specific' properties) is often justified. In the early days of semiconductor

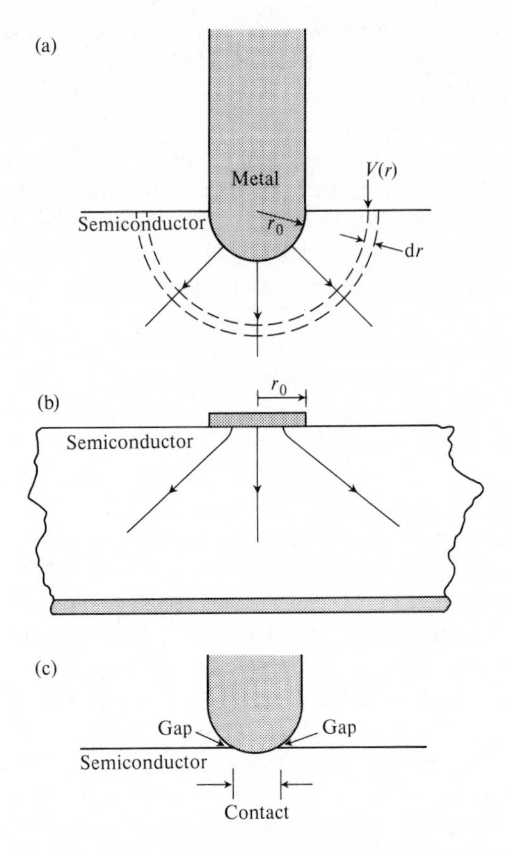

FIG. 1.8. Non-planar current flow: (a) through hemispherical contact; (b) through planar contact of a size much smaller than that of the counter electrode; (c) through a sub-hemispherical contact with edge effects.

technology, one distinguished between only two kinds of contacts:

(a) *Area* contacts, which could be assumed to be large enough to justify the neglect of all geometrical complications, and
(b) *Point* contacts, which were small (e.g. 10^{-3} cm radius) and could always be taken as hemispherical, implying purely radial current flow (Fig. 1.8(a)).

In more recent years, this distinction has become drastically blurred, since *plane* contacts of very small sizes can easily be made by photolithographic techniques, and are frequently so made in the context of integrated circuit manufacture. Correspondingly, edge effects have become more important. Of course, the assessment of what is important and what is not depends also on the thickness of the system; a very small planar contact confronting a large counter-electrode at a substantial distance can also be regarded as a radial system (Fig. 1.8(b)), even though it is manifestly not radial close to the small contact itself.

The radial flow region gives rise to a spreading resistance, and whether this is of any practical significance depends *inter alia* on the resistance of the barrier itself. For simplicity, the spreading resistance will here be considered as a separate entity, governed by a constant bulk conductivity σ_b, and in simple series connection with the barrier itself. In accordance with Fig. 1.8(a), the spreading resistance R_s is then given by

$$R_s = (1/2\pi\sigma_b)\int_{r_0}^{\infty} r^{-2}\,dr = 1/2\pi r_0\sigma_b. \quad (1.1.1)$$

Since the radius of curvature r_0 may be very small (e.g. 10^{-3} cm) the spreading resistance may attain high values. Of course, contacts encountered in practice are never hemispherical, and the assumption that they are is no more acceptable than the notion that they might be treated as infinite. However, it can be shown that

$$R_s = 1/4r_0\sigma_b \quad (1.1.2)$$

for a contact in the form of a circular disk of radius r_0 (Fig. 1.8(b)), again considered in isolation. This is done by regarding the contact interface first as a circular ellipsoid, and then allowing the axis which is perpendicular to the surface to tend to zero (Holm 1946). The result is more realistic, since contacting metals, even if originally hemispherical, flatten under the influence of ordinarily prevailing contact pressures. Thus, a contact thrust of (say) 10 g, applied to an old-fashioned wire 'whisker' contact of 10^{-3} cm radius, would generate a contact pressure of over 3000 kg cm^{-2}, under which even hard metals yield. Indeed, the final contact area would be one for which the contact pressure is just below the yield point, and which can therefore be sustained indefinitely. That such pressures can also damage the semiconductor is obvious enough, and this is one of the (many) reasons for the total demise of the 'cat's whisker' point contact in modern technology. However, point contacts of that type are still widely used during experimentation, e.g. as potential probes for the currentless measurement of the local potential (Section 1.1.1). In the absence of current, the geometrical complications are, of course, irrelevant. Conversely, when currents are involved and theoretically interpretable results are sought, great care must be taken to fabricate stable contact structures which minimize edge effects. A survey of the various possibilities has been given by Rideout (1978); three of these are shown in Fig. 1.9. The object, in each case, is the reduction of leakage currents at the electrode edges, brought about by locally higher fields. The fact that this degree of sophistication has been found necessary suggests that results obtained by simpler means should be interpreted with caution.

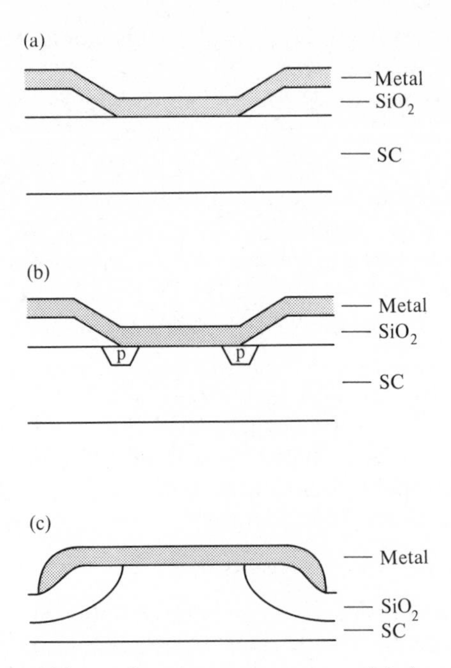

FIG. 1.9. Precautions which may be taken to prevent or minimize electrode edge effects. Based on Rideout (1978), where further references are given.

Complications of a different kind can arise in the case of contacts as shown in Fig. 1.8(c), involving peripheral air gaps in which currents can flow by cold field-emission processes. Sillars (1955) and Stratton (1956a) have concerned themselves with appropriate edge corrections for 'soft' reverse characteristics arising from this cause.

1.1.4 Geometrical aspects; non-radial systems

Equations (1.1.1) and (1.1.2) have the status of 'classical approximations' and serve well as long as the current flow pattern is at least quasi-radial. However, in many modern systems (often quite outside the context of rectifying barriers), the flow pattern is more of the type shown in Fig. 1.10a, and the spreading effects are then quite different. When experiments are performed with contacts of controlled size, the practical symptoms of such effects show themselves as departures from the proportionality between contact conductance and contact area. Moreover, the contact *shape* then enters into the argument.

There are two generically different ways of analysing such problems. One approach is exemplified by the work of Hall (1967), and Ting and Chen (1971), based on conformal transformation of the prevailing electrode geometry into a geometry for which the solution is obvious. Thus,

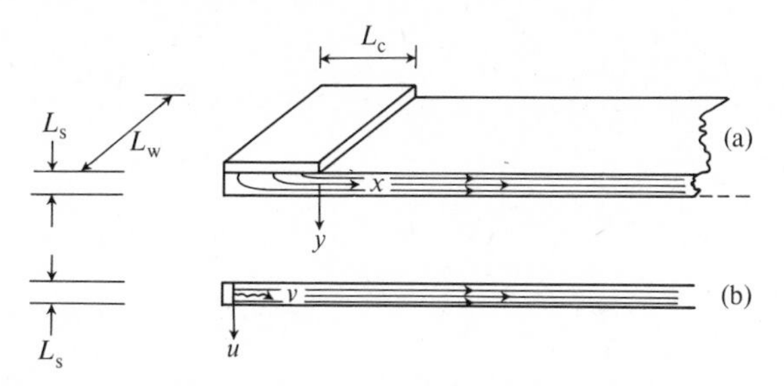

FIG. 1.10. Spreading resistance arising from non-radial current flow I; geometrical transformation through conformal mapping. (a) Actual electrode situation. Complex plane $z=x+iy$. (b) Equivalent geometry after transformation. Complex plane: $w=v+iv$. See Hall (1967).

for example, the system (a) shown in Fig. 1.10 can be 'transformed' into the shape (b) by a mathematical procedure, such that every point in the complex $z=x+iy$ plane corresponds to a unique point in the complex $w=u+iv$ plane.

The only problem is to find a relationship between w and z which will permit a transformation of the desired kind, and while this is not always easy, the method has in fact been applied to a great variety of electrode shapes (e.g. see Kober (1957) for a variety of examples, without proof, and Betz (1964) for an instructional text). The method is not only precise, as far as electrode and specimen configurations are concerned, but can also (via the reverse $w \rightarrow z$ transformation) lead to the shape of the current flow curves, and thus to the local current divergence. Spreading resistances are obtained by comparing voltage and current at equivalent points. Thus, a certain length of the column in Fig. 1.10(b) will be equivalent to the spreading resistance between $x=-L_c$ and $x=0$ in Fig. 1.10(a). For a contact of the kind shown there, Hall (1967) demonstrates the use of transformation given by

$$z=\frac{i\pi^2 L_c}{2L_s}+2\log\frac{(1+\sin w)^{1/2}+(b_z+\sin w)^{1/2}}{(b_z-1)^{1/2}} \tag{1.1.3}$$

with

$$b_z=-1+2\left(\coth\frac{\pi L_c}{2L_s}\right)^2. \tag{1.1.4}$$

L_w does not enter, because it cannot affect the nature of the current distribution as such. For the special case of $L_c=L_s$, resistance between $x=0$ and x can be shown to be

$$R=\frac{1}{L_s\sigma_b}\left(\frac{x}{L_w}+0.469\right) \tag{1.1.5}$$

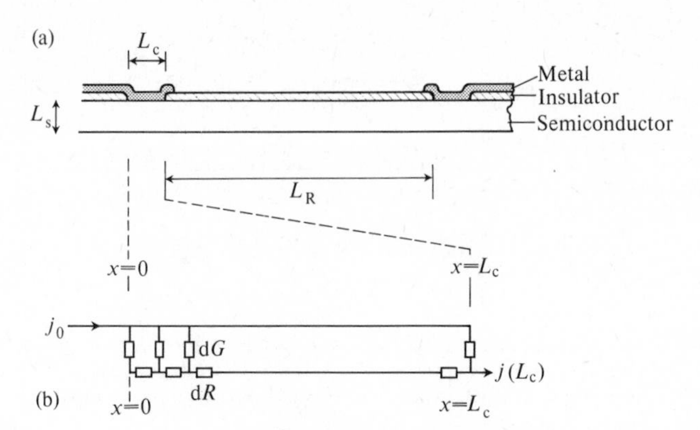

FIG. 1.11. Spreading resistance arising from non-radial current flow II; equivalence of spreading region to a lossy transmission line. (a) Actual electrode situation. (b) Equivalent circuitry; enlarged view of (a). In a practical case, the semiconductor might be Si, and the insulator SiO_2. Width of system: L_w.

as compared with $x/L_sL_w\sigma_b$ for a simple end-contact. This illustrates the order of magnitude of the spreading correction.

Again, the bulk conductivity σ_b is here taken as constant, but there are many circumstances in which its local effective value is in fact dependent on position and current density (see Section 1.2). Calculations of spreading resistance are then more complicated.

An entirely different approach, leading to approximate (but often entirely adequate) results is exemplified by the work of Murrmann and Widmann (1969), Berger (1972), Schuldt (1978), and Fang *et al.* (1979). This considers the spreading resistance as a lossy transmission line with lumped components. Figure 1.11 shows how this is done, and defines dimensions L_s, L_c, L_w, and L_R. The metal electrode itself is equipotential but, of course, there is a horizontal field within the semiconductor. As a working approximation, it may be assumed that the current flow is streamline everywhere for $x > L_c$. The analysis, e.g. along the lines of Murrmann and Widmann (but here in terms of bulk constants), can be simply outlined. Each element dx is associated with a certain vertical conductance dG and a horizontal resistance element dR. Thus,

$$j(x+\mathrm{d}x)-j(x)=V(x)\,\mathrm{d}G=\mathrm{d}j$$

and

$$V(x+\mathrm{d}x)-V(x)=J(x)\,\mathrm{d}R=\mathrm{d}V. \tag{1.1.6}$$

We also have

$$\mathrm{d}G=\frac{\sigma_b L_w\,\mathrm{d}x}{L_s} \quad \text{and} \quad \mathrm{d}R=\frac{\mathrm{d}x}{\sigma_b L_s L_w}. \tag{1.1.7}$$

Hence

$$\frac{\mathrm{d}j}{\mathrm{d}x}=\frac{\sigma_{\mathrm{b}}L_{\mathrm{w}}}{L_{\mathrm{s}}}V \quad \text{and} \quad \frac{\mathrm{d}V}{\mathrm{d}x}=\frac{j}{\sigma_{\mathrm{b}}L_{\mathrm{s}}L_{\mathrm{w}}}. \tag{1.1.8}$$

When the voltage V is eliminated, this gives

$$\frac{\mathrm{d}^2 j}{\mathrm{d}x^2}-\frac{j}{L_{\mathrm{s}}^2}=0 \tag{1.1.9}$$

of which the solution with appropriate boundary conditions is

$$j(x)=j_0\frac{\sinh(x/L_{\mathrm{s}})}{\sinh(L_{\mathrm{c}}/L_{\mathrm{s}})}, \tag{1.1.10}$$

j_0 being the total current. Correspondingly, we have for $V(x)$

$$V(x)=\frac{L_{\mathrm{s}}}{\sigma_{\mathrm{b}}L_{\mathrm{w}}}\left(\frac{\mathrm{d}j}{\mathrm{d}x}\right)=\left(\frac{j_0}{\sigma_{\mathrm{b}}L_{\mathrm{w}}}\right)\frac{\cosh(x/L_{\mathrm{s}})}{\sinh(L_{\mathrm{c}}/L_{\mathrm{s}})}. \tag{1.1.11}$$

If we put $R_{\mathrm{s}}=V(L_{\mathrm{c}})/j(L_{\mathrm{c}})$, then

$$R_{\mathrm{s}}=\frac{1}{\sigma_{\mathrm{b}}L_{\mathrm{w}}}\coth\left(\frac{L_{\mathrm{c}}}{L_{\mathrm{s}}}\right) \tag{1.1.12}$$

and this would have to be added to the resistance of the bulk to obtain the total resistance. Thus, if we consider a sheet resistor of length L_{R} between the electrodes, the total resistance

$$R_{\mathrm{T}}=\frac{1}{\sigma_{\mathrm{b}}L_{\mathrm{w}}}\left\{\frac{L_{\mathrm{R}}}{L_{\mathrm{s}}}+\coth\left(\frac{L_{\mathrm{c}}}{L_{\mathrm{s}}}\right)\right\}. \tag{1.1.13}$$

For $L_{\mathrm{c}} \gg L_{\mathrm{s}}$, $\coth(L_{\mathrm{c}}/L_{\mathrm{s}}) \to 1$. In practice, L_{c} is never likely to be smaller than L_{s}, which means that $\coth(1)=1.313$ may be regarded as an upper limit for the correction. In coarse terms, the spreading resistance adds an effective length of the same order as L_{s} to L_{R}, and is thus important only when L_{R} is small, which is what one intuitively expects. In modern microcircuitry this can, however, be a significant factor. If a point contact were applied to a thin film, the spreading resistance could be very large. The distinction between such a case and that discussed in Section 1.1.3 is, of course, that the current flow lines are here grossly non-radial.

The above analysis is formulated in terms of a transmission-line network based on a single value of y, namely L_{s}. A further sophistication (albeit of questionable value) could be achieved by assigning an appropriate network to each y-layer, and summing the currents.

Descalu and co-workers (1981) have described some of the problems arising in connection with contacts which, by virtue of their edge effects, cannot be properly described by a one-dimensional model. For an example of specific edge-effects on silicon, see Tove *et al.* (1974).

Further reading†

On the topography of contact surfaces: Cuthrell and Tipping 1973.

On spreading resistance measurements: Kramer and Van Ruyven 1972b.

On variational methods for calculating spreading resistivities: Choo *et al.* 1976, 1977, 1978, 1981.

On the design of diode geometry to minimize edge effects: Rusu *et al.* 1977.

1.2 Current-controlled non-equilibrium processes

1.2.1 Current–composition ratios

The possibility of non-equilibrium situations arising close to a current-carrying contact has already been mentioned in connection with Figs. 1.1 and 1.2. Such effects can be classified into four distinct categories, and their origin will now be be discussed. The fraction γ_n of *bulk* current carried by holes in n-type material is a useful reference parameter. It is given by

$$\gamma_n = \frac{\mu_p p_e}{\mu_p p_e + \mu_n n_e} = \frac{j_p}{j_p + j_n} = j_p/j \tag{1.2.1}$$

where the subscript e denotes equilibrium. Of course, a corresponding quantity could be defined for p-type material, namely $\gamma_p = \mu_n n_e/(\mu_n n_e + \mu_p p_e)$, the fraction of bulk current carried by electrons. However, the present discussion is formulated in n-type terms, as is customary.

Throughout the undisturbed, homogeneous bulk, γ_n is (by definition) constant. However, at the contacts and in their vicinity, the minority carrier participation in charge transport may (and, indeed, generally must) differ. Accordingly, there are various possibilities, as may be seen by reference to Fig. 1.12, which designates current composition ratios γ_1 and γ_2 for the two end contacts. (We always have $0 < \gamma < 1$, of course.)

† The inclusion of references in these lists (which follow many sections) is not intended to imply that the findings described are necessarily in agreement with those discussed here.

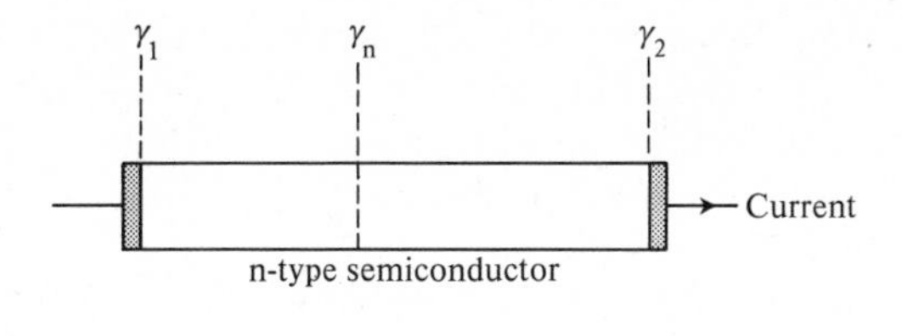

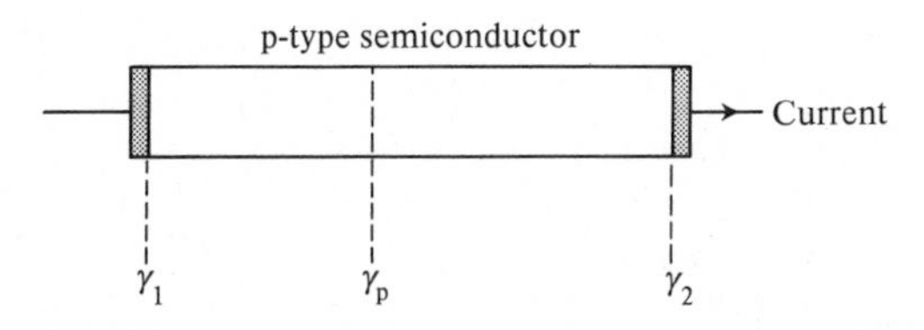

FIG. 1.12. Current composition ratios in the bulk material and at contacts. γ, in each case and location, denotes the fraction of total current carried by minority carriers.

Classification of carrier concentration disturbances in n-type material

Forward current $\gamma_1 > \gamma_n$ minority-carrier injection
Forward current $\gamma_1 < \gamma_n$ minority-carrier exclusion
Reverse current $\gamma_2 > \gamma_n$ minority-carrier extraction
Reverse current $\gamma_2 < \gamma_n$ minority-carrier accumulation

The term 'injection' was first coined by Bardeen and Brattain (1948); the last three terms are due to Low (1955). When, in the ordinary way and without further qualification, reference is made to 'injection', the term is always taken to mean *minority* carrier injection, in the sense envisaged here. The process is, of course, the basis of transistor operation in its most important form. Unfortunately, the term 'accumulation' is now more frequently used to characterize contact layers of *negative* barrier height ϕ_{ns}, irrespective of current flow, implying an excess of *majority* carriers. (See also Section 2.1.2.) In close discussion, no problem is likely to arise, but casual references to 'accumulation' may be misleading unless explained. As far as is known, current-controlled minority carrier accumulation (the concept relevant here) has no technical applications, though a possible function in electroluminescence has been suggested (Henisch and Marathe 1960). Of course, all four effects influence the behaviour of semiconductor systems with contacts, to an extent which depends on circumstances, as discussed below. When neutrality (or, at any rate, quasi-neutrality) prevails, as it generally does in a so-called 'lifetime' semiconductor (see Section 3.2), the phenomena of minority carrier injection, exclusion, extraction, and accumulation are necessarily associated with majority carrier accumulation, extraction, exclusion, and injection respectively.

The designations 'forward' and 'reverse' currents, as used in the above list, are intended to mean only that the metal contact on n-type material is positive for the former and negative for the latter, irrespective of contact resistance. For p-type material, the polarities would, once again, be reversed. In principle, the cases $\gamma_1 = \gamma_n$ and $\gamma_2 = \gamma_n$ could also be envisaged, but they are trivial, because such situations would depend on unlikely coincidences, all the more because γ_1 and γ_2 are themselves current-dependent. Contrary to popular misconception, non-equilibrium effects do *not* represent unusual situations; they may or may not make their presence felt in prominent ways, but they represent the norm.

The above classification is quite general, and leaves two important sets of questions for later discussion:

(a) What are the detailed contact parameters which govern γ_1 and γ_2?
(b) What are the *consequences* of carrier non-equilibrium in the four cases?

Question (a) is obviously concerned with the structure of the contact interface and is thus influenced by the nature of the contacting materials. It is by no means necessary for those to represent different phases; on the contrary, *any* boundary, even one that separates two regions of the same crystal (as in Fig. 1.6) can be so analysed and described. Some specific boundary configurations which arise in metal–semiconductor contacts will be discussed in Chapters 2 and 3 but, for the moment, all that matters is the existence of descriptive parameters γ_1 and γ_2 no matter how they may arise from contact structure. Question (b) is the subject of the section which follows. Another problem concerns the depth to which non-equilibrium situations are propagated into the bulk material. That depth is the distance over which the current composition ratio adjusts itself from γ_1 to γ_n and from γ_n to γ_2, and this depth can be calculated from general transport relationships (see Chapter 3). Figure 1.12 assumes that the specimen dimensions greatly exceed the depth of penetration (giving γ_n in the interior), but in practice this may not be true. Accordingly, it is possible to envisage non-equilibrium regions at the two contacts which overlap and thus interact with each other, but a full quantitative analysis of such situations is complex and a task for the computer.

1.2.2 Electrical consequences of non-equilibrium carrier concentrations

We are now concerned with the manner in which (essentially) bulk material *under* the contact reacts to the four situations specified above. As long as γ_1 or γ_2 differ from the bulk equilibrium value γ_n (Fig. 1.12) there is bound to be a departure from electronic equilibrium, to an extent which depends on the distance x from the contact and on the current density (hence 'current-controlled'). Such departures, i.e. $\Delta n = n - n_e$ and

$\Delta p = p - p_e$, affect all properties which depend on the free-carrier concentration. Thus, for instance, the optical absorption of a semiconductor is generally altered in the vicinity of a current-carrying contact, and so is the local *effective conductivity*. The term 'effective' is used here in recognition of the fact that current-controlled departures from equilibrium always generate diffusion currents. Accordingly, the effective conductivity is essentially non-ohmic. In such circumstances, the specific bulk conductivity σ_b ceases to have any operational significance, since it is a term that relates fundamentally to conduction in homogeneous bulk material. However, it still serves for purposes of comparison with the effective local conductivity, which may be larger or smaller than σ_b. It may even be larger for some currents and smaller for others. Moreover, the consequences of $\gamma_1 \neq \gamma_n \neq \gamma_2$ depend on whether the contacted material is a *lifetime semiconductor* ($\tau_0 > \tau_D$) or a *relaxation semiconductor* ($\tau_0 < \tau_D$), where τ_0 is the (diffusion length) lifetime of the carriers and τ_D the dielectric relaxation time. It is a fact that nearly (though not quite) all practical observations in the past have concerned themselves with the behaviour of injecting contacts on lifetime semiconductors, i.e. with only one of the possible cases. There are excellent reasons for this, but it also means that many gaps remain in our knowledge of contact effects, to be filled by future research.

The parameters which govern the consequences of non-equilibrium include not only the above time constants, but also n_e, p_e, μ_n, μ_p, as well as the concentration of traps, their energetic position in the band structure and their capture cross-section. It follows at once that simple *and* general predictions cannot be made; various important cases have to be analysed specifically.

Of the four current-controlled processes, minority carrier injection ($\gamma_1 > \gamma_n$, in n-type material) is technically the most important since it is the basis of transistor action. In modern technology, the injecting entity is a p–n junection, but this does not alter any fundamental relationships. In any event, the phenomenon was first discovered at contacts. According to Fig. 1.8, the potential $V(r)$ along the surface of a semi-infinite semiconductor, with current flowing through a point contact, should vary as $1/r^2$. This matter was first subjected to experimental test by Bardeen and Brattain (1948), and was duly confirmed for reverse currents. However, in the presence of substantial forward currents, the potential did *not* vary in this way; it varied less steeply than expected near the contact, suggesting an enhanced effective conductivity in the spreading region, arising from an enhanced carrier concentration. This was the first evidence of current-controlled non-equilibrium processes, and led directly to the invention of the transistor, along well-known lines. One of the key observations was, of course, the fact that the conductivity enhancement occurred only for currents in the forward direction. Had the experiments been more exhaustive, corresponding (but not equal) effects would also have been found for reverse currents. However, the results were sufficient to prove that the mechanism of current flow (i.e. the extent and manner of minority carrier participation) must be different in the two cases. At the time, this was a startling conclusion, since the electrical conductivity of a material had hitherto been regarded as a fixed parameter. The concept of current-dependent modulation was completely new.

Bardeen and Brattain at the time envisaged only lifetime semiconductors ($\tau_0 > \tau_D$) and, indeed, only germanium and silicon among those. They also worked with relatively high current densities, for which $\gamma_1 > \gamma_0$ corresponds to $\Delta p > 0$. It was at once realized that, under such conditions, one would have to take account also of an incremental concentration of *majority* carriers Δn, such that $\Delta n \approx \Delta p$, for neutrality reasons. The observed effective conductivity was clearly current-dependent, but (for all currents) approached σ_b at increasing distances from the injecting interface. Because the observation of enhanced effective conductivities led originally to the formulation of the minority carrier injection concept, it was considered for a long time thereafter that a conductivity increase is the only possible consequence of injection. It will be shown in Section 3.2 that this is not necessarily so; in some circumstances, and particularly at very low current densities (not used by Bardeen and Brattain during their original experiments), injection can actually lead to a local conductivity *decrease*. At high currents it leads to a conductivity increase which helps to diminish the bulk resistance in series with the barrier resistance, and thereby makes rectification more pronounced than it would otherwise be. In the context of transistor technology, the injecting electrode is called an *emitter*.

Though minority carrier injection is by far the most important of the four non-equilibrium processes, the others are by no means without interest. Exclusion has been the subject of a limited computer analysis by Manifacier and Henisch (1979), of which some further details are given in Section 3.4.1. Extraction has been experimentally and theoretically explored by Rahimi, Manifacier, and Henisch (1981), but very little has been done on accumulation. It is unfortunately true that purely intuitive predictions can be very misleading in this field. However, on quite general grounds, one would expect the effects of exclusion and extraction to be small, except in near-intrinsic material. Both effects involve a *diminution* of the minority carrier concentration, and that concentration is already minute in any doped semiconductor. Accumulation does involve an increase, but under geometrically unfavorable conditions which make Δp hard to detect; accumulation regions tend to be extremely thin. When $\gamma_1 > \gamma_n$, γ_1 is also called the *injection ratio* of the interface. An *exclusion ratio, extraction ratio,* and *accumulation ratio* can be defined in analogous ways. In the professional jargon, 'non-injecting' is sometimes used as if it meant 'ohmic', but the term carries no such implication, nor does it signify 'low-resistance'; see Section 1.1.2. An excellent example of the distinction is provided by some 'ohmic' low resistance contacts on n–InSb, which were found to be also highly injecting (Ancker-Johnson and Dick 1969). The assumption $\gamma_1 = 0$ ('neglecting injection') is sometimes made in the course of attempts to simplify contact analysis, but it is important to remember that zero injection automatically implies minority carrier exclusion, which may or may not amount to a simplification.

All the above comments concern bulk effects arising from non-equilibrium carrier concentrations. In addition, such departures from equilibrium also modify the properties of any *contact* placed on the non-equilibrium region. When we are dealing with injection or accumulation, the resulting modification is very similar to that produced by optical illumination, and is best discussed under that heading (Chapter 5). Under the form of non-equilibrium produced by purely electrical means, floating potentials arise at contacts; reference to those has already been made in Section 1.1.1. The difference is purely one of terminology. It should also be possible to modify contact properties in the opposite direction, via extraction or exclusion (implying $\Delta p < 0$) but it is not known whether such effects have ever been looked for or observed. In the absence of such a thing as 'negative light', they would have no photoelectric equivalent.

1.2.3 *Optical consequences of non-equilibrium carrier concentrations; injection luminescence*

Apart from the electrical consequences discussed above, the existence of non-equilibrium carrier concentrations also has optical consequences, in two senses: (a) absorption, and (b) emission.

In the long-wavelength range, the absorption coefficient of a semiconductor is governed by the absorption due to free carriers. The classical expression is

$$a(\lambda) = Zn\lambda^2 \tag{1.2.2}$$

where n is the free carrier concentration, λ (here) the wavelength, and Z a constant, inversely proportional to m_n and τ (the collision time). Z, and thus $a(\lambda)$, are therefore influenced by the nature of the prevailing scattering processes, a fact investigated and confirmed by Spitzer and co-workers (1961). Indeed, scattering processes also influence the exponent of λ, here given as 2 in the form originally proposed by Drude. More detailed analysis shows, however, that acoustic phonon scattering leads to $\lambda^{1.5}$, as found by Fan and Becker (1951), and optical phonon scattering to $\lambda^{2.5}$, as found by Visvanathan (1960). Scattering by ionized impurities yields higher exponent values still. Pankove (1971) has provided a very useful summary of the situation. However, what concerns us here is the dependence of $a(\lambda)$ on n. The equilibrium value of n, namely n_e, depends on doping, of course, but n itself is current dependent when non-equilibrium conditions prevail as a result of contact effects. Thus, Lehovec (1952) and Gibson (1953) demonstrated that the long-wavelength absorption increases with current under injection conditions, as expected. Under extraction conditions the opposite ought to occur, but the effect would be small (except in near-intrinsic material) and, as far as is known, has never been reported.

Radiative recombination (luminescence) is another phenomenon expected as a consequence of any contact process which *augments* the free-carrier concentration. This is, of course, the principle of the light-emitting diode (LED). In such systems, the inevitable competition between radiative and non-radiative recombination processes has been adjusted (by selection of the parent lattice and its built-in recombination centres) to favour the former. Direct band-gap materials are needed for high efficiency, and even in those a high degree of structural order is necessary to prevent indirect recombination through defects. (See Haynes and Nilsson (1965) for a treatment of relative recombination probabilities, and Gooch (1973) and Bergh and Dean (1976) for a general discussion of device aspects.) Minority-carrier accumulation should likewise lead to current-dependent emission (Henisch and Marathe, 1960). See Section 3.3 for a more detailed discussion of carrier injection, and Section 6.6 for one on avalanche formation. Light emission from contacts is, of course, very similar to light emission from p–n junctions, and in that context the phenomenon is now of major technical importance. (An important background literature source is *Radiation and recombination in semiconductors*, Volume 4 of the *Proceedings of the 7th International Conference on Semiconductors*, 1964. (Academic Press, New York, 1965).)

1.2.4 Space-charge-controlled conduction

All current-controlled non-equilibrium processes involve space charges in varying degree, from trivial to highly important. All are therefore 'space-charge influenced', though not necessarily 'space-charge controlled'. The latter term, when used without further qualification, has come to stand for a specific set of phenomena involving space charges which arise from *majority* carriers alone. For this to happen, minority carriers must either be absent, or present only in negligible concentrations. The whole complex of problems associated with such systems has been extensively discussed by Thredgold (1966), and by Lampert and Mark (1970), and is not among the main themes of this book. It is relevant to materials which are ordinarily classified as 'insulators' in bulk, and is analogous to space-charge-controlled conduction in vacuum. Indeed, a solid 'insulator' differs from vacuum only in three (albeit important!) ways; it has a dielectric constant, it has a band structure, and it contains carrier traps in fixed positions. The dielectric constant modifies the field, and so (in a different way) do the space charges held in traps.

It is often insufficiently recognized that 'insulators', as we know them, owe most of their characteristic properties to their trap content. Mott and Gurney (1940) were the first to show that an insulator in contact with a metal acquires a charge, which is distributed over the traps near the interface. This charge, in turn, gives rise to a field which opposes the entry of further charges, and thus prevents current flow (supported by carriers 'borrowed' from the metal) (Section 6.3.1). Were it not so, insulating layers would have the same properties as vacuum gaps, which is manifestly not the case. In solids, therefore, one has to analyse (a) the magnitude of free-carrier space charges in the presence of current flow, (b) the dynamics of the interaction between free carriers and traps, and (c) the trap concentration and the energetic position of trapping levels. These problems will be further discussed in Chapter 3.

Single carrier systems (and especially those containing traps) lack a mechanism for the neutralization of injected space charges. Space charges therefore tend to play a more important role in such materials than they do in media which contain free charge carriers of both signs. Traditionally analyses of space-charge-controlled processes in single carrier systems regard a contact simply as an 'infinite source' of carriers, and ignore all problems associated with such a concept. This may be permissible in some cases, but there is no general justification for such a procedure. (see Section 1.3.3.)

Further reading

On light emission from contacts:

with GaP: Card and Smith 1971, Miyauchi *et al.* 1969, Rosenzweig *et al.* 1969, Vasilieff *et al.* 1977;

with InP: Blom and Woodall 1970;

with diamond: Lepek *et al.* 1976, Levinson *et al.* 1973;

with II–VI compounds: Livingstone *et al.* 1973, Robinson and Kun 1969, Waite and Vecht 1971.

1.3 Origin of interface barriers

As discussed in Sections 1.1 and 1.2, contact resistances arise from contact barriers and those in turn are associated with space charges, even in the absence of third phases at the interface. The barriers can actually arise in two distinctly different ways: (a) through differences in the

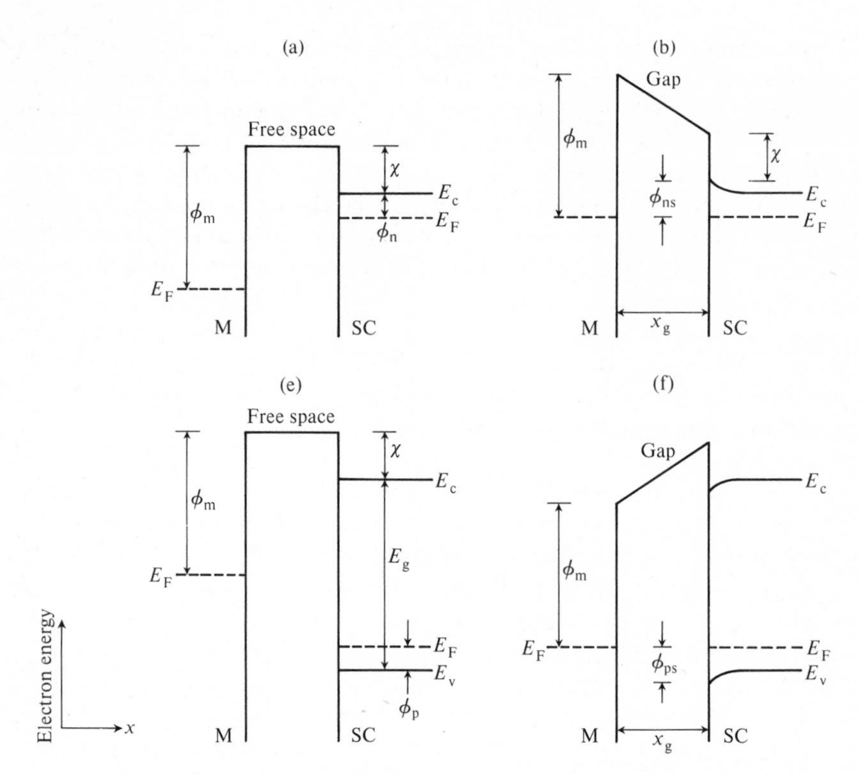

FIG. 1.13. Contact barriers due to work-function differences. ϕ_m = thermionic work function of Metal M; χ = electron affinity of semiconductor SC; E_g = band gap. (a)–(d), contact on n-type material; (e)–(h), contact on p-type material; (a) and (e), non-equilibrium situations; (d) and (h), final barrier contours. The parabolic energy contour within the space charge region in the semiconductor is explained in Section 2.1.1.

thermionic work functions of the two contacting materials, and (b) through the action of surface states. When they arise from surface states, barriers pre-exist at the semiconductor surface even before a contact is established. The presence of the metal then modifies that barrier (though often only slightly), and in that sense factors (a) and (b) may both come into play. However, it is convenient to discuss them separately.

1.3.1 *Barriers due to work-function differences*

Let us first consider the consequences which are expected to arise from differences of work functions, in the total absence of surface states. Figure 1.13(a) shows an energy profile for a metal and an n-type semiconductor not (yet) in contact with it, and hence not in equilibrium. When electronic equilibrium is established (e.g. by making contact between the metal and the semiconductor at some other surface, not shown in the diagram), the

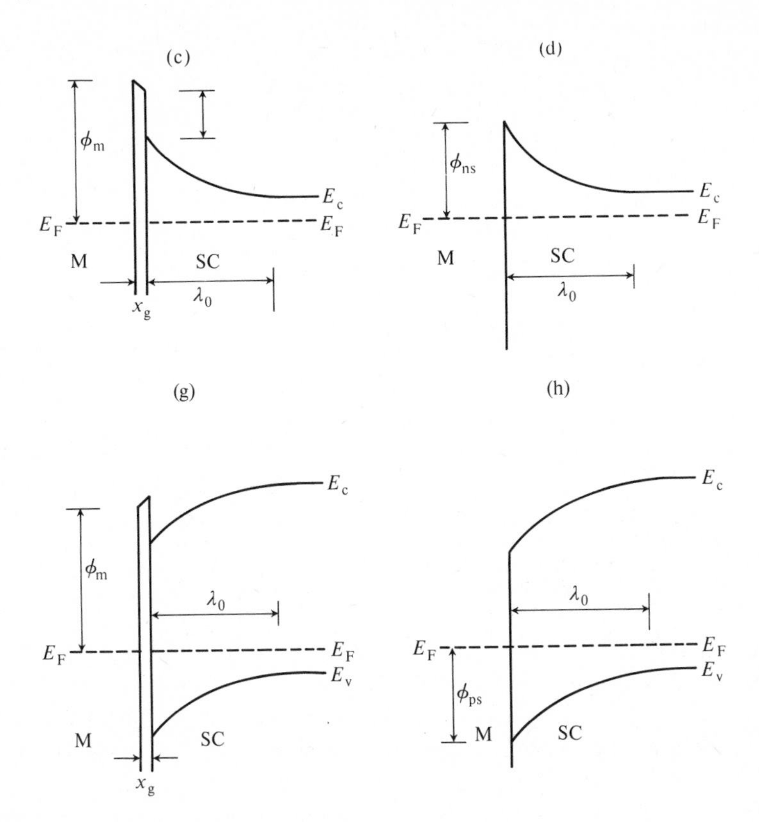

Fermi levels of the two materials must, on thermodynamic grounds, come into coincidence. The corresponding energy profile in the gap is then shown in Fig. 1.13(b). Relative to the Fermi level in the semiconductor, the Fermi level in the metal has risen by an amount equal to the difference between the two thermionic work functions. This potential difference, amounting to $\phi_m - (\chi + \phi_n)$, is called the *contact potential* (and should be clearly distinguished from the current-controlled contact and interface voltages defined in Section 1.1.1).

When the two surfaces (both regarded as plane and infinite) are allowed to approach, the capacitance of the system increases at constant voltage. Accordingly, an increasing negative charge is built up on the surface of the metal, and is distributed within the very small thickness of a metallic Debye layer. To that extent, the external field penetrates into the metal, but in most contexts this very small effect (not shown in Fig. 1.13) is totally unimportant. It does, however, make itself felt in experiments on photoemission, as demonstrated by Mead, Snow, and Deal (1966). The field penetration implies a curvature of the bands in the metal, and

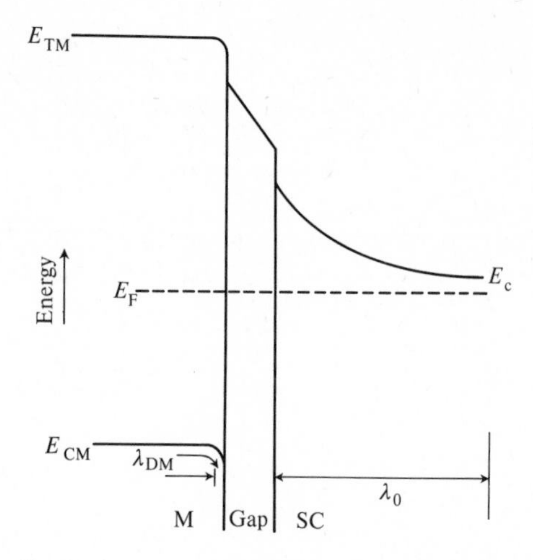

FIG. 1.14. Band bending by field penetration into the metal. The space charge extends over a Debye length λ_{DM} (ordinarily of the order of one or two atomic distances) and is negligible in many contexts. Contours here shown for zero applied voltage. E_{TM} and E_{CM} denote respectively the top and bottom of the conduction band in the metal. The dielectric constant of the semiconductor is ignored here.

its effect is to lower the barrier height by a very small amount. Figure 1.14 shows this, bearing in mind, however, that notions of band bending are very artificial when we are dealing with ultra-thin regions of the material. If we nevertheless accept them here, then the barrier lowering arises from the fact that there is a fixed energy interval between the top of the conduction band in the metal and the energy level associated with free space. The amount of barrier-lowering increases when the field is augmented by the application of external voltages (Section 2.3.5).

If there is a charge on the surface of the metal, then an equal and opposite charge must exist within the semiconductor. This is also distributed over a screening length (λ_0), but that region is thicker because the semiconductor is less conductive (Section 2.1.1). In that region most of the conduction electrons are missing, having been transferred to the metal. ϕ_{ns} is small, as long as the gap width x_g is large, and most of the potential difference then exists across the gap. As (in our hypothetical procedure) x_g diminishes, the total space charge in the semiconductor increases, and ϕ_{ns} and λ_0 thus increase likewise. The role of the semiconductor is now mixed: it acts partly as a dielectric and partly as a distributed electrode. As long as a gap exists, the electric field within it must be the same as the boundary field of the barrier, except for the adjustment arising from the difference of dielectric constants (ignored in Figs. 1.13 and 1.14).

As the gap width is (hypothetically) diminished, the smallest value of x_g

which can reasonably be envisaged is not zero, but a distance of the same order as the interatomic spacing (Fig. 1.13(c)). The thickness λ_0 thus reaches a limiting value as the two surfaces approach. Under such conditions, x_g is small enough to be transparent to electrons, and the potential barrier which the gap itself represents can thus be disregarded. The conditions which then prevail are illustrated by Fig. 1.13(d). It will be seen that the final barrier height (ignoring field penetration) will be given by

$$\phi_{ns} = \phi_m - \chi \tag{1.3.1}$$

where ϕ_m is the thermionic work function of the metal and χ the electron affinity of the semiconductor.

The particular model shown in Fig. 1.13 clearly depends on the assumptions made concerning the relative values of ϕ_m, ϕ_n, χ, and E_g, where ϕ_n is the Fermi energy below the conduction band, and E_g the band gap of the semiconductor. Of course, various other models could be analyzed in the same manner. However, inspection shows that none of these implies the existence of high barriers, a condition which will be shown to be essential for pronounced contact rectification. Correspondingly, energy profiles of the only case which implies high and thick p-type barriers are shown in Figs. 1–13(e), (f), (g), and (h). The polarity of the space charge is now reversed, and the final value of the barrier height (now measured downward, of course) is

$$\phi_{ps} = \chi + E_g - \phi_m. \tag{1.3.2}$$

The barrier height thus depends again on the thermionic work function of the metal, but now in the opposite sense to that at an n-type contact.

These relationships are simple enough but to test their validity in practice is actually a complex task because one would have to operate with totally clean surfaces, free even of adsorbed gas, and the choice of the contacting members would have to be restricted to materials which are chemically inert towards one another. A great deal of confusion was caused in the early years by failures to ensure this last requirement (e.g. see Schmidt 1941).

It will be clear that barriers that are found to exist at contacts between identical semiconductors cannot arise from the causes described above; they must be ascribed to the action of surface states alone. (See also Section 2.1.7.)

1.3.2 Barriers due to surface states

The familiar band picture that serves so well for the explanation of many bulk properties applies in principle only to an infinite solid or, in practice, to portions of a finite solid which are far from any boundary. For several reasons, the picture does not apply at the surface itself, at which even the lattice constant may differ from that characteristic of the bulk. In that sense, the band structure is expected to be *distorted* at (and within close range of) the surface. We must expect additional states to make their appearance there, and these are called *surface states*. In as much as they arise from the discontinuity of the lattice, their existence was first recognized by Tamm (1932), who analysed the wave equation for an idealized one-dimensional crystal in which the local potential at the surface was represented as a step function. The detailed results of such calculations depend actually on the precise manner in which the crystal is assumed to terminate (e.g. see Rijanow 1934, Maue 1935, and Goodwin 1939. In the course of more detailed analyses, Shockley (1939) and Statz (1950) were able to show how the surface states arise from the atomic levels of the constituent members. [Extensive bibliographies can be found in the books by Many *et al.* (1965) and by Frankl (1967), and also in an important status report by Davison and Levine (1970).] In practice, the surface state spectrum is, of course, influenced by absorbed matter, notably oxygen. However, for present purposes (only), the precise origin of the surface states is not important, only their existence, energy spectrum, and occupancy.

Let us then assume that surface states exist, whatever their origin, and that at least some of them are capable of acting as electron traps on n-type material. Even before any metallic contact is established, these states will capture electrons from the bulk, giving the (two-dimensional) surface a negative charge. Conversely, the adjoining border region of the semiconductor will become positively charged, thus forming a 'double layer'. The positive charge in the semiconductor is distributed over a small thickness of material which we call the *barrier thickness* (λ_0). Figure 1.15(a) shows these relationships. (Compare Fig. 1.13). The exact shape of the barrier, and a more precise definition of what is meant by its thickness will be discussed in Chapter 2. Here it will be sufficient to note that the charge density arises from ionized donor centres within the barrier which are no longer compensated by a corresponding electron density. The barrier thickness thus depends on the donor concentration, as well as the barrier height, and this, in turn, depends on the energetic position and number of surface states. A steady state is reached when the negative charge in surface states is equal to the positive charge in the barrier, both being determined by the same Fermi level. It will be shown in Chapter 2 that the positive charge within the barrier is roughly proportional to the barrier width which, in turn, increases as the square root of the barrier height. On the other hand, the greater the barrier height, the *smaller* is the amount of negative charge that can be accommodated in surface states below the Fermi level. This situation is

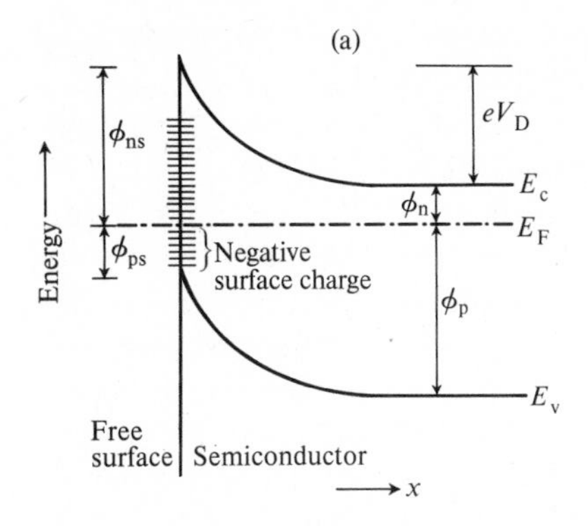

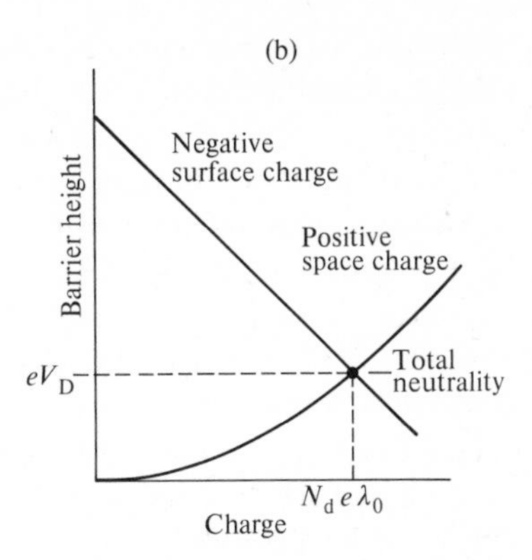

FIG. 1.15. Barrier on a free surface due to homogeneously distributed surface states; zero current. (a) Band picture. (b) Condition governing total neutrality and hence equilibrium. Barrier height 'seen' from the semiconductor: eV_D. E_c denotes the bottom of the conduction band of the semiconductor, E_v the top of the valence band (or, at any rate, the top of the first mostly occupied band).

schematically represented by Fig. 1.15(b). The barrier height will increase to the point where the two curves intersect. In general, this will leave some surface states unoccupied (above the Fermi level). The resulting barrier height will very much depend on how the occupied states are distributed over the energy spectrum. Figure 1.15(b) applies to a uniform distribution, an idealized situation, but if the levels were densely bunched in a small energy interval, the Fermi level would be bound to come to rest somewhere within that interval. The Fermi level is then said to be 'pinned', since its position cannot vary much. (Fig. 1.16). In the absence

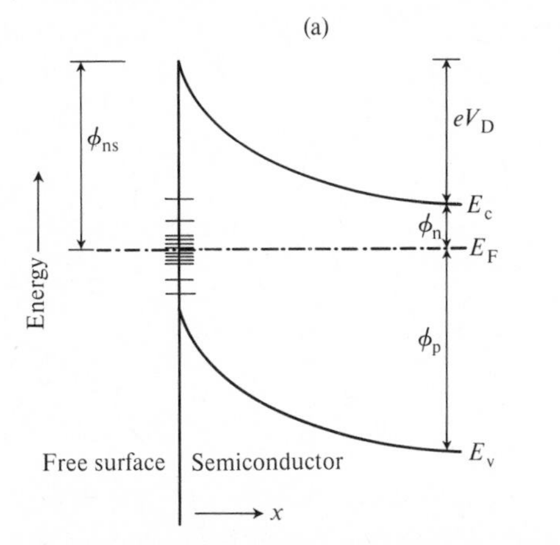

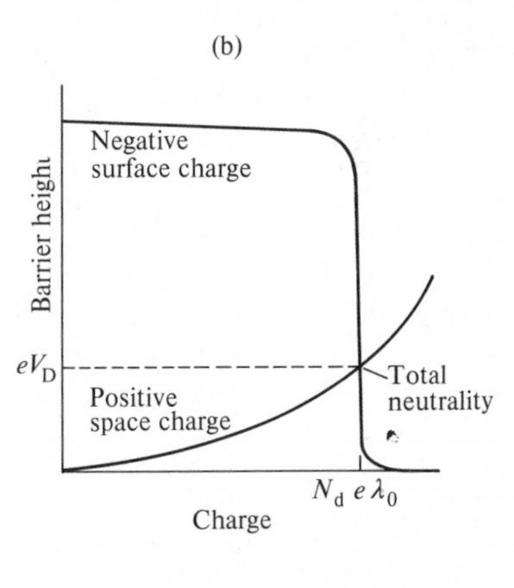

FIG. 1.16. Barrier on a free surface due to bunched surface states; zero current. (a) Band structure. (b) Conditions governing total neutrality and hence equilibrium. Barrier height 'seen' from the semiconductor: eV_D.

of additional complications, this must be true for p-type as well as n-type material. Pinned Fermi levels therefore imply $\phi_{ns}+\phi_{ps}=E_g$. In either case, the barrier height 'as seen from the semiconductor' is eV_D, where V_D is called the *diffusion potential.* (See also Section 1.1.2.)

We shall now examine what happens when a metal makes contact with such a system, and will assume (for simplicity only) that the surface state distribution is thereby unaffected (see, however, Section 2.1.6). The problem is one of interaction with the surface states in the first instance, and only secondarily one of interaction with the bulk semiconductor, since the latter comes into play only when the surface states fail to provide complete screening. Before contact, we will again have a difference in the position of the Fermi levels relative to free space, e.g. as shown in Fig. 1.17(a). When the contacting members are brought into equilibrium, the Fermi levels must come into coincidence, which involves a charge exchange, just as it did in the work-function case above. However, this exchange now takes place between the metal on the one hand and the surface states on the other, always assuming that the density of surface states is high enough to allow this. In the particular case envisaged, the metal would gain electrons from the surface states, thereby increasing the charge in *its* Debye layer which, as before, is exceedingly thin; we can again neglect it here. If the surface states were densely bunched somewhere in the energy spectrum, the barrier in the semiconductor would remain entirely unaffected by the diminishing gap x_g, irrespective of the thermionic work function of the metal. Figures 1.17(a)–(d) show what is then expected to happen (again neglecting

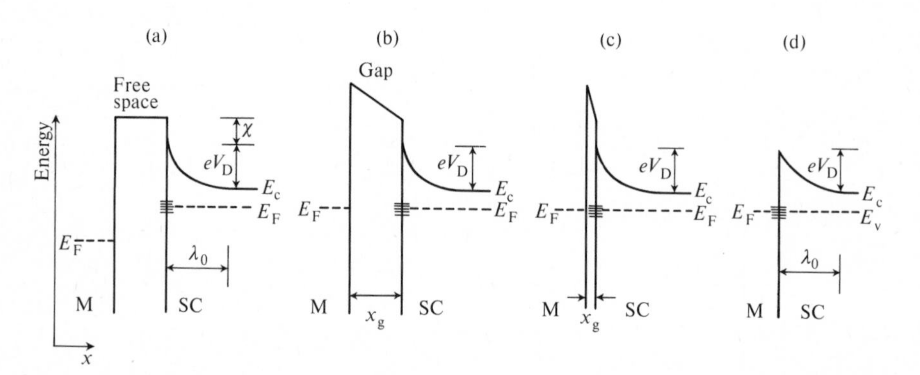

FIG. 1.17. Contact between a metal and a surface barrier controlled by a high concentration of tightly bunched surface states. (a) Non-equilibrium situation; (b) after charge exchange; (c) final situation. Note λ_0 and V_D unchanged after contact is established. Field differences due to the dielectric constant of the semiconductor are ignored in this representation, and the interface states which exist in (d) are for simplicity assumed to be identical with the surface states in (a), (b), and (c).

differences of field arising from the dielectric constant of the semiconductor). The fields in the gap and in the semiconductor are not the same now, even after allowing for the dielectric constant, because of the two-dimensional surface charge. If the surface states were tightly bunched somewhere near the middle of the forbidden band, then contacts on n-type and p-type material would be associated with symmetrically reversed barriers, similar to those in Fig. 1.13; otherwise not (but see below).

Note (by reference to Fig. 1.17) that λ_0 and V_D remain unchanged while the field in the gap increases during the approach. Of course, this can happen only when the surface states are capable of accommodating the entire charge exchange; in general, they are not. As x_g diminishes, the barrier within the semiconductor remains unchanged only up to a point (when the surface states are either completely full or completely empty, depending on circumstances); after that, all further adjustments involve the barrier and change it to an extent that must again depend on the thermionic work-function of the metal. This form of partial screening is the norm, rather than the exception.

Instances of Fermi-level pinning are well documented in the literature; see below. As a rule, experiments are performed by contacting n-type and p-type material of a given type, measuring the barrier heights by one of the methods available for this purpose, and testing whether $\phi_{ns}+\phi_{ps}$ do indeed add up to E_g. However, there is generally some doubt as to whether identical interface conditions have been achieved. It is therefore of interest to examine what happens at a given contact on initially n-type material when that material is converted (*in situ*) to p-type by stages of fast-neutron bombardment (and subsequent annealing). Such an experiment has been reported by Rahimi and Henisch (1981) and, while the barrier height on Ge was not actually measured, it was confirmed that the barrier remained essentially unchanged (because pinned) during the conversion. Taylor *et al.* (1952) reported that an n-type barrier is 'destroyed' by neutron bombardment, but this is true only in the sense that its resistance is lowered.

If Fermi-level pinning were indeed total and universal, the situation would be very simple, but there are in fact a number of complications. For instance, an excellent example of pinning is provided by the contrasting behavior of hafnium contacts on n-type and p-type silicon. On p-type material, the measured barrier height is $\phi_{ps}=0.9\,\text{eV}$. On n-type material, hafnium contacts have very low resistances, signifying a very low value of ϕ_{ns}, and suggesting strongly that the sum $\phi_{ns}+\phi_{ps}$ is indeed constant (Saxena, 1971). However, the very same experience indicates that the interface state distribution of the hafnium–silicon system is unusual, with tight-bunching closer to E_c than (as is much more common) to E_v. The fact is that the states which exist on a free surface cannot be expected to survive unchanged when a second phase is brought within an atomic distance of the first. We shall return to this matter in Sections 1.3.3 and 2.1.6. A further complication arises from the formation of third phases at the interface, sometimes with unexpected consequences. Thus, for instance, Card (1976b) has shown for Al contacts on n-type and p-type Si that $\phi_{ps} < \phi_{ns}$ before oxide formation, but $\phi_{ps} > \phi_{ns}$ afterwards, the $\phi_{ns} + \phi_{ps} = E_g$ relationship being miraculously maintained at all times.

1.3.3 Surface and interface state distributions

When the subject of surface states first came into practical prominence through the work of Bardeen (1947), the material under discussion was Ge, for which surface state densities within the forbidden band (i.e. between E_c and E_v) tend to be high. Until that time, all barriers were believed to be work-function determined, partly because nobody had suggested anything else, and partly because some of the early work on selenium rectifiers had given that indication (Schmidt, 1941). However, Bardeen was confronted by measurements of barrier heights on germanium, involving different contact metals, heights that varied very much less than expected on the basis of work-function arguments. Indeed, the barrier was almost independent of the contacting metal, implying screening by surface states. The same was found to be true for Si (Meyerhof 1947), and a great deal of work was subsequently done to confirm this conclusion, e.g. by Lepselter and Andrews (1969) and by Kircher (1971). (See also Andrews, 1974.)

The observations provoke three questions:

(a) Is the surface barrier always screened in this way?
(b) If the screening is less than perfect, what is the residual sensitivity of the barrier to the nature of the metal?
(c) How does the height of the (metal-insensitive) barrier depend on the exact composition of the semiconductor?

Under the last heading Mead and Spitzer (1964) reported observations on 14 different materials (Group IV and Groups III–V semiconductors) and showed that the Fermi level comes to rest at some distance from the valence band that is a constant fraction of the band gap. Thus, as the band gap increases (e.g. from InSb to BN in their series), barrier height also increases and does so linearly, irrespective of ϕ_m. Seven different metals were used as contacts, and the semiconductor band gaps varied widely. In all significant cases, the Fermi level ended up at an interval of about $E_g/3$ from the top of the valence band. Barrier heights varied from 0.17 eV (InSb) to 3.1 eV (BN), as determined by capacitance measurements (Section 4.1). [See also Mead and Spitzer 1963a,b, Spitzer and Mead 1963,1964, Mead 1965,1966a.)

During the early stages of this research, the belief gained ground that all semiconductors would eventually be found to exhibit these properties, but there was no real foundation for this view, and later work showed indeed that it is not so. Thus, Mead (1966b) and Kurtin and co-workers (1969) demonstrated that ϕ_m plays an increasing role as the *ionicity* of the semiconductor increases (see Fig. 2.9(a)). Whereas contacts on Ge and Si were relatively insensitive to the choice of metal, contacts on near-insulating ZnS, ZnO, Al_2O_3, etc, exhibited very considerable sensitivity. It was certainly tempting to interpret this as meaning that ionic solids have no surface states, but this was not actually a permissible conclusion. If Tamm states arise from the discontinuity of the lattice (as, by definition, they do), then *all* solids of finite size must have them, and this turns out to be correct. In a classic paper, Shockley (1939) showed that ionic solids do, indeed, have Tamm levels, but that their energetic position is very close to the band edges. In that position, they are harmless, since their occupancy is almost the same as that of the bands themselves. In contrast, materials which form their valence and conduction bands as a result of more substantial wave-function overlap (i.e. more covalent materials) support surface states which are situated closer to the middle of the forbidden gap, and are bunched there. Mead (1965, 1966a) has subjected this notion to an elegant experimental test, by measuring barrier heights on CdS crystals with varying amounts of Se substitution for sulfur. Selenium promotes wave-function overlap and, thereby, the formation of surface states far from the band edges. As expected, these states render the contact properties less sensitive to the nature of the contacting metal. To that extent, the picture is consistent, and we conclude, *inter alia*, that impurity content is one of the factors that can influence surface states in general. For the case of (111)-surfaces on diamond, a detailed calculation is available (Pugh, 1964) which confirms the $E_g/3$ position for the host material as such, irrespective of impurities. Pugh considered the distortion which a diamond lattice is expected to suffer in the immediate vicinity of a surface, and predicted a surface state distribution with a strong maximum just below the middle of the band gap. However, there are also materials (e.g. InAs, InP, GaSb, CdTe and CdSe) for which different (though equally constant) ratios prevail.

The correlation between surface states and ionicity arises naturally from the character of the bonding process, but there is no reason for expecting the ionicity to be the sole independent factor in this relationship. Indeed, Lindau and co-workers (1978) have established a very similar relationship by using the *heat of formation* of various semiconductor compounds as the plotting variable, a quantity which likewise measures the tightness of bonding and chemical stability. Undoubtedly, there is such a factor and we shall return to this matter also in Section 2.1.7. As far as experimental results are concerned, Mead (1966a) has given an extensive table of barrier heights, complete with primary source references.

There remains the fact, already touched upon in Section 1.3.2, that the very process of establishing a contact changes the surface (then 'interface') states in number and energetic distribution. Indeed, there is a viewpoint which maintains that the surface states on a free surface are only marginally relevant once a contact is in place. Heine (1965), Andrews and Lepselter (1970), and others have pointed out that there is an interpenetration of the electronic wave functions contributed by the contacting metal and, because these states are partly occupied, their presence sets up a thin region of negative space charge that decays rapidly (but not abruptly) into the semiconductor. Indeed, these wave-function tails *are* the interface states. (See also Parker *et al.* 1968.) The depth of penetration (a few atomic distances) determines the 'thickness' of the interface charge. One consequence of the non-zero thickness is to make the barrier contour rounded and continuous, rather than pointed and discontinuous (as in all the barrier representations discussed so far). Figure 1.18 shows this. The rounding implies a small diminution of barrier height, dependent, of course, on the density of penetrating states and their depth. Heine (1965) and Pellegrini (1974) have given distribution profiles and have shown that the density of such states is indeed expected to have a maximum (albeit not a sharp one) somewhere nearer the middle of the forbidden band than to either E_c or E_v. Heine also investigated the extent to which such states would become attenuated by oxide formation, and has shown that, despite thin oxide layers at the interface, the peaked distribution is essentially maintained. This explains why observed relationships between barrier height and the nature of the contacting metal

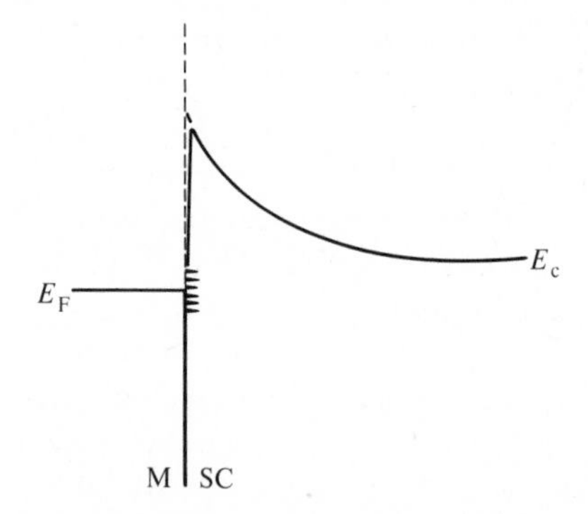

FIG. 1.18. Schematic representation of barrier rounding due to the penetrating tails of wave functions in the metal. A negative space charge resides in the thin region of semiconductor in which these states (here symbolized by thin triangular features) are attenuated. As always, notions of band bending on such a small scale must be handled with great caution, because the bands are not well defined over these intervals.

remain at least partly intact when interface oxide layers are formed. Pellegrini (1974) has shown that the density of penetrating states decreases as the thermionic work function of the metal increases. We thus have once again a relationship between ϕ_m and ϕ_{ns}, but via a mechanism very different from that discussed in Section 1.3.1, and not as strong. (See also Pellegrini 1974). Nevertheless, the barrier rounding which results from the penetrating tails will be ignored in the diagrams which follow. The effect is in any event overwhelmed by a more immediately important one due to image forces, as will be seen in Section 2.1.5.

Of course, all the present arguments assume that the contacting members are totally separated by the boundary, but it is now known that they actually interdiffuse to a significant extent. This complicates matters, and is an excellent example of the way even sophisticated models can come to be out of harmony with reality; neither a fundamental starting point, nor a rigorous development guarantee that a model reflects the actual situation in hand. The interdiffusion itself has been investigated by a variety of methods, including powerful synchrotron techniques (Chye *et al.* 1978; Lindau *et al.* 1978; Brillson *et al.* 1980a). The diffusion can be 'inert', but it often results in chemical reaction, forming reaction products (e.g. between semiconductor anions of III–V compounds and the contacting metal) from 3 to 20 Å thick (Brillson, Margaritondo, and Stoffel 1980).

The consequences tend to invalidate all simplistic interpretations of barrier heights and work-function differences. In the very least, such comparisons should distinguish between four important cases, as Andrews and Phillips (1975) have pointed out:

(a) Non-metal is a near-insulator and metal is physically adsorbed, corresponding to the case of the true work-function determined Schottky barrier. (Section 1.3.1.)
(b) Non-metal is covalent (e.g. like Ge and Si) and metal is physically adsorbed, corresponding to Bardeen's model of the work-function independent barrier, entirely controlled by surface states. (Section 1.3.2.)
(c) Non-metal is covalent and tends to form one or more chemical compounds with the contacting metal, corresponding, for instance, to the silicide contacts described in Section 2.1.6. (See also Brillson 1982.)
(d) Non-metal is covalent, and a thin oxide film tends to exist between it and the contacting metal, corresponding, for instance, to metal contacts on chemically etched silicon. (Section 1.3.4 and Section 2.4.)

Of course, there are other cases, but the above are frequently encountered in connection with technologically important systems and materials (see also Section 2.1.7).

Evidence of barriers due to surface states can also be obtained directly by measurements on clean contacts between identical, homogeneous semiconductors (Section 8.4). The first systematic experiments with this objective in mind appear to be those of Benzer (1947, 1949) on germanium. Since work-function differences cannot be involved (and assuming that foreign contaminants have been successfully avoided, e.g. by experimenting with freshly cleaved surfaces) surface states provide the only possible explanation for any barrier effects observed. Such barriers tend to be lower than those associated with metal contacts on the same semiconductor, suggesting that surface states may be suppressed by the symmetry of these systems. A residual density of interface states would remain due to local imperfections and lattice non-alignment.

1.3.4 Barriers due to third phases

Barriers of the kind discussed in Sections 1.3.1 and 1.3.2 are, of course, abstractions. Mostly they presuppose a contact between two perfectly clean materials or, at any rate, contacts with no more than a monatomic gas layer at the interface. In fact, as already mentioned, metals and semiconductor can interact with one another to form third phases, which may be insulators (in bulk), or else semiconductors, as suggested under (c) above. Even when this does not happen, both materials may in practice be associated with oxide layers. Indeed, in one form or another, the metal–oxide–semiconductor (MOS) system is the most generally encountered configuration, whereby the interface layers may be crystalline, polycrystalline, or amorphous. Quite generally, we must expect 'real-life contacts', as distinct from contacts specifically made for research purposes, to be associated with third phases whose exact composition, structure, and bulk properties are ill-defined and, in practice, often unknown. On some occasions, interface matter is deliberately created to augment or otherwise improve specific aspects of contact behaviour. Such layers tend to be materials of wide band gap, but specific band structures are rarely considered. Indeed, these barrier regions are often too thin and too disordered to allow standard band considerations to be applied. Most of the models found in the literature therefore deal with the problem on a schematic basis, taking their cue from an early and much discussed analysis by Dilworth (1948). Accordingly, we have configurations as shown in Fig. 1.19, here drawn on the assumption that space-charge effects within the interface layer are absent or negligible and $\epsilon = \epsilon_i$. (See however, Sections 2.3.8 and 3.5.5). The missing reference to a valence band may often be justified, but the neglect of trapping states is inherently implausible; few models include them. Third-phase interface

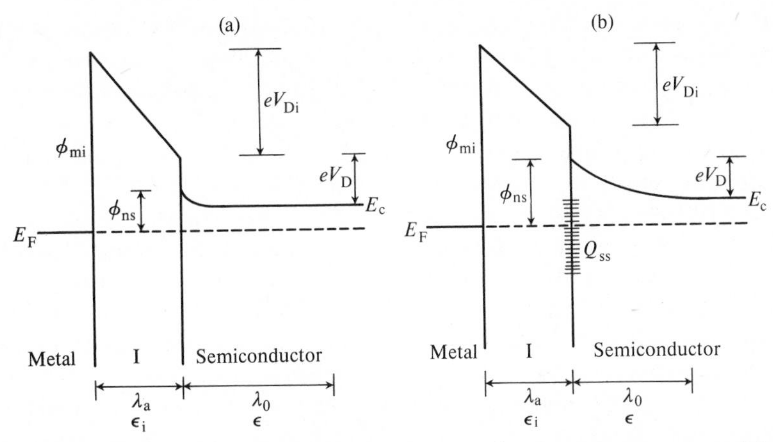

FIG. 1.19. Artificial barrier system; profiles in the quiescent state. ϵ_i, ϵ = permittivities ($\epsilon_i = \epsilon$ assumed here). (a) Without surface states. (b), (c) With surface states containing charge Q_{ss}; ϕ_m kept constant for comparison. The small amount of barrier-lowering due to the non-zero thickness of the interface charge (Section 1.3.3) is neglected here and, indeed, in all other barrier representations that follow.

layers are sometimes called 'artificial' barriers, to distinguish them from barriers which arise naturally at contacts between totally clean and inert materials.

The field in the insulating layer arises, of course, from the difference of thermionic work functions between metal and semiconductor. In the ordinary way, the semiconductor has a much higher conductivity and (in the absence of interface states) its share of the voltage drop is therefore relatively small (Fig. 1.19(a)). Interface states trap negative charge which, other things being equal, leads to a higher barrier within the semiconductor and, correspondingly, to less voltage drop across the insulating layer (Fig. 1.19(b)). How much less depends, of course, on the density of interface states, on their distribution, on λ_a, and on the donor concentration N_d.

Figure 1.19 is based on the conventional assumption that the thermionic work function of the metal is larger, but the opposite situation is conceivable. Thus, in unlikely but not impossible circumstances, the standing fields in the semiconductor and insulator might be of opposite sign when interface states are present in sufficient concentration. Figure 1.19(a) also assumes that $Q_{ss} = 0$, which makes the boundary fields in the insulator and semiconductor equal. This is, of course, a highly improbable state of affairs. In general terms, $\epsilon < \epsilon_i$ leads to lower, and $\epsilon > \epsilon_i$ to higher fields (and barriers) in the semiconductor. (See Section 2.3.8.)

The fields prevailing in oxide barriers tend to be high and, under such conditions, the structure and properties of the layer can be modified by ionic drift of metal and semiconductor constituents ('dry electrolysis'). Even in the absence of fields, ionic and atomic diffusion can play such a role, and it is possible to determine the resulting compositional profiles by sputter profiling. Migration effects can be promoted or retarded by externally superimposed fields, but tend to be only partially reversible, and their influence on barrier characteristics is highly complex. Of course, dry electrolysis can also occur within the Schottky barrier itself, but modern device technologies have succeeded in reducing these effects to a minimum. In selenium rectifiers of old, they used to be one of the major sources of instability and concern (Henisch 1957).

In principle, artificial barrier layers involve *two* contacts, of course, one between the metal and the interface material, and one between that material and the semiconductor. Each may be expected to make its distinctive contribution to the system as a whole. Conventional treatments which regard such a layer as a single non-complex entity are all based on the simplifications which are believed to follow from the small thickness of the third phase. The notion is that neither contact can develop its full characteristic properties because the two contact systems interact within the thickness λ_a, and this is bound to be true. Whether it justifies extreme simplification is another matter.

Though many aspects of this field are not yet clearly understood, the importance of artificial barrier layers has been recognized for a long time. In the course of early and, as yet, somewhat 'informal' experiments, de Boer and van Geel (1935), Hartmann (1936), and Klarmann (1939), for instance, showed that intermediate layers of such utterly diverse materials as shellac, mica, sulfur, potassium fluoride, and quartz can augment rectification. A great deal of early work was also done on the artificial barriers associated with selenium rectifiers (e.g. see Clark and Roach (1941), Sato and Kaneko (1950), Nagata and Agata (1951), and Yamaguchi and co-workers (1950, 1951)). Though the detailed context is now different, problems related to third phases, and especially to oxide films, continue to be of great importance in modern integrated circuit technology.

Currents through 'insulating' barriers can be carried in two ways, by electron (or hole) activation over the barrier or by wave-mechanical tunnelling. In most (but not all) contexts, the latter is the predominating mechanism. (See also Sections 2.1.5, 2.2.8, 2.3.3, and 6.5.3.) Attempts to calculate tunnelling characteristics date from the earliest days of semiconductor theory (Sommerfeld and Bethe 1933; Holm and Kirchstein 1935), and received additional impetus in the context of metal-metal contacts with (unwanted) intermediate layers (Holm 1951). For a general overview of basic problems and some of the more likely complications, see Simmons (1963).

Further reading

On the measurement of interface state density: Borrego, Gutman, and Ashok (1977).

On barrier heights at contacts:

with InP, GaAs and Si: Grinolds and Robinson (1980); Hökelek and Robinson (1981).

with Si: Barret and Vapaille (1976); Card and Singer (1975).

On the effect of oxide thickness variations: Barret, *et al.* (1975).

2
SINGLE CARRIER SYSTEMS

CONTACTS on semiconductors can be classified into three generically different groups:

(a) *Blocking contacts*, interpreted here as contacts which permit *no* current to pass.
(b) *Semi-permeable contacts*, which do allow current to pass in varying degree, but constitute a localized source of high resistance.
(c) *Low-resistance contacts*, often (but misguidedly) referred to as 'ohmic', which represent a negligible impediment to carrier flow, in comparison with other impediments present. Whether a low resistance contact does indeed obey Ohm's Law, matters hardly ever. More precisely, it matters only in the context of high-current pulse experiments.

The present chapter, and Chapter 3, will concern themselves only with contacts of category (b). Contacts of type (c) receive consideration in Chapter 6. Contacts of type (a) will not be discussed in any detail, but passing references to them will be found in Sections 2.2.9, 2.3.2, and 4.1.3. We can think of such contacts for present purposes as having artificial barrier layers of such height or thickness (or both) as to prevent all charge carrier transitions. See also Section 6.1.

Contacts in general need not, of course, be associated with barriers at all (see reference to *accumulation layers* in Section 2.1.2), but they usually are, and barriers do indeed represent configurations which are technologically more important than accumulation systems; hence the extensive treatment which they receive here.

Many of the considerations which follow are concerned with space charges. In general, the charge density ρ in a trap-free n-type semiconductor is given by

$$\begin{aligned}\rho/e &= (N_d - n_d) + p - n \\ &= (N_d - n_d) + (p_e + \Delta p) - (n_e + \Delta n) \qquad (2.0.1)\end{aligned}$$

where N_d is the concentration of donor centres and n_d the concentrations of electrons remaining at these centres. $(N_d - n_d)$ thus represents the number of ionized donors per unit volume. The two other terms refer to the concentrations of free holes and free electrons respectively. In homogeneous material, N_d, p_e, and n_e are, of course, constants; the other quantities may be functions of x. The systems discussed below are all assumed to be non-degenerate in bulk. For a relaxation of this simplifying assumption, see Section 6.5.

In this chapter we shall consider systems which are entirely controlled by majority carrier transport; minority carriers (i.e. holes in the n-type material under consideration) are totally ignored ($p = p_e = 0$). This is a realistic model for barriers on wide band-gap materials with a sufficient number of donors to deserve the designation 'n-type', and it will also serve as an excellent introduction to the more complex systems in which minority as well as majority carriers play a role. To be sure, the single carrier system is theoretically a special case of the more general situation, but that situation has too many parameters to serve as a convenient beginning. It is therefore desirable to simplify first and to generalize later. All the systems considered will be deemed to be one-dimensional, unless otherwise stated.

A variety of voltage–current relationships will be developed, each appropriate to its own set of initial assumptions. This is the textbook approach. However, when we are faced with the practical need to interpret experimental results, the problem is reversed: we confront a set of measurements and would like to know of what model they are a logical consequence. As a rule, we do not know on *a priori* grounds what assumptions to make and, indeed, look to the measurements themselves for clues and guidance. In the circumstances, the chance of pursuing circular arguments is greater than we tend to think, but it can be minimized by adhering to some important procedural priorities:

(a) What is the mechanism of current flow? Does it call for a single carrier or a double carrier model?
(b) Which model, in the appropriate category, gives absolute magnitudes that compare plausibly with experimental results? Only if such an order-of-magnitude can be reached is the pursuit of a more detailed analysis worthwhile.
(c) Which model gives the best agreement as regards shape of the voltage–current relationship?
(d) Can any structural feature be inferred from the results and confirmed by independent means?

All too often, one finds in the literature an exclusive concentration on step (c) which, by itself, is not capable of yielding conclusions with any degree of reliability.

2.1 Static barriers

2.1.1 The classic Schottky barrier

We are in the first instance, concerned here with the current-less case, which came to be analysed long before the days of semiconductor interest, in the context of electrolysis. Extensive discussions of such pioneering work have been provided in papers by Macdonald (1954, 1958, 1959), where further references may be found.

In the ordinary way, the barriers here discussed are, of course, much too thin to lend themselves to direct optical observations, but there is just one case in which their existence can be *visually* demonstrated. This arises when the semiconductor is ferroelectric, as described by Wemple (1969), and by Kurtz and Warter (1966). When the contact is illuminated with monochromatic laser light, the resulting fringe distribution provides evidence of the way in which the local birefringence varies and that, in turn, is proportional to the square of the polarization. The existence of a linear field distribution was ascertained in this way, in accurate confirmation of the Schottky model described below.

Once we are convinced that a metal–semiconductor contact is the seat of a potential barrier, we need to know something about the relationship between barrier shape (height, thickness, contour) and various semiconductor parameters (concentration of active centres and their activation energies). In their idealized form (but not necessarily in practice) these relationships are very simple. They are in any event 'contained' in the solution of Poisson's equation under suitable boundary conditions (Schottky 1942). For this purpose, the space charge density is ordinarily considered as *continuous*, despite the fact that it is in fact composed of integral charges in fixed positions, namely at the donors (in n-type material), or at acceptors (in p-type material). This amounts to the assumption that the distance between such centres is much smaller (preferably orders-of-magnitude smaller) than the barrier thickness. In practice, this is hardly ever true, but this simplifying assumption still manages to yield acceptable solutions in many cases. The matter receives more detailed discussion in Section 2.1.3.

For n-type material of permittivity ϵ, we thus have

$$\left.\begin{aligned} \mathrm{d}^2V/\mathrm{d}x^2 = -\rho/\epsilon, \qquad E = -eV, \\ \frac{\mathrm{d}^2E}{\mathrm{d}x^2} = \frac{e\rho}{\epsilon} \end{aligned}\right\} \tag{2.1.1}$$

where E is the electron energy. (Many treatments use E to denote the electric field, and a curly form of E (often difficult to distinguish from ϵ) to denote energy. Each convention has its devotees.) The charge density ρ has already been defined above. For present purposes, it is simplest to consider the donors to be completely ionized ($n_\mathrm{d} = 0$), which also means $N_\mathrm{d} = n_\mathrm{e}$ in the bulk. The 'classic' Schottky barrier is one in which not only minority carriers, but also majority carriers are neglected. This is equivalent to putting $\rho = eN_\mathrm{d}$ within the barrier, and $\rho = 0$ outside it, and thus involves a space-charge discontinuity which is wholly artificial. However, in many situations which are of practical interest, the consequences of this approximation are not actually drastic, which is why they have acquired the status of a classic convention. The assumptions will be relaxed in Section 2.1.2.

Meanwhile, we have

$$\frac{d^2E}{dx^2}=\frac{N_d e^2}{\epsilon} \tag{2.1.2a}$$

and, after one stage of integration,

$$dE/dx=(N_d e^2/\epsilon)(x-\lambda_0) \tag{2.1.2b}$$

where λ_0 is the barrier width, *defined* as the value of x for which the field is zero. The implied fields $F=(dE/dx)/e$ elsewhere (of the order of 10^5 volts cm^{-1} at $x=0$), may be sufficiently high to cause drift velocity saturation, an effect ignored here, but discussed in Section 2.2.7.

Further integration of eqn (2.1.2) yields

$$E(x)=(N_d e^2/\epsilon)(\tfrac{1}{2}x^2-\lambda_0 x)+\text{constant}. \tag{2.1.3}$$

When $x=\lambda_0$, then $E(x)=E(\lambda_0)=\phi_n$ where ϕ_n is the energy interval between the bottom of the conduction band and the Fermi level. ϕ_n, and the corresponding ϕ_p, are by definition positive quantities. The Fermi level will be regarded as the reference level for all other energies: $E_F=0$, as shown in Fig. 2.1(a). The integration constant is thus given by

$$\phi_n+N_d e^2\lambda_0^2/2\epsilon=\text{constant},$$

and, by substitution into eqn (2.1.3), we have

$$\begin{aligned} E(x)-\phi_n&=(N_d e^2/\epsilon)(\tfrac{1}{2}x^2-\lambda_0 x+\tfrac{1}{2}\lambda_0^2) \\ &=\tfrac{1}{2}(N_d e^2/\epsilon)(x-\lambda_0)^2. \end{aligned} \tag{2.1.4}$$

When $x=0$, we have

$$E(0)-\phi_n=\phi_{ns}-\phi_n=N_d e^2\lambda_0^2/2\epsilon \tag{2.1.5}$$

and this is the barrier height as 'seen' from the semiconductor side. We often write $\phi_{ns}-\phi_n=eV_D$, giving

$$V_D=N_d e\lambda_0^2/2\epsilon \tag{2.1.6}$$

where V_D is called the *diffusion potential* (and is always positive). It represents a built-in potential difference in the absence of current, and this has sometimes been a point of difficulty. However, it need not be so: the potential difference is actually *essential* for maintaining the zero current in the presence of the prevailing carrier concentration gradient.

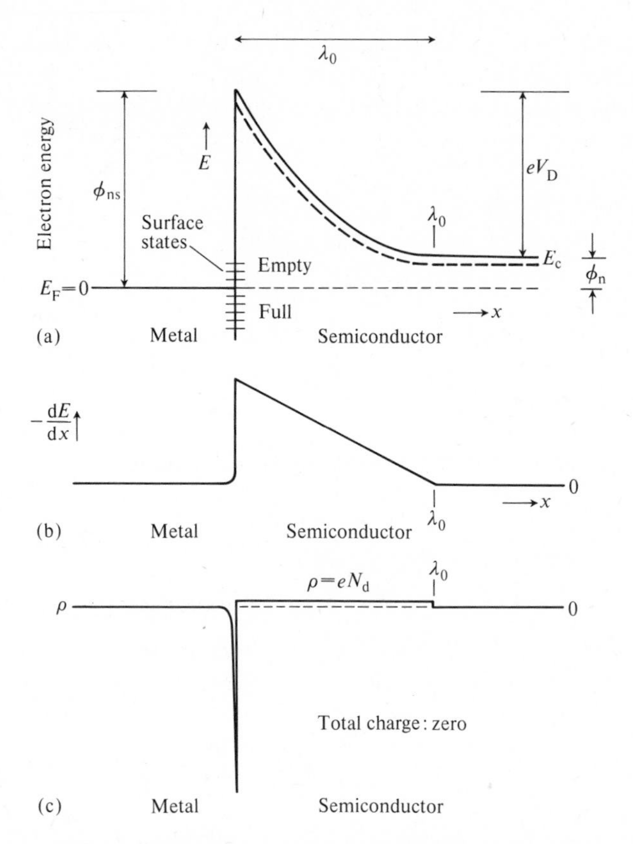

FIG. 2.1. Classical Schottky barrier on n-type material; quiescent condition (zero current). (a) Electron energy versus distance profile through the metal-semiconductor interface. (b) Field versus distance profile (idealized, in as much as $F=0$ at a definite point given by $x=\lambda_0$). (c) Space-charge density profile. The surface charge on the metal is numerically equal, but spatially much less extended than the charge of opposite polarity on the semiconductor.

Thus, a field current and a diffusion current (both x-dependent) are exactly balanced at every point within the barrier. Because the diffusion potential is part of the equilibrium situation, it cannot be measured by external means, i.e. by applying a voltmeter to the system. A voltmeter measures only the difference between two Fermi levels and no such difference is involved here. Spenke (1955) has discussed an interesting geophysical analogy, namely that provided by the earth's atmosphere. There is a pressure gradient with altitude, but no vertical wind, because that pressure difference is exactly compensated by gravitational forces. Pressure gradient *and* gravitational forces maintain the whole system in equilibrium.

Equation (2.1.4) gives the shape of the barrier (parabolic), and eqn (2.1.6) shows how barrier height and barrier thickness are related to one another. A convenient way of expressing this relationship makes use of the fact that

$$(kT\epsilon/N_d e^2)^{1/2} = \lambda_D \qquad (=\text{here } \lambda_{D_n}), \tag{2.1.7}$$

which is the *Debye length* under the (single-carrier, total ionization, $N_d = n_e$) conditions assumed here. [Note that, by analogy with the diffusion length $L_n = (D_n \tau_n)^{1/2}$, where τ_n is the carrier lifetime, the Debye length λ_{Dn} of electrons can alternatively be expressed as $\lambda_{Dn} = (D_n \tau_{Dn})^{1/2}$, where τ_{Dn} is the dielectric relaxation time $\rho\epsilon$, since $(D_n \rho\epsilon)^{1/2} = (D_n \epsilon / N_d e \mu_n)^{1/2} = (kT\epsilon/N_d e^2)^{1/2}$.] This makes it possible to write eqn (2.1.6) in the form

$$\left(\frac{eV_D}{kT}\right) = \frac{1}{2}\left(\frac{\lambda_0}{\lambda_D}\right)^2, \tag{2.1.8a}$$

and eqn (2.1.4) in the form

$$\{E(x) - \phi_n\}/kT = \tfrac{1}{2}(x - \lambda_0)^2/\lambda_D^2 \tag{2.1.8b}$$

which gives the barrier energy in units of kT, referred to the bottom of the conduction band in the bulk material.

The last equation can also be written in the form

$$\left\{\frac{E(x) - \phi_n}{kT}\right\} = \left(\frac{\phi_{ns} - \phi_n}{kT}\right)\left(\frac{x}{\lambda_0} - 1\right)^2 \tag{2.1.8c}$$

which is sometimes convenient. In Fig. 2.2, values of eV_D/kT are plotted against the quiescent barrier width λ_0, normalized to the Debye length (eqn 2.1.8a) for use in conjunction with Fig. 2.3.

For reference purposes, and because a knowledge of λ_D is useful in many other contexts, Fig. 2.3 relates the room-temperature value of λ_D to N_d and the dielectric constant κ. These relationships apply as long as the carrier concentration in the bulk is sufficiently low to make the free-carrier gas non-degenerate. For higher concentrations, an expression based on Fermi–Dirac statistics would have to be used. Moreover, the screening lengths then become so small as to make the 'collective' treatment of the solid (as exemplified by the use of a dielectric constant) inappropriate. In such circumstances, the real screening lengths are up to one order-of-magnitude larger than eqn (2.1.7) would suggest. The above formulation assumes complete donor ionization, but the situation is not radically different when that ionization is incomplete. The free-carrier concentration in the bulk is then given by $n_e = N_d - n_d$, and that quantity, rather then N_d, then governs the Debye length. Of course, n_d must vary throughout the barrier, though it eventually becomes a constant in the bulk, governed by the statistical occupancy of the donor level.

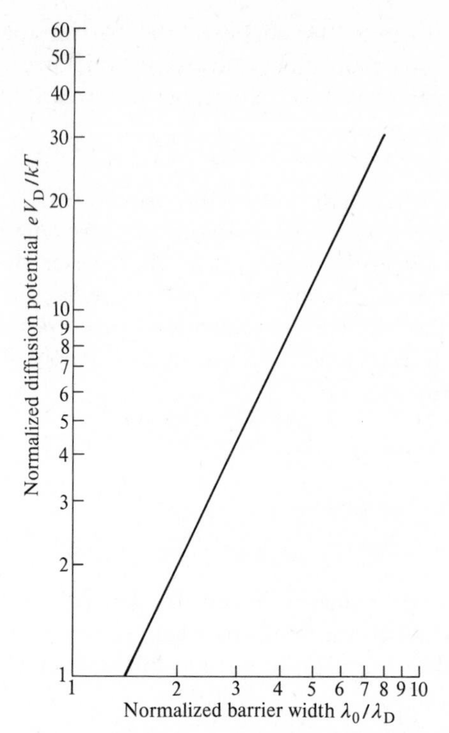

FIG. 2.2. Barrier height (normalized to kT), as seen from the semiconductor, versus barrier thickness (normalized to the Debye length); idealized Schottky system. (See also Fig. 2.3.)

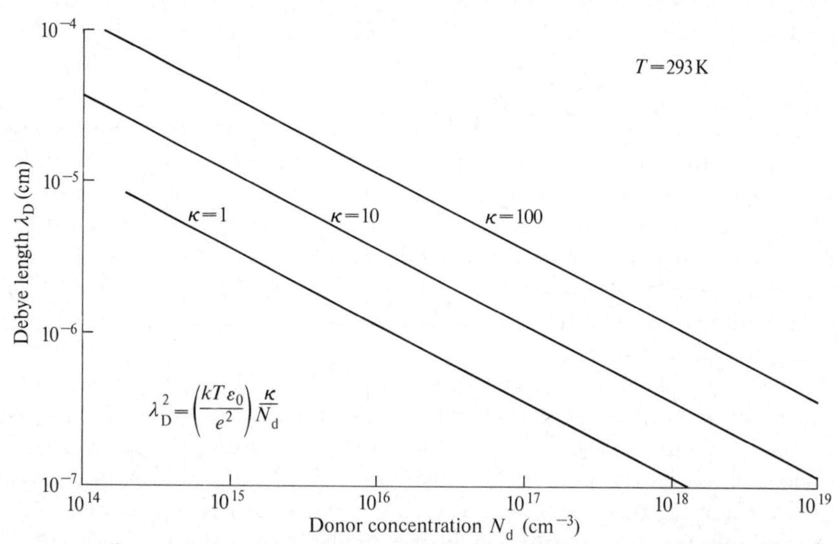

FIG. 2.3. Debye length in a non-degenerate semiconductor versus donor concentration, for three values of the dielectric constant. All donors assumed to be ionized ($N_d = n_e$).

As one would expect, the conductive barrier properties which are of interest here are more sensitive to changes of V_D than they are to changes of λ_0. This means that a great deal depends in practice on the lateral uniformity of the barrier height (Section 2.3.6), which in turn may be entirely controlled by surface contamination.

For a metal contact on p-type material, *other things being equal*, the barrier profile would be a mirror image of that shown in Fig. 2.1(a), with the barrier height (as seen from the metal side) then called ϕ_{ps}, and a bulk parameter ϕ_p in place of ϕ_n, such that $\phi_p + \phi_n = E_g$. This dualism between contacts on n-type and p-type material appertains throughout the present treatment but, for the sake of brevity, (and because they are somewhat easier to visualize) only n-type contacts are explicitly discussed.

The barrier width λ_0 has quite obviously the character of a screening length, and it is therefore important to discuss how it *differs* from the Debye length λ_D, since λ_D is also a screening parameter. As such, it governs the decay of electric fields in a screening region of carriers which are mobile and free. It does so in accordance with

$$F = F_0 \exp(-x/\lambda_D). \tag{2.1.9}$$

Within the decay region, the local space charge density is variable. In contrast, the Schottky barrier field decays linearly with distance (eqn (2.1.2b)) in a manner governed by the fixed space charge in ionized donor centers. Thus, under the present set of assumptions, ρ cannot exceed eN_d anywhere. That is also why λ_0 is larger than λ_D, under the conditions assumed; but there are situations in which the relationship is reversed. This can happen in amorphous semiconductors containing a high concentration of states within the forbidden gap. Those states hold very substantial numbers of carriers (and their space charge) which are not available for conduction. As a result, the true screening length may be orders-of-magnitude *smaller* than λ_D. The matter will receive more detailed consideration in Chapter 5.

In all the above considerations, N_d is considered constant. If it were not a constant, but were still a continuous function of x, then eqn (2.1.3) would have to be replaced by

$$E(x) - \phi_n = \frac{e^2}{\epsilon} \int_{\lambda_0}^{x} \int_{\lambda_0}^{x} (N_d)_x \, dx \, dx \tag{2.1.10}$$

(see Lubberts and Burkey 1975).

The contacts here considered are planar, and it will be clear on *a priori* grounds that contact *curvature*, as envisaged in Section 1.1.3, cannot affect the barrier in any material way, unless the radius of curvature is small enough to be comparable with the barrier width. This can apply only to the very thickest (highest) of barriers and, as far as is known, effects arising from interface curvature have never been experimentally confirmed. In principle, of course, there is no difficulty. The equivalent of eqn (2.1.2) in spherical coordinates is

$$\frac{1}{r^2}\cdot\frac{\mathrm{d}}{\mathrm{d}r}\left(r^2\frac{\mathrm{d}E}{\mathrm{d}r}\right)=\frac{N_\mathrm{d}e^2}{\epsilon} \tag{2.1.11}$$

and the integration of this equation yields a barrier thickness λ_{s0} which compares with the planar value λ_0 as follows:

$$\lambda_0=\lambda_{s0}\left(1+\frac{2}{3}\frac{\lambda_{s0}}{r_0}\right)^{1/2}\approx\lambda_{s0}(1+\lambda_{s0}/3r_0) \tag{2.1.12}$$

(Henisch 1957). Corresponding to eqn (2.1.6), we now have

$$V_\mathrm{D}=\left(\frac{N_\mathrm{d}e\lambda_{s0}^2}{2\epsilon}\right)\left(1+\frac{2\lambda_{s0}}{r_0}\right) \tag{2.1.13}$$

which likewise shows that $\lambda_{s0}<\lambda_0$.

Barrier heights eV_D are *temperature dependent* for two reasons: because the Fermi level changes relative to the band structure, and because the band gap itself is temperature dependent. The relationships which control ϕ_n and ϕ_p are well known (arising, as they do, from specimen neutrality). The behaviour of ϕ_ns is generally more complex. In a system governed entirely by thermionic work function differences, or else by densely bunched surface states, ϕ_ns should be virtually constant. In other circumstances, it may vary with temperature in a manner that depends, *inter alia*, on the surface state spectrum.

The existence of a built-in diffusion potential is at the very core of all explanations of current rectification, because external voltages can be superimposed in two directions which are not equivalent to one another. In one direction they augment V_D, in another they diminish it. (See Section 2.2.1.) Any diffusion potential can serve in this capacity, even though it may arise in ways very different from that described above, e.g. in association with a bulk impurity gradient or a graded heterojunction, as described by Allyn *et al.* (1980).

We have dealt here with isolated barriers, but there are practical situations in which surface barriers overlap, e.g. on semiconducting layers of thickness less than $2\lambda_0$. The consequences of such overlap will be further described in Sections 2.2.9 and 2.4.

2.1.2 *The free-carrier contribution; accumulation layers*

Equation (2.1.2) neglects the electronic contribution to the space charge, and this is clearly permissible for barriers which are very high: $eV_\mathrm{D}\gg kT$, since the carrier concentration diminishes exponentially with increasing

E. However, even then the conventional approximation introduces some difficulties, as will be seen in the course of considerations which involve current flow (Section 2.4). For low barriers, i.e. barriers for which eV_D/kT is between zero and (say) 4–5, the neglect of the free carriers introduces a significant error. When the electronic contribution is taken into account, the appropriate form of Poisson's equation is

$$\frac{d^2E}{dx^2} = \frac{e^2}{\epsilon}\{N_d - n(x)\} \tag{2.1.14}$$

where, in the quiescent case,

$$n(x) = n_e \exp\left\{-\frac{(E-\phi_n)}{kT}\right\}$$

as long as the electron gas is non-degenerate. In this form the equation cannot be fully integrated by analytic methods, though approximations can be devised (Goldsmid 1968). In any event, a step-by-step integration can always be achieved numerically, easily implemented on a programmable calculator. In pre-computer days, the process was far more tedious (Davidov 1939). (Numerical solutions of the transport equations receive detailed discussion in Section 2.4 and Chapter 3.)

An alternative approach may be patterned after a suggestion by Jonscher (1968). That makes use of the fact that the exponential part of eqn (2.1.14) can be expanded when the space-charge density is very small. With $N_d = n_e$ and higher terms neglected, this yields

$$\frac{d^2E}{dx^2} = \frac{N_d e^2}{\epsilon}\left[\frac{E(x)-\phi_n}{kT}\right] \tag{2.1.15}$$

and hence

$$E(x) - \phi_n = C\exp(-x/x_0) \tag{2.1.16}$$

where x_0 is easily ascertained by differentiating, as

$$x_0 = (\epsilon kT/N_d e^2)^{1/2} = \lambda_D, \tag{2.1.17}$$

the Debye length, and this is what one would expect. The most important conclusion is that the tail is exponential. Since C is arbitrary, it can be used as a fitting constant. One could, for instance, accept eqn (2.1.4) as valid up to some point $x = x_f$, and then allow the exponential decay to take over. At the fitting point, eqns (2.1.4) and (2.1.16) would have to coincide as regards values, as well as the first and second derivatives, i.e. the field and its curvature, must be continuous. Moreau, Manifacier, and Henisch (1982) have shown that the point which satisfies this condition is given by

$$\frac{x_f}{\lambda_D} = \left(\frac{2eV_D}{kT} - 1\right)^{1/2} - 1 \tag{2.1.18}$$

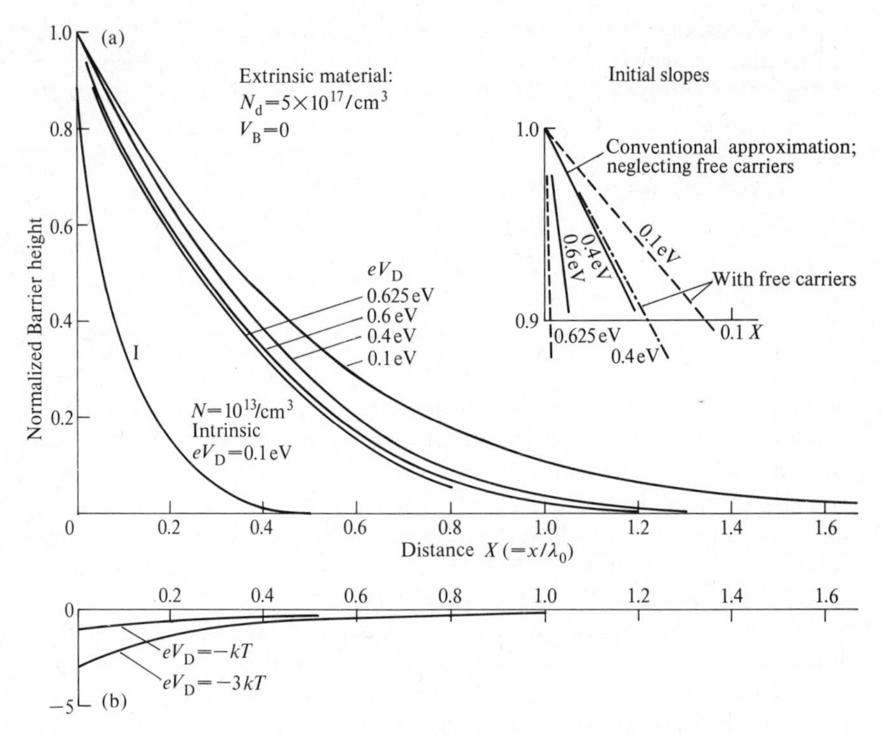

FIG. 2.4. Effect of free carriers (a) on the shape of normal exhaustion barriers and (b) accumulation layers. Wide band gap and non-degeneracy assumed. Curve I in (a) refers to a barrier ($eV_D=4kT$ at room temperature) on intrinsic germanium. Note high initial field due to free holes of x-dependent concentration. Other curves show barrier contours for various diffusion potentials; insert: initial slopes. (See also Section 3.1.) Zero current.

Typical results of numerical calculations (in the absence of current) are shown in Fig. 2.4(a) for barriers of several different heights. The results are here normalized to facilitate comparison. Thus, all ordinates express fractions of the total barrier height attained at various distances from the interface, the distances being themselves expressed as fractions of the classic Schottky barrier width λ_0. The following observations may be made:

(a) High barriers (e.g. $eV_D \approx 0.6$ eV) give rise to almost parabolic contours, just as they do when free carriers are neglected. However, now that free carriers are taken into account, the barriers no longer 'end' at $\mathbf{X}=1$. They continue further, but the overall shape remains almost 'classical'.
(b) As barrier heights diminish, the contribution of the free electrons makes itself increasingly felt. Thus, for $eV_D=0.1$ eV, the shape differs very substantially from that of an ideal Schottky barrier, and the point

$x = \lambda$, ($X = 1$ on the diagram), no longer denotes a 'barrier width'.

From a capacitive point of view (Chapter 4), the barrier functions as a dielectric; however, as we have just seen, the conductivity is not actually zero in any particular interval $0 < x < \lambda_0$. An improved assessment comes from regarding the barrier as non-conductive only where $E(x) - \phi_n > kT$, to allow for the effect of free carriers. In that sense, the quiescent barrier, considered as a dielectric, might be said to end not where $E(x) = \phi_n$ (since this is true only at infinity) but at a value of x which makes $E(x) = \phi_n + kT$. The correction is equivalent to reducing the barrier height by kT. Thus, eqn (2.1.6), for instance, should be replaced by

$$eV_D - kT = N_d e^2 \lambda_0^2 / 2\epsilon \tag{2.1.19}$$

where λ_0 is now an *effective* width. The validity of this formulation is easily demonstrated, taking advantage of the fact that eqn (2.1.14) can be explicitly integrated *once*, yielding the local field as a function of the (alas, unknown) local potential. However, a simple appeal to Gauss's Law then gives

$$\{F(0)\}^2 = \frac{2eN_d}{\epsilon}\left(V_D - \frac{kT}{e}\right) \tag{2.1.20}$$

as shown by Atalla (1966) and Rhoderick (1978). Compare eqns (2.1.2b) and (2.1.6). It is thus possible to consider $eV_D - kT$ as an *effective diffusion potential.* It is the value *experiments* are expected to yield. It may be noted in passing that since no model ever includes *all* the features of reality, every experiment yields 'effective' values in one sense or another, whether we explicitly say so or not.

Equation (2.1.19) applies to a semiconductor with a dilute electron gas, one to which Boltzmann statistics is applicable. Goodman and Perkins (1964) have shown that the equation must be modified for a degenerate electron gas. It then becomes

$$eV_D + 2\phi_n/5 = N_d e^2 \lambda_0^2 / 2\epsilon. \tag{2.1.21}$$

By definition, ϕ_n is always positive. The sign of the correction is now different, because the Fermi level is now above E_c.

All the above equations can, of course, be made more sophisticated and more general by including a provision for incomplete donor ionization, but this further complication is avoided here.

In the situations discussed above ($V_D > 0$), the free carrier contribution is most important when V_D is small, but this is not the only configuration of interest. V_D can actually be negative. This can happen for appropriate positions of bunched surface states, or even in the absence of surface states, if the thermionic work function of the semiconductor were greater than that of the metal. The space charge would then consist entirely of free electrons. Its density would still be given by eqn (2.1.14), except that $E-\phi_n$ would now be negative, and $n(x) > N_d$ (for full ionization in bulk). Accordingly, one would expect a screening distance equal to a Debye length. Such a majority-carrier space-charge is shown in Fig. 2.4(b) (for two non-degenerate cases). The region over which it extends is called a (majority-carrier *accumulation layer*,† to distinguish it from the *depletion layer* which is associated with positive diffusion potentials. Accumulation layers constitute majority-carrier reservoirs which can support *lateral* conduction, and are therefore of great importance in connection with a variety of experiments on surface behaviour; see Many and co-workers (1965), Eger *et al.* (1976), and Frankl (1967) for details. In the context of interfaces and *transverse* conduction (i.e. conduction at right angles to the metal–semiconductor interface), accumulation layers are of interest only in connection with low-resistance contacts (see also Chapter 6).

2.1.3 *Problems of space-charge discontinuity*

In all the above considerations, the barrier space-charge was taken as *continuous*, whereas we know that it consists in fact of ionized centres in fixed, randomly spaced locations. An assumption of continuity is reasonable when there are many such centres within the (barrier) region under discussion, but this condition is often unfulfilled. Take a barrier of (say) 5×10^{-5} cm width, on a material of (say) 10^{15} donors cm^{-3}. If these donors were arranged on a superlattice, they would be 10^{-5} cm apart. Even in a random system, 10^{-5} cm would be the average distance between them, which means there would be room, on average, for only 5 such centres within one barrier thickness. In these circumstances, the continuity assumption is hard to justify. Schottky knew this very well, but was unable to resolve the issues arising from this fact. Not much has happened since 1942 to change this situation, which is almost universally ignored, even (elsewhere) in this book. The complications involved would seem to be very serious, but it is true that no specific observations are known which link barrier properties to this feature. One could therefore hold that the appearances are more disturbing than the facts.

If, for instance, the donors were located on a superlattice (or, at least, on planes perpendicular to x), then the barrier contour would have no curvature *between* these planes, and would therefore look as shown in Fig. 2.5(a). This would be a relatively unimportant departure from the 'ideal' shape, and may account for the absence to date of any theoretical

† The terminology, though deeply rooted by now, is nevertheless unfortunate, since the term 'accumulation', as used here, has nothing whatever to do with the *minority-carrier accumulation* which is briefly discussed in Section 1.2.

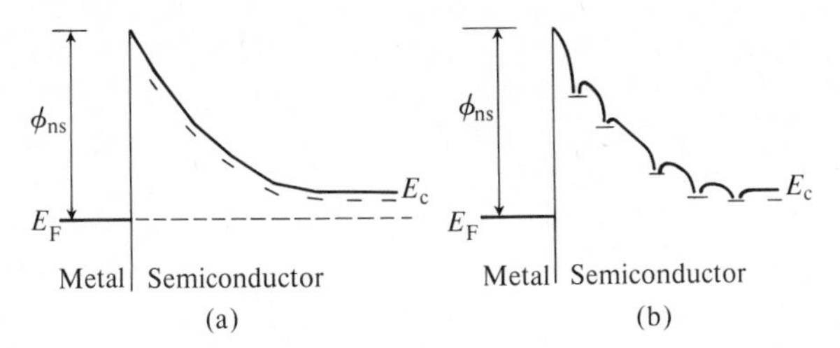

FIG. 2.5. Profile of barriers with discontinuous charge distributions; schematic representation. (a) Charge in planes parallel to the interface. (b) Charge at point centres, each associated with radial field components.

attempt to include space-charge discontinuity in barrier models. However, the arguments which support the traditional picture were not dictated on Mount Sinai, and a more pessimistic view suggests itself with comparable force. Thus, the assumption of charges arranged in parallel planes is highly artificial, and the situation is much more complicated when randomly spaced donors are under consideration. The barrier contour would then look somewhat like that shown in Fig. 2.5(b), which takes account of local coulomb attractions. Over substantial distances, the effects should average out, but the initial barrier peak (near $x = 0$) may well be transparent to electrons at some locations in the y–z plane, for the reasons discussed in Section 2.1.5. This would lead to highly unequal current densities ('hot spots') when external voltages are applied, and the current flow geometry would cease to be one-dimensional on a small scale. In practice, it would be exceedingly difficult to distinguish the consequences from a variety of complications arising in other ways, and the problem therefore remains, in theory and experiment, one of the outstanding gaps in our understanding of these matters.

Data of illustrative (if highly schematic) interest can be derived by a computer simulation of randomly spaced donors, and their effect on the boundary field $F(0)$. Figure 2.6 shows such results, in which the coulomb fields at the interface are summed over donors, which are distributed in an array of cells, 20 cells long and 6 cells deep. In this particular case, the cells are assumed to be one unit apart in the direction parallel to the electrode, and half-a-unit apart in the perpendicular direction. Of course, a great deal depends on the (arbitrary) nature of these assumptions, and one would in any case expect the field fluctuations to be diminished (by a factor of about 4) in a more realistic three-dimensional calculation. Figures 2.6(a) and (b) refer to different percentages of occupancy. Complete occupancy leads to excellent averaging, as shown by the horizontal portions of Fig. 2.6(a), whereas, even one missing donor in the first row has appreciable consequences in terms of the local field, and leads to a local increase of barrier resistance. More drastic field fluctuations are

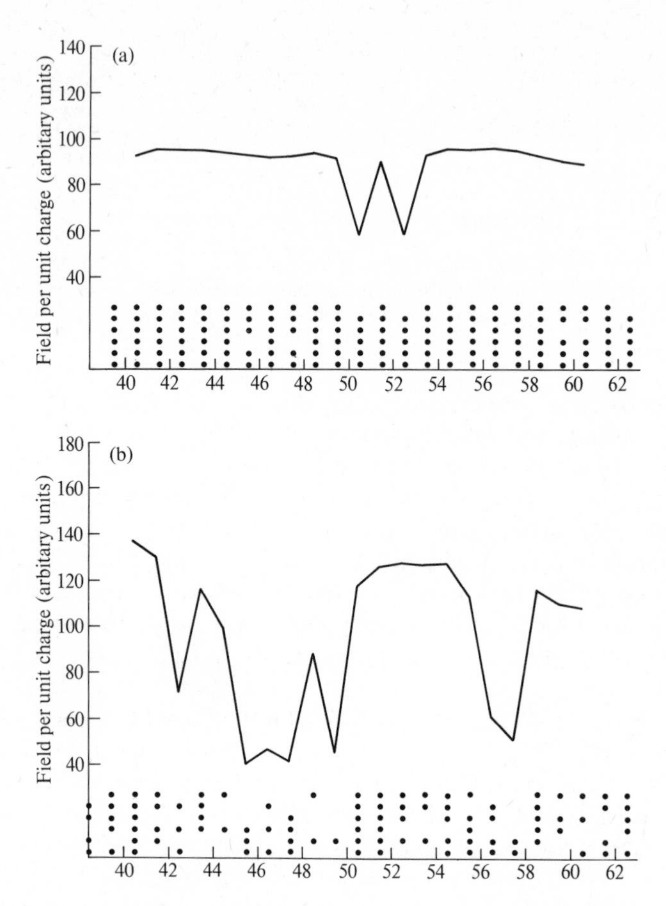

FIG. 2.6. Simulation of discontinuous barrier charge, with random occupancy of possible donor sites; boundary field pattern resulting from: (a) almost complete (96.0%) occupancy; (b) 63.0% occupancy. Abscissa: arbitrary count of box numbers.

expected for lower site occupancy rates, as shown in Fig. 2.6(b). Further complications must in practice be envisaged as a consequence of donor clustering (not part of the model). Since the problem is analytically intractable, simulation methods may in fact be the best way to attack it, but very little has been done along those lines.

Meanwhile, the prevailing hope is that the complicating effects will somehow average out. It will be shown in Section 2.2.4 that the barrier resistance at very low currents should be proportional to $\exp(eV_D/kT)$ and, because very thin barriers are transparent to electrons (Section 2.1.5), we must expect the *effective* value of V_D to fluctuate in sympathy with the fluctuating field. The consequences would be amplified by the exponential term. (See Section 2.3.6.)

2.1.4 *The Mott barrier*

When theories of rectifying barriers first came into prominence, the materials which were of practical interest were mostly non-stoichiometric oxides (except for selenium, which was widely used in rectifier manufacture). Such materials tend to be influenced by the surrounding atmosphere, a fact which sometimes limits their long-term stability. An n-type oxide, for instance, which owes its conductivity to a stoichiometric excess of metal, is for that very reason also a reducing agent. Oxygen in the air interacts with it and tends to restore the stoichiometric balance. It is therefore plausible to expect a thin surface layer which is less conductive than the bulk by virtue of having a lower concentration of active donor centres. This situation 'leads to the barrier model proposed by Mott (1939).

Let us then assume that a surface layer of thickness λ_M has a donor concentration $N_{dM} \ll N_d$, where N_d is the bulk value. Assuming complete ionization, as before, the built-in conduction band step between these regions would be given by

$$V = (kT/e)\log(N_d/N_{dM}) \qquad (2.1.22)$$

which is always of the order of kT, and thus small compared with common barrier heights. The situation is as shown in Fig. 2.7(a). Upon contact, there must be a further charge exchange, which means that other

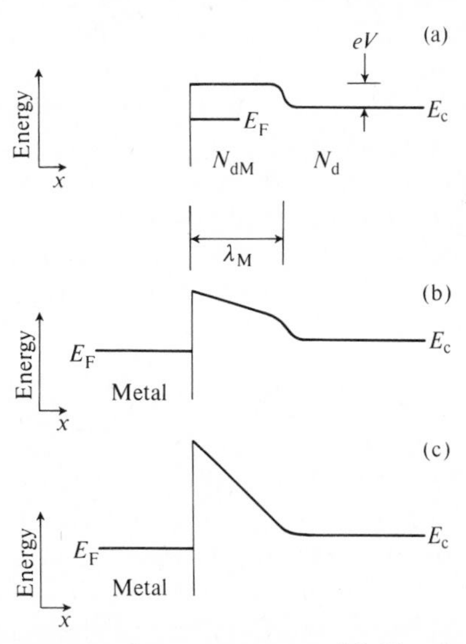

FIG. 2.7. Formation of a Mott barrier on n-type material. N_d = donor concentration in the semiconductor bulk; N_{dM} = donor concentration within a distance λ_M from the metal interface. Energy contours: (a) before the establishment of contact, (b) for a low contact barrier, (c) for a high contact barrier.

fields come to be superimposed. Figures 2.7(b) and (c) show qualitatively how this development will proceed as the barrier is built up, whether it is basically due to work-function differences or surface states. If the barrier were high enough the original diffusion step would become negligible (Fig. 2.7(c)). If N_{dM} were sufficiently small (in the limit, tending to zero), then the field would be linear within the barrier and, apart from small boundary problems, the barrier width would never be very different from λ_M.

In the *ideal* Mott barrier there is an abrupt transition from $N_{dM}=0$ to N_d at $x=\lambda_M$; in practice, hybrid cases may be expected to arise, in which N_a/N_{dM} is closer to unity, and such situations are more complex. The positive charge arising from ionized donors would then exist partly within the Mott barrier region and partly in the diffusion region underneath.

2.1.5 High field boundary effects; image forces

The high field of the Schottky barrier in the vicinity of $x=0$ gives rise to two further effects, not yet considered in the above analysis, one arising from image forces ('Schottky effect') and one from electron tunnelling. Both influence the effective barrier height. In the conventional analysis, the barrier is assumed to be rather high ($eV_D \gg kT$), in the belief that this will enable each electron-transit across the interface to be considered a rare event, uninfluenced by other electrons which support the total current flow. Image force effects are then superimposed upon the band picture as if they were additive. This is certainly an oversimplification but, without it, the formulation of a more general model incorporating image forces would be much more difficult. In effect, we here use the notion of electron transition as a 'rare event' to justify the neglect of all screening processes. The present models are thus hybrids and deserve to be viewed with caution, despite their universal popularity.

We begin with the fact that an electron at a distance x from the interface between a metal and a dielectric of permittivity ϵ_∞ is subject to an image force $\mathcal{F}$,

$$\mathcal{F} = -\frac{e^2}{4\pi\epsilon_\infty(2x)^2} = -\frac{e^2}{16\pi\epsilon_\infty x^2} \tag{2.1.23}$$

which attracts it to the metal (hence, by reference to Fig. 2.8, the negative sign). ϵ_∞ is the high-frequency value of the permittivity, since the electron spends very little time in the image-force region during one barrier transition. The idea is that the semiconductor does not have time to become polarized, a matter which will be further discussed below.

In the first instance, let us forget about the barrier as such, and its space charge. The electron energy at a distance x from the metal would then be simply ϕ_{ns}, and independent of x. The image forces changes the situation,

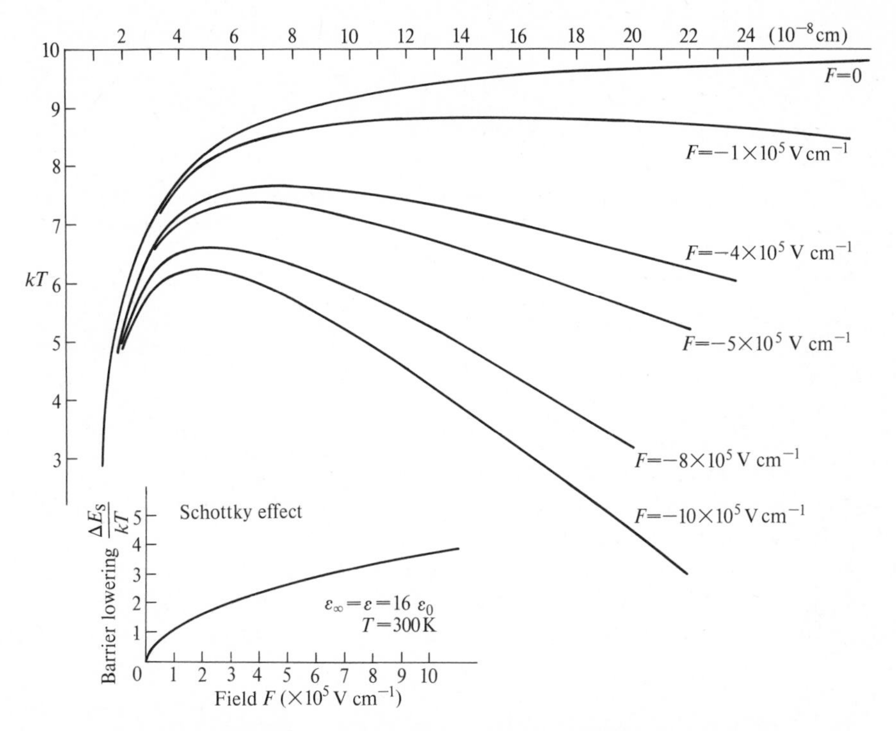

FIG. 2.8. Energy profile of a boundary region subject to the influence of image forces, in the presence of various externally applied fields F (Schottky effect). Insert: barrier lowering by Schottky effect; summary.

and makes the energy

$$E(x) = \phi_{\text{ns}} - \int_{\infty}^{x} \mathcal{F}\,\text{d}x = \phi_{\text{ns}} - \frac{e^2}{16\pi\epsilon_\infty x}. \qquad (2.1.24)$$

The top line in Fig. 2.8 shows this situation, and the fact that the energy goes to $-\infty$ as $x \to 0$ is sufficient to show that some more sophisticated assumptions are needed. When x is very small, the metal is no longer a continuum, and various kinds of provisions have been made to assess that region, of which the most important are due to McColl and Thomas–Fermi. (See Crowell and Sze 1966c for a more detailed discussion.) Fortunately, that region is not of great importance in the present context; we are concerned here with the formation of an energy maximum, and with its height. The lower lines in Fig. 2.8 show how the contours are modified by the presence of various fields, superimposed so as to make the metal negative with respect to the semiconductor ($F<0$). For all such fields, the $E(x)$ contour has a maximum. At that point, $x = x_{\text{m}}$, the image field and F compensate for one another exactly. The maximum therefore occurs where

$$e(-F) = e^2/16\pi\epsilon_\infty x_m^2 \qquad (2.1.25)$$

since the field is negative. Thus,

$$E(x_m) = \left[\phi_{ns} - \frac{e^2}{16\pi\epsilon_\infty x_m}\right] - x_m e(-F) = \phi_{ns} - \frac{e^2}{8\pi\epsilon_\infty x_m} \qquad (2.1.26)$$

or

$$\phi_{ns} - E(x_m) = e^2/8\pi\epsilon_\infty x_m = \Delta E_s. \qquad (2.1.27)$$

Substituting for x_m from eqn (2.1.25), we obtain, for the lowering of the barrier height,

$$\Delta E_s = e^{3/2}(-F)^{1/2}/2\pi^{1/2}\epsilon_\infty^{1/2}. \qquad (2.1.28)$$

This phenomenon is well known in the context of thermionic emission from metals into vacuum. On that context, the barrier-lowering is often referred to as the 'Schottky effect'.

Equation (2.1.28) is general, in the sense that it applies to any negative field. When we are concerned with a Schottky barrier, then F is, of course, the barrier field, given by eqns (2.1.2), namely

$$F(x) = (N_d e/\epsilon)(x - \lambda_0). \qquad (2.1.29)$$

Here we use the low-frequency permittivity ϵ, since the barrier field is static. The barrier field is a function of x, of course, but we have every justification for assuming $x_m \ll \lambda_0$, and therefore take $F(x_m) \approx F(0) = -N_d e\lambda_0/\epsilon$ for use in eqn (2.1.28). Accordingly, for zero current,

$$\Delta E_s = e^2 N_d^{1/2}\lambda_0^{1/2}/2\pi^{1/2}\epsilon_\infty^{1/2}\epsilon^{1/2}. \qquad (2.1.30)$$

Alternatively, we could express the barrier lowering as a function of the initial barrier height eV_D, using eqn (2.1.6). Thus

$$\Delta E_s = \frac{e^2 N_d^{1/2}\lambda_d^{1/2}2^{1/4}(eV_D/kT)^{1/4}}{2\pi^{1/2}\epsilon_\infty^{1/2}\epsilon^{1/2}}$$

$$= \left(\frac{e^7 N_d V_D}{8\pi^2\epsilon_\infty^2\epsilon}\right)^{1/4}. \qquad (2.1.31)$$

In Section 2.2.8, this expression will be used to calculate an approximate image-force correction in the presence of current flow. The correction evidently increases with N_d.

Rideout and Crowell (1970) have shown that the use of $F(0)$ in place of $F(x_m)$ in eqn (2.1.31) is still a worthwhile approximation, even when the barrier lowering ΔE_s becomes comparable with the barrier height itself!

When ΔE_s is actually equal to V_D, giving effectively zero barrier height, we have from eqns (2.1.6 and (2.1.31)

$$N_d = 8\pi^2 \epsilon_\infty^2 \epsilon V_D^3 / e^3 \qquad (2.1.32)$$

which gives the critical donor concentration for which this condition is satisfied. The concept is purely notional, since the discreteness of the donor charge would make itself felt very forcibly long before that state of affairs is reached. It is also very artificial, in as much as we are faced with a 'barrier' which is no barrier to the only entity that is capable of attempting a crossing, namely an electron. In such circumstances, electron transitions cannot be considered 'rare events', the key idea underlying the image-force concept. In varying degree, this constitutes a problem even at lower fields. Nevertheless, the image-force correction, as conventionally applied to barrier models, has proved its worth, qualitatively and at times quantitatively, in the analysis of reverse voltage–current characteristics, which is why it is reproduced here. A more profoundly satisfactory theory remains to be formulated, and beginnings in that direction have been made; e.g. see Crowell (1974). Meanwhile, it is worth noting that the classic Schottky effect can indeed be observed on free semiconductor surfaces in vacuum (Howorth *et al.* 1973; see also Inkson 1971).

Several attempts have been made to obtain experimental evidence for the entry of ϵ_∞ into eqn (2.1.31), and early work by Kahng (1963) on selenium did indeed suggest that the permittivity, as evaluated from voltage–current characteristics, was very low. However, such evaluations (especially on Se) involve many uncertainties, and it is therefore better to establish the field-dependent barrier height by measuring the photoelectric threshold. This was done by Sze, Crowell, and Kahng (1964) for gold contacts on n-Si, and the measurements yielded $\epsilon_\infty = 12.0 \pm 0.5$, the same value as that measured by purely optical methods in the infrared (3–15 μm). The optically measured dielectric constant does not in fact diminish until the wavelength is below about 1 μm, when band-to-band excitation sets in.

For a discussion of image forces and their effect on voltage–current characteristics, see Section 2.2.8 and 2.3.9.

2.1.6 *The wavemechanical tunnel effect*

A second important boundary effect arises from electron tunnelling, and is in practice augmented by high fields. The apex of the barrier is transparent to electrons, but this is not ordinarily the most important aspect, because so few electrons are available at that energy to take advantage of the transparency. The most important penetration is usually well below the apex. It is easy enough to show from first principles (see,

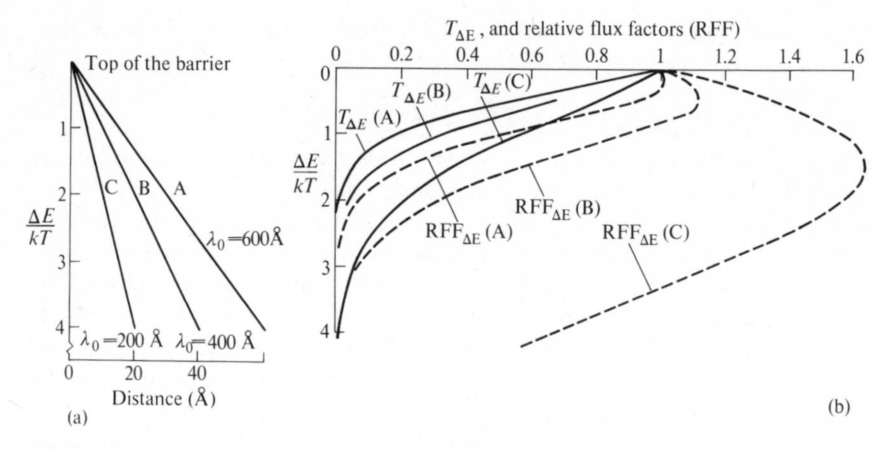

FIG. 2.9. Tunnelling probabilities at various heights of the barrier for three different barrier thicknesses. (a) Energy contours near top of barrier. (b) $T_{\Delta E}$ (full lines) and relative flux factor $T_{\Delta E}\exp(\Delta E/kT)$ (broken lines).

for instance, Mott and Sneddon (1948)) using the familiar WKB approximation, that the transmission coefficient (i.e. the wavemechanical tunnelling probability) for a barrier is given by

$$T_E = \exp\left[-2\int_0^{x_t}\left\{\frac{2m}{\hbar^2}(E(x)-E)^{1/2}\right\}dx\right] \tag{2.1.33}$$

assuming free electrons on either side. Here $E(x)$ describes the shape of the barrier contour, and E the energy of the incoming electron. x_t is the tunnelling distance of that electron. For a triangular barrier of the kind shown in Fig. 2.9(a) this becomes

$$T_{\Delta E} = \exp\{-(8\pi x_t/3h)(2m)^{1/2}(\Delta E)^{1/2}\} \tag{2.1.34}$$

where $\Delta E = E(0) - E$. Taking x_t in Angstrom units, this can be written as

$$T_{\Delta E} = \exp\{-0.11x_t(\Delta E/kT)^{1/2}\}. \tag{2.1.35}$$

For a triangular barrier, x_t itself depends on ΔE (being proportional to it), and this means that $T_{\Delta E}$ actually depends on $(\Delta E/kT)^{3/2}$.

The WKB approximation can be safely used only when the potential is a slowly varying function of distance, and in connection with contact barriers this is not always the case. The limits of validity have been carefully explored by Stratton and Padovani (1967). Moreover, calculations which relate to actual situations must come to terms with the three-dimensionality of systems, with the fact that different densities of states prevail on opposite sides of the barrier, and with back-tunnelling, not forgetting the possible anisotropy of the k-vector. Realistic calculations of $T_{\Delta E}$ have been made by Crowell and Sze (1966c) and Chang and

Sze (1970). (See also Section 2.3.4.) However, even the present coarse picture yields an important result.

Let us first consider a typical quiescent Schottky barrier and envisage there different thicknesses, as in Fig. 2.9(a). The corresponding values of $T_{\Delta E}$ are shown by the full lines in Fig. 2.9(b); they decrease rapidly as ΔE increases. However, the number of electrons which are in a position to attempt tunnelling increases with ΔE, and it is therefore interesting to consider how the relative flux factor $T_{\Delta E} \exp(\Delta E/kT)$ behaves. It is represented by the broken lines, and we assume here, for simplicity, that ΔE is a sufficiently small energy interval for the density of states to be considered constant. A maximum transmission develops at an energy below the apex, and electrons close to that energy make the maximum contribution to tunnelling currents when a potential difference is applied. In the absence of such a difference, the tunnelling fluxes in the two directions are, of course, balanced. The fact that the energy range over which the transmission takes place is comparatively narrow permits various simplifications to be introduced into more sophisticated analyses, as Padovani and Stratton (1966b) have shown. The entire process is referred to as *thermionically assisted tunnelling*, or *thermionic field emission* to distinguish our situation from that which prevails when the only important tunnelling takes place at the Fermi level. See Crowell and Rideout (1969a) for a more extensive treatment of thermionic field emission (but neglecting image-force and space-charge effects), Murphy and Good (1956) for a general treatment of the combined image-force and tunnelling effects, Rideout and Crowell (1970) for an analysis of tunnelling through Schottky barriers, and Stratton (1969) for a general overview.

The total electron flux must, of course, be found by integration over the total height of the barrier, and allowance must be made for the reflection of electrons with energies *above* the barrier height. It is tempting to assess the last item (intuitively) as small, but calculations by Crowell and Sze (1966c) show that the transmission probability of electrons with energies of (say) 0.1 eV in excess of the barrier height is typically only 50–70%. The total tunnelling transmission depends critically on the shape of the barrier, and Fig. 2.9(b) shows schematically how sensitive these relationships are to small changes. The barrier itself will in practice look more like the image-force modified contours of Fig. 2.8, rather than the simple outlines of Fig. 2.9(a). Even so, precise formulations for the transmission are now available (see papers cited above). Whether they are directly useful is another matter, for whereas the general principles are not in doubt, the detailed consequences are difficult to verify. We never have sufficient information about practical systems; as a result, corrections for tunnel effects introduced into other theories are always schematic to some extent, and beset by major uncertainties. There is therefore little point in pursuing the corrections themselves to high precision and in lavishing mathematical refinement upon them, all the more because uncertainties arising from quite different factors prevent the interpretation of experimental results in such detail. These 'other factors' include electrode geometry, surface shape and structure, accidental contamination, structural and compositional non-homogeneities, strain, temperature gradients, etc. Approximations therefore carry the day.

In the ordinary way, it is assumed that the tunnelling barrier is free from intermediate steps, but this is not necessarily justified. The semiconductor may contain deep traps which can act as tunnelling bridges, and thereby affect the total tunnelling probability. A good deal of work has been done on this problem (e.g. see Sarrabayrouse *et al.* (1977), and Feuchtwang *et al.* (1977)), but the results are (once again) difficult to apply to practical cases, because experimental systems are rarely defined with sufficient precision. Under extremely high fields (which may, at times, prevail within barriers), the band structure itself changes. The field cannot then be superimposed in any simple way on any band structure calculated in its absence; it must be introduced into the wave-equation from the beginning. (e.g. see Eger *et al.* (1976)). As self-consistent field contours can then be calculated by methods which are beyond the present scope (see Quinn and Styles (1976)). The outcome varies from semiconductor to semiconductor, and is therefore not predictable by any standard equation. For present purposes we shall assume that the boundary fields are high enough to warrant corrections of the type discussed above, but not high enough to lead to changes of band structure.

2.1.7 *Barrier height, bulk structure, and interface relationships*

As outlined in Section 1.3, semiconductors without surfaces states are expected to form barriers which depend directly on the difference of thermionic work functions, whereas barriers formed on surfaces with a high density of surface states should be much less dependent on the contacting metal. Many practical cases are intermediate, providing for some but not for complete screening by surface (interface) states. The significance of the ionicity and the heat of formation of the semiconductor has already been pointed out (Section 1.3.3), and some further details will now be discussed.

For the comparison of interface behaviour, one frequently uses a parameter denoted by S, which represents $d\phi_{ns}/d\chi_m$, where χ_m is the electronegativity of the contacting metal. Figure 2.10(a), for example, represents the relationship between S and ionicity, after Kurtin and co-workers (1969). Pioneering work along those lines was done by Mead (1965, 1966a,b) and Kurtin and Mead (1968) in America, and by Turner

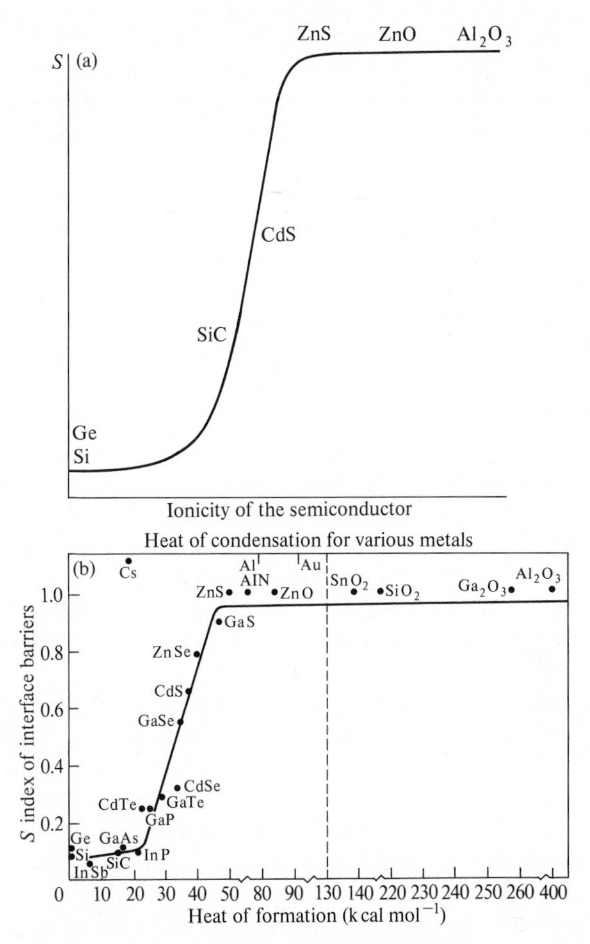

FIG. 2.10. Sensitivity of the barrier height to the bulk parameters of the contacting metal: (a) correlated with the ionicity of the semiconductor, simple data after Kurtin, *et al.* (1969), (b) correlated with the heat of formation of the semiconductor, after Lindau *et al.* (1978). The heat of formation may be regarded as an indirect measure of binding strength.

and Rhoderick (1968) in England. The results derived from electrical measurements were in close agreement with optically determined interface properties. Correspondingly, Fig. 2.10(b) shows the relationship between S and the heat of formation, after Lindau and co-workers (1978), who have provided *inter alia* an extensive review of experimental data in the literature; see also Brillson (1978a,b).

All simple models of the barrier assume that the surfaces involved are totally clean but, of course, this is an ideal which one can only hope to approach. One can do so by working with freshly cleaved surfaces in ultra-high vacuum (e.g. see Crowell (1965), Banbury *et al.* (1962) and

Brillson (1982)) but the technique is difficult. Measurements are further impeded by the fact that they must be performed quickly (e.g. within 20–30 minutes, depending on the quality of the vacuum and one's faith in the vacuum gauge) before appreciable gas adsorption can take place. Moreover, freshly cleaved surfaces, though clean, have a step-like structure (e.g. see Noble and Henisch (1967)), which severely limits the available regions of lateral uniformity. The same is, of course, true for surfaces cleaved under oxygen-free liquids, a technique which has been explored as a simpler (if limited) alternative to cleavage in ultra-high vacuum (Henisch and Noble 1966; Noble *et al.* 1967). Because of the inherent difficulties, experimentation on surfaces cleaved under controlled conditions is rare; a more common procedure is to prepare chemically clean surfaces, and to 'etch' them by argon sputtering. This avoids the step-like structure, but does so at the cost of substituting a uniform disorder caused by the argon bombardment itself.

All experimentation thus involves compromise, sometimes slight, often severe. Oxide layers, rather than 'foreign contamination' usually constitute the main problem. These layers, whether accidental or deliberately made, tend to be ill defined, and can therefore be safely blamed for a variety of discrepancies between experimental expectations and experimental results.

Because oxide films can vary in thickness and structure, superficially similar systems may exhibit a great variety of properties. Fortunately, the converse can also be true. Thus, for instance, Varma and co-workers (1977) observed that gold contacts on (111)-silicon prepared under very different conditions gave similar results. Turner and Rhoderick (1968) reported that though chemical cleaning and etching of silicon surfaces led to progressively lower barrier heights, the *final* heights were insensitive to further treatment, and had to be ascribed to the growth of a stable oxide layer. Along similar lines, Williams and co-workers (1977) found that barrier heights on (110)-InP are ultimately constant, apparently uninfluenced by the thickness of the thin (10–15 Å) oxide layers associated with them. Corresponding experiments on *cleaved* silicon after various stages of oxidation have been carried out by Crowell (1965). It is clear from all this that stable surfaces *can* be made with perseverance and care, but with which barrier model any particular surface should be associated is often difficult to decide. This explains also why agreement between barrier heights measured by independent methods is comparatively rare (Kahng 1963).

The many complications which can ordinarily arise at contact interfaces have generated a strong motivation to achieve simpler and more predictable systems. There is, of course, no general solution to this problem, but there is one that is specifically applicable to silicon, namely contacts made

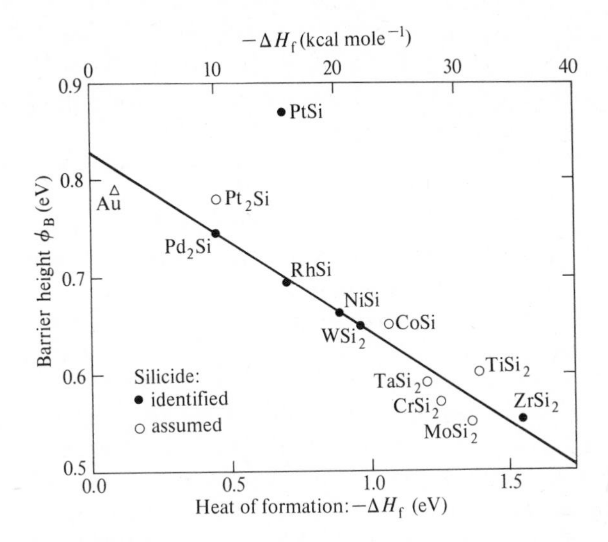

FIG. 2.11. Relationship between the barrier height at silicide–silicon interfaces and the heat of formation of the silicide involved. After Andrews and Phillips (1975).

of silicides grown *in situ*. In the course of silicide formation, the interface moves into the semiconductor, and away from surface (physical and chemical) imperfections as Andrews and Phillips (1975) have pointed out. (See also Section 2.2.8.)

A great deal of work has been done on such interfaces, and many further references will be found at the end of this section. Andrews (1975) and Andrews and Phillips (1975) for instance, have reported on the heights of barriers between Si and a remarkable series of twelve transition-metal silicides. The barrier heights varied linearly with the heat of formation (of those silicides), as shown in Fig. 2.11, demonstrating convincingly that interface (bonding) relationships are involved. The same is, in all likelihood, true for other types of contacts, as yet less well defined. As far as Fig. 2.11 is concerned the suggestion is that the barrier height (0.8 eV) at zero heat of formation corresponds to that of the free-surface barrier on silicon, and that, as one might expect, is not very different from the barrier of a gold–silicon interface which is free from compound formation. It was also shown that ordering of the silicide can lead to changes in the nature of the bonding, and hence to changes of barrier height.

Alloyed contacts on GaAs have also been widely studied, e.g. by Robinson (1975), Benenz and co-workers (1976), Guha and co-workers (1977), and Wittmer *et al.* (1977). For an extensive review of experimental work in this field, see Lindau and co-workers (1978). Chye *et al.*

(1978) studied GaAs, GaSb, and InP by photoemission and found that, in contrast to the observations on Si, gold interacts strongly with these semiconductors. The Fermi level is 'pinned' when the surface is covered with as little as one-fifth of a monolayer of gold.

The motivation for many of these researches is highly practical: the barrier height is one of the most important parameters of the system; small differences of ϕ_{ns} lead to order-of-magnitude differences in barrier resistance (Andrews 1974). One of the most powerful tools is the photoemission of electrons excited by monochromatic synchrotron radiation. In this way, Spicer and co-workers (1982) have been able to analyse the composition of semiconductors within 3 atomic layers of the surface. GaAs, InP, and GaSb have been investigated by this method, and it was found that properly cleaved crystal surfaces of these materials initially have no detectable surface states within the band gap, and that surface states are subsequently generated by the addition of metals and oxygen in astonishingly small quantities. Another unexpected result was the fact that the surface states created were tightly bunched (the cause of 'Fermi-level pinning') within an energy range that was found to be independent of the contaminating surface atoms. The conclusion was that foreign surface atoms ('adatoms') actually create defects in the semiconductor surface which then, in turn, give rise to the (always equal) surface states. See Spicer *et al.* (1980, 1982) for references to earlier work of the Stanford University group.

X-ray photoemission spectroscopy, low-energy-loss spectroscopy, and Kelvin probe measurements are some of the other methods that have been brought to bear on these problems (e.g. see Brillson (1977, 1978c, 1982). Brillson, Margarito, and Stoffel (1980) have shown that the presence of reactive metals, e.g. Ti, Al, and Ni, at the interfaces of compound semiconductors with *unreactive* metals (e.g. Au) has a sensitive effect on the electromigration of semiconductor constituents, and also on the height of the barrier measured by electrical means after establishment of a steady state. They conclude that the strength and character of microscopic interfacial binding are critical factors in determining the barrier properties. See Brillson (1982) for extensive references to previous papers.

A detailed review of this rapidly developing field is not feasible here, but it will be clear that these investigations are concerned with key problems relating to barrier structure and stability. In the rest of this book, various barrier structures will be regarded as 'given', without further enquiry into their origin, and principal attention will be devoted to the logical consequences of these structures in terms of electrical properties.

2.1.8 *Deep donors; incomplete ionization*

In all the above discussions it was assumed that $n_e = N_d$ within the semiconductor bulk, which is true enough for (say) germanium at room temperature or above, but cannot be said to be generally valid. The assumption owes its popularity only to the obvious simplicity of the resulting relationships. It implies donor levels which are within kT or so of the conduction band edge (or, of course, acceptor levels within kT or so of the valence band), whereas actual donors are often much deeper, and therefore incompletely ionized. Partial ionization is generally (but not necessarily) associated with *non*-degenerate electron gases, but band bending at contacts may bring some of the donors close to the Fermi level, which means that *their* occupancy cannot always be calculated on the basis of Boltzmann statistics, even though the free-electron concentration in the bulk might be so calculated. This complication affects the space charge, and thus the barrier profile, but the qualitative picture is immediately clear. Up to a certain distance, the donor levels are far above the Fermi level and the local space charge density is eN_d, as for the fully ionized case. This region is followed by an interval in which the charge density is $eN_d - en_d(x) - en(x)$, where $n_d(x)$ is the concentration of electrons remaining in donors. Of course, $n_d(x) + n(x) \rightarrow N_d$ as $x \rightarrow \infty$. The barrier space-charge density is evidently reduced by incomplete ionization, which means that the barrier has to be thicker to fulfil its screening function. On the other hand, if the bulk material is to have a useful conductivity despite the fact that the donors are deep, then a high donor concentration is needed, and that reduces the barrier width. The two factors are thus in opposition, and the extent to which they compensate for one another varies from case to case.

The equilibrium occupancy of donor levels is subject to considerations slightly different from those that govern extended band states. This is because any particular donor can hold an electron of either spin, but never two electrons (as an extended level can). Guggenheim (1953) has shown that this leads to a concentration of electrons at bulk donor sites equal to

$$n_d(\infty) = n_d = \frac{N_d}{1 + \frac{1}{2}\exp\{(E_d - E_F)/kT\}} \tag{2.1.36}$$

where E_d denotes the energy of the donor levels, and E_F the Fermi level. Under the conditions here assumed (holes absent, and all donors characterized by the same E_d), the Fermi level must be equidistant from E_c and E_d (see Fig. 2.12), so that $\phi_n = \Delta E_d/2$, which gives for the bulk

$$n_d = \frac{N_d}{1 + \frac{1}{2}\exp(-\Delta E_d/2kT)}. \tag{2.1.37}$$

When ΔE_d increases, $n_d \rightarrow N_d$, as expected. (Correspondingly, the concentration p_a of *holes* remaining at acceptor sites, and thus the concentration of unionized acceptors, is given by

$$p_a = N_a/[1+\tfrac{1}{2}\exp\{(E_F - E_a)/kT\}]$$

where N_a is the concentration of acceptors and E_a their energy.)

To calculate the barrier shape, we must now modify eqn (2.1.14), which becomes

$$\frac{d^2E}{dx^2} = \frac{e^2}{\epsilon}\{N_d - n_d(x) - n(x)\} \tag{2.1.38}$$

where $n_d(x)$ and $n(x)$ can be simply calculated, as long as we know where the Fermi level is situated. That will depend on the band gap. For a very wide forbidden-band, a simplification is possible, because we know that the Fermi level is then (and only then) half-way between E_c and E_d in the bulk.

As long as Boltzmann statistics can be used for the free electrons,

$$n(x) = (N_d - n_d)\exp\left\{\frac{-\left(E(x) - \frac{\Delta E_d}{2}\right)}{kT}\right\}. \tag{2.1.39}$$

From eqn (2.1.36), with $E_d(x)$ now a variable, we have

$$n_d(x) = \frac{N_d}{1+\tfrac{1}{2}\exp\left[\frac{E(x) - \Delta E_d}{kT}\right]} \tag{2.1.40}$$

since E_F is taken as the zero reference level.

Check (1): As $x \rightarrow \infty$, $E(x) \rightarrow \Delta E_d/2$, we have $n_d(x) \rightarrow n_d(\infty) = n_d$, by eqn (2.1.37). At the same time: $n(x) \rightarrow N_d - n_d = n_e$, and the space charge disappears.

Check (2): As $E(x)$ becomes very large, $n_d(x)$ tends to zero, i.e. all the donors tend to become ionized.

Equation 2.1.38 can in any event be numerically integrated. Results for a wide forbidden-band and a non-degenerate electron population are shown in Fig. 2.12 for three activation energies, of which the lowest ($\Delta E_d = 2kT$) is close to the very limits of permissibility. For comparison, a contour for complete ionization (of the same carrier concentration) is also given. The assumed barrier height is substantial here. For small values of x the contour is then almost unchanged, since the donors in that region are all ionized as usual. The change shows itself most drastically in terms of ultimate barrier thickness. The consequences of incomplete ionization are even greater for barriers of lower height.

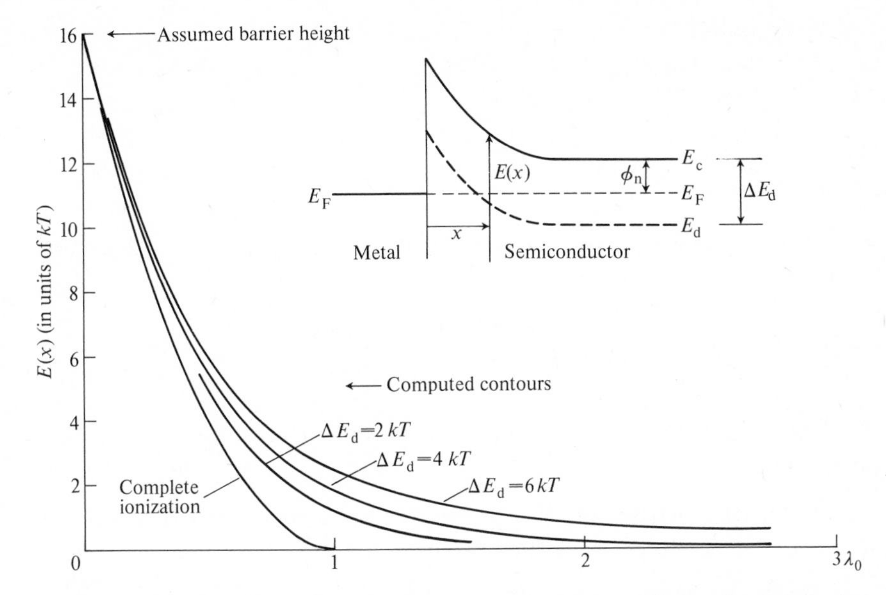

FIG. 2.12. Barrier profiles with incomplete donor ionization for different donor activation energies; computed results. Local energy plotted against distance in units of λ_0. Zero current.

In the sections and chapters which follow, these matters are almost completely ignored, not at all because they are necessarily unimportant, but because their consequences have not yet been fully worked out. (See, however, Section 4.1.3.) Thus, pending more numerical computation, we do not know how such a barrier distorts in the presence of current flow, and how that distortion affects its conductive properties. In as much as incomplete donor ionization is an additional complication, it makes the interpretation of experiments more uncertain. On the other hand, when monochromatic light is used as an additional tool, effects associated with deep donors can sometimes be successfully identified and interpreted, e.g. see Grimmeis (1974), Braun and Grimmeis (1973), and Marfaing and co-workers (1974). The purpose of such work is generally to find ΔE_d, either as a discrete value, or else as a distribution over a small energy range. Another way is to employ the techniques of thermally stimulated current (TSC) measurement, e.g. as described by Smith (1972, 1974) for gold contacts on GaP.

In the ordinary way, we are rarely concerned with degenerate electron gases in the semiconductor, but such cases do arise when ΔE_d is less than $2\,kT$. However, the resulting barrier is then very thin, and it is more convenient to consider such problems in connection with the topics of Chapter 6.

Further reading

On two-dimensional space-charge barriers: Brown and Lindsay 1976.

On the charge distribution in metal surfaces: Rhoderick 1975.

On interface wave functions and states: Barret and Vapaille 1976, Gossick 1969b.

On barrier heights and ellipsometric surface characterization: Adams and Pruniaux 1973.

On tunnelling through an insulating film: Simmons 1963.

On absorption at semiconductor–metal interfaces: Brillson *et al.* 1981.

On atomic redistribution at compound semiconductor interfaces: Brillson *et al.* 1980b.

On chemical mechanisms of Schottky barrier formation: Brillson 1979, Spicer, *et al.* 1980, Brillson 1983.

On modifying barrier height by ion implantation: Shannon 1974, 1976.

On modifying barrier-height by thermal treatments: Card 1975a, Reith and Schick 1974.

On the relationship between barrier height and the (metal) atomic number: Heime 1970b.

On diffusion effects at contacts: Grinolds and Robinson 1980, Sinha *et al.* 1976.

On silicon–silicide contacts: Andrews and Koch 1971, Andrews and Lepselter 1970, Buckley and Moss 1972, Coe and Rhoderick 1976, Deneuville and Chakraverty 1972, Fertig and Robinson 1976, Grinolds and Robinson 1977, Hosack 1972, Kircher 1971, Nakamura *et al.* 1976, Ohdomari *et al.* 1978, Ottaviani and Tu 1980, Parker and Mead 1969, Parker, *et al.* 1976, Pellegrini 1975a, Robinson 1974, Terry and Saltich 1976, Van Gurp 1975, Van Gurp and Langereis 1975, Walser and Bené 1976, Wittmer *et al.* 1981.

On interface chemistry at contacts with GaAs: Sinha and Poate 1973, Waldrop and Grant 1979.

On barrier heights at contacts:
with Si: Ali *et al.* 1979, Basterfield *et al.* 1975, Cowley 1970, Jäger and Kosak 1969, Ponpon and Siffert 1978, Reith 1976, Smith and Rhoderick 1971, Szydlo and Poirier 1973;
with Ge: Jäger and Kosak 1969;

with GaAs: Aydinli and Mattauch 1982, Cho and Casey 1974, Kajiyama *et al.* 1973, Muraka 1974, Sleger and Christou 1978, Uebbing and Bell 1967;
with GaP: Lei *et al.* 1978, 1979;
with InGaAs: Kajiyama *et al.* 1973;
with $GaAs_{1-x}P_x$: Neamen and Grannemann 1971, Rideout 1974, Schade *et al.* 1971;
with InAs: Walpole and Nill 1971;
with InSb: Ancker-Johnson and Dick 1969, Korwin-Pawlowski and Heasell 1975;
with InP: Thiel *et al.* 1977, Wada *et al.* 1982;
with CdS, CdSe, CdTe: Courtens and Chernow 1966, Fuchs and Heime 1966, Learn *et al.* 1966, Parker and Mead 1969;
with PbSe, PbTe: Hohnke and Holoway 1974, Walpole and Nill 1971;
with SiC: Dunlap and Marsh 1969, Wu and Campbell 1974;
with PbO: Heijne 1962;
with TiO_2: Davis and Grannemann 1963, Szydlo and Poirier 1980;
with diamond: Ihm *et al.* 1978;
with Cu_2O: Assimos and Trivich 1973;
with Se: Champness *et al.* 1968, Schmidlin *et al.* 1968;
with tetracyanoquinodimethane: Bickford and Kanazawa (1976);
with tetracene: Ghosh and Feng (1973).

2.2 Current-carrying barriers; analysis of basic models

Because the transport equations cannot be solved in closed analytic form, the only 'exact' solutions are those derived by numerical methods. These will be discussed in Section 2.4. However, for many purposes, numerical solutions are inflexible and inconvenient. Amongst other things, they make it difficult to see how the various structural parameters of the system influence the outcome. In that connection, analytic solutions are more helpful, even when based on approximations (as they always are). The problem is to select approximations which are inherently plausible, and to remember what assumptions have been made. The use of analytical models *outside* the range in which the underlying assumptions are valid is a widespread transgression. In as much as the present chapter is limited to single-carrier systems it is itself a grand approximation, but an extremely plausible one when applied to materials of relatively wide band gap. The higher the doping level and the lower the temperature, the smaller is the permissible band gap for which the present considerations are valid. However, *general* treatments must certainly be based on two-carrier considerations, solved numerically. Indeed, even the single-carrier system calls for numerical methods, if the contribution of the *free* charge carriers is taken into account. That contribution is actually small, but very interesting in special circumstances (see Section 2.3). In the present section, it is neglected, in accordance with convention, and for purely didactic reasons.

Since minority-carrier injection and free charge-carrier effects are not contemplated here, the treatments given apply equally to lifetime and relaxation semiconductors; but in Chapter 3 that distinction will again have to be made. Despite occasional references to band theory, most topics of the present section and, indeed, nearly all the models described elsewhere, are essentially classical, applicable to barriers of a height substantially in excess of kT, and thick enough everywhere to make wave-mechanical tunnel penetration impossible. This restriction is relaxed in Sections 2.2.8 and 2.3. Wave-mechanical considerations which apply especially to low barriers will be the subject of Section 6.5. Unless otherwise stated, conditions are assumed to be isothermal.

It will be recalled (Section 1.2.4) that in single-carrier systems as discussed here, departures from equilibrium ($\Delta n = n - n_e$) involve space charges, which means in practice that such departures are relatively difficult to establish. In double-carrier systems, departures from the electron equilibrium and from hole equilibrium ($\Delta p = p - p_e$) can provide partial or complete compensation for one another, and this means that Δn and Δp can coexist over long distances under conditions of quasi-neutrality.

2.2.1 Barrier contours in the presence of current flow

The imposition of external voltages causes a current to flow by disturbing the internal equilibrium relationships discussed above. The barrier contours are thus distorted in the presence of a current, and since the barriers are non-symmetrical to begin with, the distortion is non-symmetrical also; Spenke (1955) refers to 'shearing'. The result is a current that is both voltage-dependent and polarity-dependent. As before, the polarity corresponding to higher currents will be called the *forward direction*, that corresponding to lower currents the *reverse direction*.

It will be recalled (Section 1.1.2) that the analysis of barrier systems depends on the assumptions made concerning the magnitude of the mean-free path in relation to the barrier width. When the mean-free path is relatively small, charge transport is governed by drift and diffusion, and this is here assumed, unless otherwise stated. When the mean-free path is large, then 'diode models' analogous to theories of thermionic emission into vacuum are in some ways more appropriate as, for instance, in Section 2.2.6. Both kinds of systems can be perturbed by quantum mechanical tunnel effects, e.g. Sections 2.1.6, 2.2.8, 2.3.3, and 6.5.

In a qualitative way, the Schottky barrier profiles under forward and reverse voltages are represented by Fig. 1.7, which explains the general principle of rectification. However, it is important to note that when we are dealing with drift-diffusion systems, there is more involved than a change of current magnitude: the very mechanism of current flow is polarity dependent. In equilibrium (zero current), diffusion and field currents balance. An externally applied voltage can cause the field component to increase, in which case field currents will predominate over diffusion currents. This is obviously the case in the reverse direction (Fig. 1.6(c)). Conversely, external voltages can cause the internal field to decrease, allowing diffusion currents to predominate, as in Fig. 1.6(a). This asymmetry is clearly and directly associated with the existence of the built-in field, and the non-linearity with the existence of diffusion currents. The different character of forward and reverse currents is made even more pronounced by changes of carrier concentration under current flow, as discussed in Section 2.2.2. Figure 2.13 shows these relationships schematically. Note that the field current changes sign at a particular value of x in the forward direction. This must be so, since the bulk field (small) and the barrier field (large) have opposite signs.

Equation (2.1.4) applies to zero current, and envisages a definite barrier width λ_0. Within the framework of the present approximation (free carriers neglected), we shall assume that the same arguments are valid, subject only to a modification of the barrier width from λ_0 to λ_B. Accordingly, we can put, for $0<x<\lambda_B$:

$$E(x)-\phi_n=(N_d e^2/\epsilon)(\tfrac{1}{2}x^2-\lambda_B x+\tfrac{1}{2}\lambda_B^2) \tag{2.2.1a}$$

and, corresponding to eqn (2.1.6), we now have

$$V_D+V_B=N_d e\lambda_B^2/2\epsilon. \tag{2.2.1b}$$

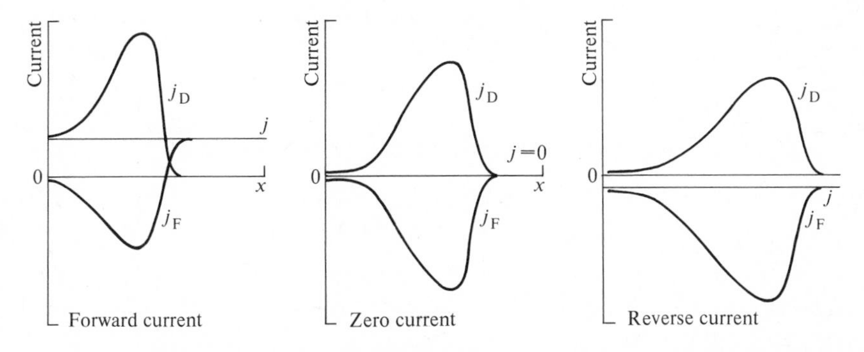

FIG. 2.13. Schematic representation of field and diffusion currents in a Schottky barrier. (a) Forward direction: diffusion current dominates. (b) Quiescent condition: diffusion and field currents are in balance everywhere, giving zero net current. (c) Reverse direction: field current dominates.

Total ionization is assumed. Equation (2.2.1b) is reflected by the contours of Fig. 1.7. Here, V_B is the *barrier voltage*, i.e. the voltage which appears across the barrier itself as a result of applying an *external* voltage V_T to the system as a whole. V_B and V_T are, of course, related by

$$V_T = V_B + V_b \tag{2.2.2}$$

where V_b is the voltage across the bulk material in series with the barrier. It is obviously desirable that it should be of the same sign as V_B for a given current. In Fig. 1.7, V_B has already been taken as positive in the reverse direction, in harmony with the fact that it adds itself graphically to the diffusion voltage V_D. Therefore, V_b should likewise be positive, but since the current density j is negative in the reverse direction, we must here put

$$V_b = -j \cdot R_b \tag{2.2.3}$$

where R_b is the total bulk resistance per unit area. We shall return to the problem of sign conventions in Section 2.2.2. In the absence of current-controlled non-equilibrium processes, V_B itself is numerically equal to the contact voltage V_c, as used in Section 1.1.1, because the extrapolation from experimental results on which V_c is based then yields a simple answer. In the presence of minority carrier injection or exclusion (Chapter 3), this identity is not maintained, because R_b is current-dependent. This explains the need for maintaining the distinction. In the present context, however, R_b is constant. In the reverse direction, it tends to be negligible compared with the barrier resistance, which is why the energy profile in Fig. 1.6(c) is horizontal outside the barrier itself.

However, for increasing currents in the forward direction, the series component cannot remain negligible, because the resistance of the barrier itself diminishes. This is why the energy profile in Fig. 1.7(a) shows a slight gradient outside the barrier. V_B is then defined at a distance from the interface which corresponds to the energy minimum. Note that the field is zero at that minimum, a fact which will be of major importance for the considerations in Section 2.4.

For a high Mott barrier, the relationship corresponding to (2.2.1a) is

$$E(x) = \phi_{ns} - e(V_D + V_B)x/\lambda_M \tag{2.2.4}$$

with λ_M constant, of course.

The above considerations apply explicitly to planar contacts. If contacts were hemispherical, as envisaged by Fig. 1.8(a), then eqn (2.2.2) would have to be replaced by

$$V_D + V_B = (N_d e \lambda_{SB}^2 / 2\epsilon)(1 + 2\lambda_{SB}/3r_0) \tag{2.2.5}$$

where r_0 is the contact radius, and λ_{SB} the new barrier thickness.

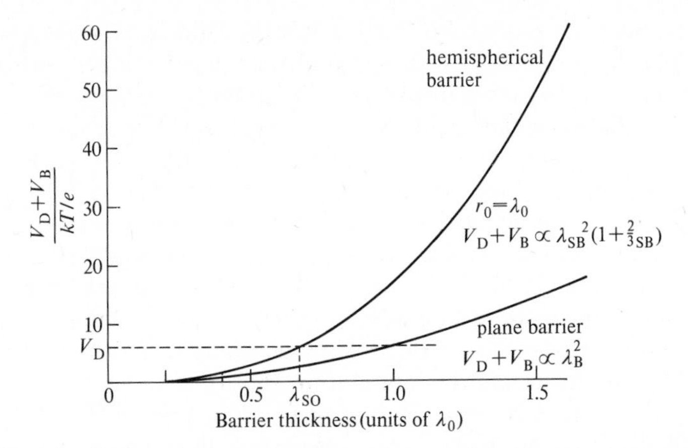

FIG. 2.14. Barrier thicknesses as a function of the applied voltage V_B, for planar and curved ($r_0=\lambda_0$) space charge regions. Abscissa normalized to λ_0. $V_D=6kT/e$.

($\lambda_{SB}=\lambda_{SO}$ when $V_B=0$) See Section 2.1.1. Comparison with the planar system shows that

$$\lambda_B=\lambda_{SB}(1+2\lambda_{SB}/3r_0)^{1/2}\approx\lambda_{SB}(1+\lambda_{SB}/3r_0) \quad (2.2.6)$$

(see Henisch 1957). Curvature thus reduces the barrier width in principle, but makes no material difference as long as $r_0 \gg \lambda_{SB}$ which is ordinarily the case. In view of this, contact curvature is neglected elsewhere in this book, except in connection with the topics of Sections 1.1.3, 3.1.1, 5.1.2, and 6.2.1. The curvature is, however, relevant to every discussion of *edge effects*, since $r_0 \gg \lambda_S$ cannot be satisfied at the lateral boundaries of the metal contact. In these locations, we must expect a barrier-lowering due to high field effects (even under static conditions), and a reduction of barrier thickness due to contact curvature, as the above equations show. When external voltages are applied, the curved barrier is not distorted at the same rate as the plane barrier, which introduces further complications. Figure 2.14 shows schematically how the thicknesses of a curved and a plane barrier are expected to vary with V_D.

2.2.2 *Voltage–current relationships and concentration profiles; general*

We will again consider an n-type semiconductor. With the appropriate change of symbols and signs, the same considerations apply also to p-type material. The first step is to solve the diffusion equation which covers the voltage–current characteristic, namely (for a one-dimensional system under isothermal conditions):

$$j=en(x)\mu_n F(x)+eD_n(dn/dx)_x \quad (2.2.7)$$

where j is the current density. This equation was first applied to the rectifying barrier problem by Wagner (1931), who used it essentially on the strength of an analogy with the known behaviour of ions in solution. To solve the equation, we need to relate μ_n and D_n to one another, and the accepted way of doing this is to assume *Einstein's Relation*, namely $\mu_n kT/e = D_n$. Einstein's own work (during the first few years of this century) actually referred to the diffusion of uncharged colloid particles and their drift in a gravitational field, and it was by no means obvious that electrons would obey a similar law. Wagner (1930) made it *plausible* that they should without, however, providing a rigorous proof. Ever since, the conventional textbook derivation of Einstein's Relation has leaned heavily on the existence of a Boltzmann distribution. Indeed, the relation follows directly from that distribution, in conjunction with eqn (2.2.7) for $j = 0$. In the static case, this derivation is unobjectionable, but it would not be appropriate to adopt such a procedure here automatically, since the electron distribution ($n(x)$ versus $E(x)$) can depart substantially from its equilibrium contour within a current-carrying barrier. Fortunately, Landsberg (1952) has shown that D_n/μ_n can be derived from fundamental conduction theory *without* assuming any particular electron distribution, as long as that ratio is the same for all carriers present. Under such conditions, Landsberg showed that the common ratio must be kT/e. Even when we are dealing with a degenerate electron gas, Landberg's findings indicate that Einstein's Relation remains significant, as long as D_n and μ_n can be regarded as *averages* over the carrier population. Except for the small additional reservations given below, the Einstein relationship can therefore be employed with confidence in the present context. Certainly, its use is now theoretically buttressed to an extent far greater than it was when Schottky first applied the diffusion equation to his newly proposed space-charge barrier (1938). Thus,

$$j = eD_n\left\{-\frac{en(x)}{kT}\left(\frac{dV}{dx}\right)_r + \left(\frac{dn}{dx}\right)_r\right\}. \tag{2.2.8}$$

Even then, Einstein's Relation is, strictly speaking, valid only at low fields and for small concentration gradients, but its use is justified as long as the current densities involved are not too large. In search for an explicit solution in terms of j and V, we use an integrating factor:

$$j\int_0^{\lambda_B} \exp(-eV/kT)\,dx = eD_n\{n(x)\exp(-eV/kT)\}_{x=0}^{x=\lambda_B} \tag{2.2.9}$$

where V is, of course, a function of x. To solve the equation, we need some boundary conditions. On the right-hand side of the barrier, these are relatively simple, and correspond to the bulk parameters of the semiconductor in equilibrium ($n(x) = n_e$). The question is exactly *where* (i.e. for what value of x) this condition is to be applied, considering that the barrier is known to be *infinitely wide* when the free carrier space charge is taken into account (Section 2.1.2). However, in the classic Schottky theory, the free carrier space charge is neglected, which makes it possible to say that the barrier comes to an end at $x = \lambda_0$ in the absence of current, or at $x = \lambda_B$ when a current is flowing. Along those lines, we therefore put

$$\begin{aligned} eV(\lambda_B) &= -\phi_n + eV_B \\ n(\lambda_B) &= n_e. \end{aligned} \tag{2.2.10}$$

This approximation is the price that has to be paid for the convenience of achieving analytic solutions. We shall later discuss how the exact solutions, obtained by numerical methods, differ (Section 2.4).

The boundary conditions at the metal–semiconductor interface are a very different problem. Astonishing as it may be, we have no clearcut information about them. Instead, we have a variety of 'options', of which the simplest is

$$eV(0) = -\phi_{ns} = \text{constant},$$

and

$$\begin{aligned} n(0) &= n_e \exp(-eV_D/kT) = \text{constant} \\ &= \mathcal{N}_c \exp(-\phi_{ns}/kT) \end{aligned} \tag{2.2.11}$$

where $\mathcal{N}_c$ is the effective density of states in the conduction band, given by

$$\mathcal{N}_c = 2(2\pi m_n kT/h^2)^{3/2} \tag{2.2.12}$$

(For p-type material, the relevant quantity would be the effective density of states in the valence band, namely $\mathcal{N}_v = 2(2\pi m_p kT/h^2)^{3/2}$.)

Other assumptions could have been made, and some of these will be further discussed in Sections 2.3 and 2.9. The crux of the matter is, of course, whether $n(0)$ is really independent of current density or not. Certainly, if it were, the constancy would have to be maintained by the recombinative and regenerative processes in the metal. For small mean free paths in the semiconductor and at low current densities, this is likely to be true. For long mean free paths, the processes in the metal are, of course, unchanged, but it will be shown (Section 2.3.2) that the value of $n(0)$ loses the simple significance here assigned to it. (See also below.)

With the above boundary conditions, eqn (2.2.9) yields

$$\begin{aligned} j\int_0^{\lambda_B} \exp(-eV/kT)\,\mathrm{d}x &= en_cD_n \exp(\phi_n/kT)\{\exp(-eV_B/kT) - 1\} \\ &= e\mathcal{N}_cD_n\{\exp(-eV_B/kT) - 1\} \end{aligned} \tag{2.2.13}$$

What remains is only to find the value of the integral on the left, which clearly depends on the shape of the barrier profile (which the right-hand side of the equation does not). Within the framework of the present assumptions, eqn (2.2.13) is thus the solution of the most general form. Quick check of polarities: with V_B positive, j turns out to be negative. This is not due to any exotic mutation of Ohm's Law, but arises directly from the (conventional) way in which the boundary conditions (2.2.10) have been formulated, i.e. by reference to a fixed Fermi level *in the metal.* With the semiconductor positive relative to the metal, electrons go towards the right (e.g. on Fig. 1.6(c)) and constitute a negative current. Of course, V_B could have been defined with opposite polarity to avoid the association of a positive voltage with a negative current. However, because V_B and V_D are truly additive in the reverse direction (see Fig. 1.7), there are good reasons for making V_B positive in the reverse direction, as has been done. This choice having been made, we shall then maintain appearances and real-world consistency by writing $J=-j$ in some contexts, while remembering that j reflects the real polarity of the current (defined as positive when electrons go towards the left). This gives

$$J=\frac{e\mathcal{N}_c D_n\{1-\exp(-eV_B/kT)\}}{\displaystyle\int_0^{\lambda_B}\exp(-eV/kT)\,\mathrm{d}x}. \tag{2.2.14}$$

This is a logical outcome of the diffusion equation (2.2.7) which, in turn, is based on the assumption that bulk conduction laws are valid within the barrier. Whether they really are, depends on the barrier field and the carrier mobility. When both are high, *hot carrier effects* must be envisaged, arising from drift velocity saturation, as discussed in Section 2.2.7. In any event, eqn (2.2.14) represents the total current density only as long as electron tunnelling can be neglected (as it here is). It can be solved for various barrier profiles, as outlined in Sections 2.2.3, 2.2.4, and 2.2.5. Meanwhile, and by way of a purely transient excursion, it is desirable to consider the carrier concentration relationships in somewhat greater detail.

The object of eqns (2.2.8) to (2.2.14) is to arrive at a solution of eqn (2.2.7), giving j in terms of V_B, the applied voltage. Alternatively, eqn (2.2.7) can be 'solved' to give $n(x)$ as a function of $V(x)$, even though $V(x)$ is not explicitly known. This makes it possible to examine to what extent $n(x)$, for a given x, differs from the value prescribed for it by Boltzmann statistics when a current is flowing. To do this, we shall keep $n(\lambda_B)$ severely constant (n_e), and allow $n(0)$ to 'float' in accordance with the requirements of the transport equation. We can then put

$$n(x)=n_e\exp\left\{\frac{eV(x)-eV_B+\phi_n}{kT}\right\}+\Delta n \tag{2.2.15}$$

where Δn measures the departure from the Boltzmann concentration. Δn will be a function of x and j. Note that when $x = \lambda_B$ we have $eV(\lambda_B) = -\phi_n + eV_B$ which makes $\Delta n = 0$. Moreover, Δn must be zero for all values of x when $j = 0$. Substituting for $n(x)$ into eqn (2.2.8) gives

$$\frac{d(\Delta n)}{dx} - \frac{e}{kT}\left(\frac{dV}{dx}\right)\Delta n - \frac{j}{\mu kT} = 0 \tag{2.2.16}$$

but this cannot be fully integrated without an explicit knowledge of $V(x)$ as a function of x. However, Spenke (1955) has shown that as long as $x \ll \lambda_B$, the most important contribution to Δn comes not from the general but from the particular solution, at any rate in the reverse direction, for which $d(\Delta n)/dx$ is small. Then

$$\Delta n(x) = \frac{-j}{\mu_n e(dV/dx)_x}, \quad \text{and} \quad \Delta n(0) \approx \frac{j}{\mu_n eF(0)} \tag{2.2.17}$$

remembering that j is negative in the reverse and positive in the forward direction of current flow. The field F is, of course, negative in all cases. We see, therefore, that reverse currents call for carrier concentrations near the boundary which are larger than the values expected from the Boltzmann equilibrium distribution of carriers. This reduces the local concentration gradients, and causes the field currents to dominate even more than they would otherwise have done. Conversely, in the forward direction, diffusion theory predicts boundary concentrations lower than those suggested by the Boltzmann equilibrium distribution. This implies higher concentration gradients, and causes diffusion currents to play an even more prominent role, and though eqn (2.2.17) is then no longer a faithful approximation, it is still qualitatively valid.

Figure 2.15 illustrates some of these relationships. The full lines correspond to the simplifying assumption of a fixed boundary concentration $n(0)$, the broken lines to the Boltzmann equilibrium distribution, based on a fixed $n(\lambda_B) = n_e$. Equation (2.2.17) calls for a compromise between these two limiting cases (dotted lines). Over much of the barrier thickness, the contours practically (though never quite) coincide. In view of this, the above simplifying assumption $n(0) = \text{constant}$, while not in perfect harmony with expectations, can be deemed to represent an acceptable basis for the calculations pursued below. Billig and Landsberg (1950) have made such calculations of carrier density for a slightly different barrier configuration, namely a Mott barrier (see Section 2.2.3), and the results were essentially similar.

In contrast, it should be kept in mind that concentration contours in accurate agreement with the Boltzmann relationship would automatically imply *zero* current. This can easily be shown to be so, because the solution of the diffusion equation (2.2.8) for $j = 0$ is indeed the Boltzmann

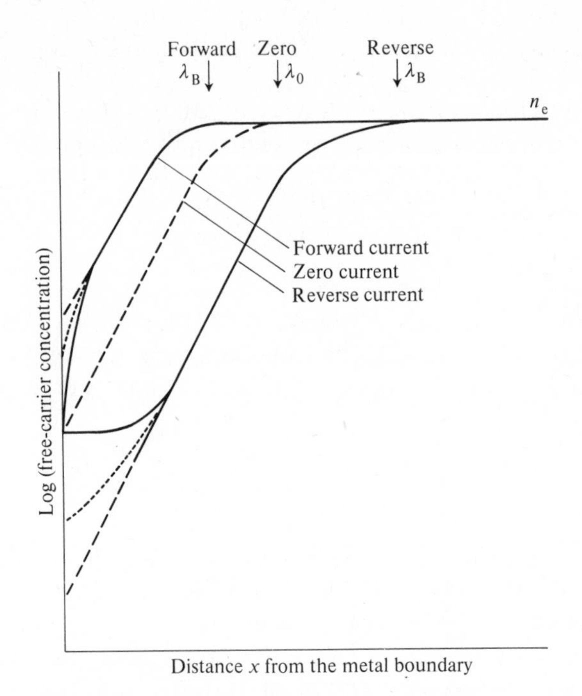

FIG. 2.15. Electron concentration contours within a Schottky barrier—schematic representation. Full lines: calculated concentrations based on a fixed value of $n(0)$. Dotted lines: expectations derived from eqn (2.2.17), and obtained by numerical methods. Broken lines: contours based on the Boltzmann equilibrium relationship alone.

equation. Thus, *every* current leads to a Δn and, conversely, a non-zero Δn can be regarded as an essential condition for current flow. Additional computed results will be presented in Section 2.4.2.

Carrier concentration contours are often graphically represented by means of quasi-Fermi levels, but these levels do not by themselves add anything new. Their position is never known from independent evidence, and their use serves only to introduce a measure of carrier concentration into band diagrams which would not otherwise contain any such reference. In the present context, and as a direct result of the above considerations, it follows that the quasi-Fermi level for electrons cannot possibly be constant *throughout* the current-carrying barrier, despite occasional claims that it is. It can, of course, be approximately constant over a part of the barrier thickness, and this is reflected by Fig. 2.15.

2.2.3 *Voltage–current relationships; Mott barrier*

Because of its simple shape, the Mott barrier lends itself most readily to the complete evaluation of eqn (2.2.14). We have, by substitution for V

from eqn (2.2.4):

$$\int_0^{\lambda_M} \exp(-eV/kT)\,dx = \exp\left(\frac{\phi_{ns}}{kT}\right)\int_0^{\lambda_M} \exp\left\{\frac{-e(V_D+V_B)x}{\lambda_M kT}\right\}dx$$

$$= \left\{\frac{-\lambda_M kT}{e(V_D+V_B)}\right\}\exp\left(\frac{\phi_{ns}}{kT}\right)\left[\exp\left\{\frac{-e(V_D+V_B)}{kT}\right\}-1\right] \quad (2.2.18)$$

which, with (2.2.13) and (2.2.14), yields:

$$J = \frac{\sigma_b(V_D+V_B)}{\lambda_M}\exp\left(-\frac{eV_D}{kT}\right)\left[\frac{1-\exp(-eV_B/kT)}{1-\exp\{-e(V_D+V_B)/kT\}}\right] \quad (2.2.19)$$

as obtained by Mott (1939). Here $(V_D+V_B)/\lambda_M$ is clearly the barrier field, and σ_b is the bulk conductivity. Thus, $\sigma_b \exp(-eV_D/kT)$ could be deemed to be the effective conductivity at the top of the barrier, a notional entity which goes under the name of *Randschichtleitfähigkeit* (σ_R) in the German literature. All non-linearity arises from the term on the right. Except for the highest forward voltages (V_B negative and comparable with V_D), the exponential term in the denominator is negligible, giving under more normal conditions

$$J = \frac{\sigma_b(V_D+V_B)}{\lambda_M}\exp\left(-\frac{eV_D}{kT}\right)\{1-\exp(-eV_B/kT)\} \quad (2.2.20a)$$

which could also be written as

$$J = \sigma_R F(0)[1-\exp(-eV_B/kT]. \quad (2.2.20b)$$

The shape of this voltage–current characteristic is easily understood. In the reverse direction ($V_B>0$), the exponential term with V_B quickly disappears, leading to a constant differential resistance. In the forward direction ($V_B<0$), the linear and exponential terms in eqn (2.2.20) are in competition with one another, the latter 'winning' by an enormous margin within the voltage range in which the present assumptions have any chance of being correct. The result is an almost exponential forward characteristic. Figure 2.16 shows typical relationships, as calculated from eqn (2.2.20a) and before correction for some of the high field effects already mentioned in Section 2.1.5. (See also Section 2.2.8.) All resistances in series with the barrier have been neglected here, but will be discussed in Section 2.2.5.

The ideal Mott barrier is rarely encountered in practice, but it remains a historically and didactically important limiting case for contacts on semiconductors with one-dimensionally non-uniform impurity content. Because the model makes no allowance for free carrier space charges and other complicating factors, actually observed characteristics must be expected to depart somewhat from the simple behaviour shown in Fig. 2.16.

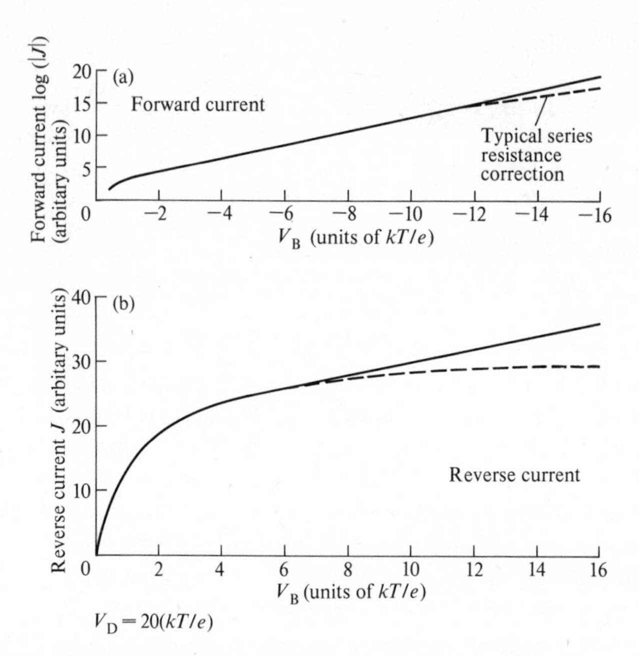

FIG. 2.16. Isothermal voltage–current relationships of a simple Mott barrier (full line), as calculated from eqn (2.2.20a). Note the linear regions in (a) and (b). Broken line in (b); reverse characteristic of a Schottky barrier, fitted at $V_B = 5kT/e$; eqn (2.2.25).

2.2.4 *Voltage–current relationships; Schottky barrier*

The non-uniform field of the Schottky barrier makes the evaluation of the integral in eqn (2.2.14) more complicated. From eqns (2.1.3–2.1.5) we have for the shape of the barrier profile:

$$eV(x) = -\phi_{ns} + (N_d e^2/2\epsilon)(2\lambda_B x - x^2). \tag{2.2.21}$$

On this basis, the integration can be achieved in two stages of accuracy. [See Section 2.2.5 for a treatment of the high forward current range.]

It will be seen that, over most of the integration range ($0 < x < \lambda_B$) we shall have

$$2\lambda_B x \gg x^2. \tag{2.2.22}$$

The interval in which this is true actually contributes most to the integral, and it is therefore permissible to neglect the x^2 term, as long as we are not concerned with high forward direction. The result is:

$$\int_0^{\lambda_B} \exp(-eV/kT)\,dx = \exp\left(\frac{\phi_{ns}}{kT}\right)\int_0^{\lambda_B} \exp\left\{\frac{-N_d e^2 \lambda_B x}{\epsilon kT}\right\} dx$$
$$= \left\{-\frac{\epsilon kT}{N_d e^2 \lambda_B}\right\}\exp\left(\frac{\phi_{ns}}{kT}\right)\left\{\exp\left(\frac{-N_d e^2 \lambda_B^2}{\epsilon kT}\right) - 1\right\}. \tag{2.2.23}$$

This may now be substituted into eqn (2.2.14). When λ_B is expressed in terms of V_B, this yields

$$J = \sigma_b \left\{ \frac{(V_D + V_B) 2 N_d e}{\epsilon} \right\}^{1/2} \exp\left(\frac{-eV_D}{kT} \right) \left[\frac{1 - \exp(-eV_B/kT)}{1 - \exp\{-2e(V_D + V_B)/kT\}} \right]. \tag{2.2.24}$$

Under the conditions assumed, the exponential term in the denominator is completely negligible, which makes it permissible to write the equation in the form

$$J = \sigma_b \{(V_D + V_B) 2 N_d e/\epsilon\}^{1/2} \exp(-eV_D/kT)\{1 - \exp(-eV_B)/kT\} \tag{2.2.25}$$

Comparison with eqn (2.2.20) for the Mott barrier shows that only the pre-exponential terms are different and, even then, the consequences are astonishingly small. Remarkable as it may seem, eqn (2.2.25) can also be expressed in a form corresponding to eqn (2.2.20b), namely

$$J = \sigma_R \{F(0)\}^{1/2} \{1 - \exp(-eV_B/kT)\} \tag{2.2.26}$$

and this fact appears to be independent of barrier profile, as long as high field corrections are neglected. However, a general proof is unavailable, and eqn (2.2.2b) should be used with caution, remembering that σ_R, $F(0)$, and V_B are not in fact independent variables.

Because the pre-exponential term is only slightly dependent on V_B, compared with the exponential term, convention (unsupported by circumspect caution) has introduced another stage of simplification, namely

$$J = J_s \{1 - \exp(-eV_B/kT)\} \tag{2.2.27}$$

thereby achieving an expression that is identical in form with one that applies to the p–n junction. It signifies total saturation at $J = J_s$ in the reverse direction, whereas the true characteristic should vary with $(V_D + V_B)^{1/2}$, and should thus fail to attain the linearity shown by the full line in Fig. 2.16(b). The broken line on this figure shows this. However, the differences are not always clear enough in practice to establish the nature of the barrier on these grounds alone. For Schottky as well as Mott barriers (in the absence of high boundary field corrections) the value of V_D can be obtained from the temperature-dependence of j_s and a knowledge of the Fermi level in the bulk material. As eqns (2.2.25) and (2.2.27) show, the temperature-dependence of J_s is governed by $\sigma \exp(-eV_B/kT)$, i.e. by $\exp(-\phi_{ns}/kT)$, since $\sigma = \sigma_0 \exp(-\phi_n/kT)$ and $\phi_{ns} = \phi_n + eV_D$. A corresponding treatment for incompletely ionized donors is not yet available.

It will be clear that Schottky barriers and Mott barriers represent limiting models, to which practical systems may or may not conform. Intermediate cases can certainly be envisaged in great variety, and for interpretational purposes the choice between models is best made on the basis of capacitance relationships (Chapter 4). Contrary to popular assumption, geometrical fit of calculated voltage–current relationships is generally not a sufficiently discriminating criterion for this purpose. Indeed, it can be grievously misleading. Thus, Landsberg (1951a) has discussed a case in which the Schottky model yielded a better *fit*, but the Mott model yielded more plausible descriptive constants of the barrier! It is precisely for such reasons that some of the hybrid models of Section 2.3 have been devised.

2.2.5 Barriers in the high forward direction

For a number of reasons, the above treatment is unsuitable for an analysis of contact behavior in the high forward direction. Among those are:

(a) The approximation involved in eqn (2.2.22).
(b) The neglect of the exponential term in the denominator of eqn (2.2.24).
(c) The neglect of free carrier space charges.
(d) The neglect of all series resistances.
(e) The complications arising from non-constancy of the pre-exponential term. Indeed, simplest inspection of eqn (2.2.25) shows that things go disastrously wrong as V_B approaches $-V_D$; the model fails to reflect the intuitive and utterly reasonable expectation that the resistance should go to zero when the applied forward voltage is large enough to make the barrier vanish.

As long as the pre-exponential term *is* taken to be constant, $\log |J|$ should vary linearly with V_B in the high forward direction, with a slope of e/kT. This expectation has come to be an article of faith among researchers in this field, but is in fact subject to challenge on all the grounds given above and, indeed, on other grounds. Most prominent among those is the open question, in any practical case, as to whether one is really dealing with a single-carrier system, as assumed here. See Section 3.5.

Some improvement upon the above treatment could be achieved by using a more accurate evaluation of the integration factor. For this purpose, we write the barrier profile as

$$eV(x) \approx -\phi_n - (N_d e^2/2\epsilon)(\lambda_B - x)^2 \qquad (2.2.28)$$

and assume that this is valid even in the presence of current flow. We substitute this for V in eqn (2.2.14). It is then convenient to change the variable to $y = (N_d e^2/2\epsilon kT)^{1/2}\ (\lambda_B - x)$, and this yields

$$J = \sigma(kT/N_d/2\epsilon)^{1/2}\{1 - \exp(-eV_B/kT)\}G^{-1}(z) \qquad (2.2.29)$$

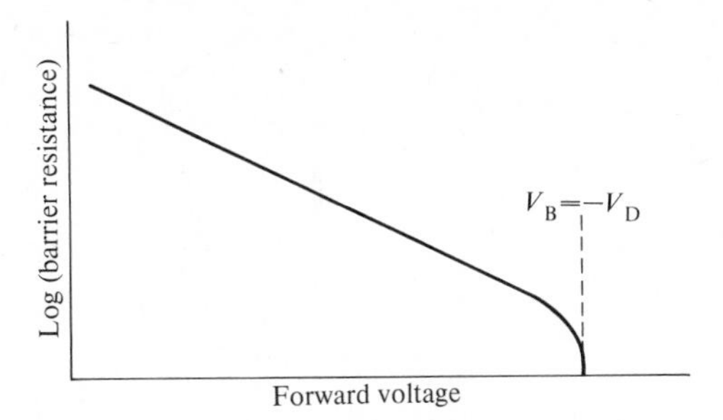

FIG. 2.17. Resistance–voltage characteristic of Schottky barriers in the forward direction; corrected analysis along the lines of Rose and Spenke (1949).

where

$$G(z) = \int_0^z \exp(y^2)\,dy \tag{2.2.30}$$

and

$$z = \{e(V_D + V_B)/kT\}^{1/2}$$

The complete solution can then be obtained by numerical methods, as demonstrated by Spenke (1949), Rose and Spenke (1949), and by Landsberg (1951a, 1952). Corresponding results are schematically shown in Fig. 2.17. They demonstrate that it is, indeed, the resistance that goes to zero as $V_B \rightarrow V_D$, and not the current, as eqns (2.2.20a) and (2.2.25) suggest. However, the method is now rarely used, and for good reason: once numerical integration is embarked upon, it might as well be applied to the diffusion equation as a whole (e.g. as in Section 2.4), and not merely to a correction of the explicit analysis.

Recognition of the fact that the barrier resistance must disappear when $V_B = -V_D$ gave rise to the notion that V_D might be evaluated in a simple way from the forward characteristic, as suggested by Fig. 2.18. However, there is no practical way of judging at what stage the composite $j - V_T$ characteristic becomes linear enough for extrapolation purposes. The method thus yields values of V_D which tend to be too low. The general fact is that we have *direct* experience only of V_T, not of V_B. In the reverse direction, $V_T \approx V_B$ tends to be a good approximation; under all other conditions an extrapolation is involved, and is often made on the basis of uncertain facts. With increasing forward voltages applied, the barrier resistance rapidly diminishes, and neglect of the bulk series resistance is less and less justified. Amongst other things, this affects the slope of the log $|J|$ versus V_B relationship, as qualitatively indicated in Fig. 2.16(a). In practice, such departures from linearity are frequently misinterpreted, and this will be further discussed under the heading of

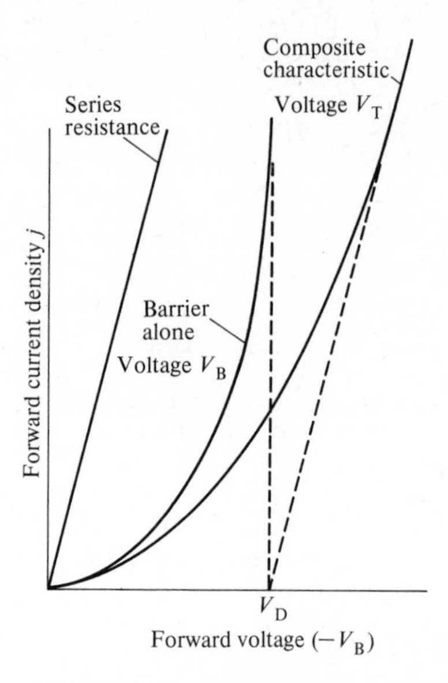

FIG. 2.18. Forward characteristics with and without series resistance. The composite characteristic is obtained by adding voltages for every current, yielding V_T.

'non-ideality'. Non-linearity, even though somewhat 'local', is often represented as a change of average slope and expressed by a modified form of eqn (2.2.27), of purely empirical character, namely

$$J = J_s\{1 - \exp(-eV_B/kT/\eta kT)\} \tag{2.2.31}$$

where η is the *non-ideality factor.* Rhoderick (1978) has pointed out that the presence of an artificial barrier layer (see Sections 2.5 and 5.2 as well as Section 1.3.4) is one of the many circumstances which can cause η to exceed unity. Crowell and Beguwala (1971) have examined the image-force correction as a cause of $\eta \neq 1$, and in Section 2.3.3 we shall see that the tunnel effect is likewise a possible mechanism. This is qualitatively easy to understand. At relatively high forward voltages, the effect of tunnelling will be very small, even negligible. At lower forward voltages, the effect will cause the current to be larger than otherwise expected. The slope of the log $|J|$ versus V_B relationship must therefore be diminished.

Thus, $\eta \neq 1$ can have a variety of interpretations, and some models even yield a non-constant (bias-dependent) η; see Wilkinson *et al.* (1977). Ordinarily, η differs from unity by a small fraction, e.g. 1.28 for Crowell and Beguwala's image force model. Instances of $\eta < 1$ will be further discussed in Section 3.5.3.

From all this, it will be clear that the actual values of η cannot be directly used for the interpretation of current flow mechanisms, being far too circumstantial and implicit. Even the conventional term 'ideality factor' is misleading, since it implies departures from a largely arbitrary 'norm'. The profound diagnostic significance with which observed values of η are often invested in the literature is therefore misplaced. There are also instances (McColl and Millea 1976 or Korwin-Pawlowski and Heasel (1975) in which the observed forward characteristic refuses spectacularly to be represented by eqn (2.2.31) at all, and this should not come as a surprise to anyone.

Among the many possible explanations of $\eta \neq 1$ the effect of series resistance is by far the simplest, particularly when one is dealing with contacts on high-resistivity material. If V_T is the total voltage and V_b the voltage dropped across the bulk material, then the plot of $\log |J|$ versus V_T would yield values of η such that $V_T/\eta = V_T - V_b$. Evidently, $\eta > 1$, even though V_b is current dependent. (See Section 2.4.2 for further discussions relating to single-carrier systems, and Chapter 3 to minority-carrier injection.) Manifacier and Fillard (1976) have provided a particularly instructive set of results on Ag–Ge contacts in which $\eta > 1$ is associated with injection. For non-injecting contacts, our principal concern here, an interesting method for separating the barrier and bulk contributions to the system resistance has been devised by Gútai and Mojzes (1975). It involves measuring the angular dependence of the magnetoresistance. High accuracy has been claimed.

2.2.6 *Thermionic emission; diode theory*

According to eqns (2.2.20) and (2.2.25), the reverse current can increase indefinitely, depending only on the barrier voltage V_B. This result is closely linked with the assumed boundary conditions and, in particular, with the last part of eqn (2.2.11). The notion there was that the carrier concentration $n(0)$ remains constant, no matter how great the externally imposed trends are to change it. We have already seen (Section 2.2.2) that forward currents tend to diminish $n(0)$ compared with Boltzmann *equilibrium* expectations, but it is equally clear that they increase $n(0)$ compared with the simplifying assumption $n(0)_{j=j} = n(0)_{j=0} = \text{constant}$. The converse is true in the reverse direction; we then face an electron depletion at the boundary, arising from the rapid removal of electrons by the field. This represents one of the most significant departures from the simplest model, and is one of the contributory needs for the development of diode theories. It will be further discussed in Section 2.3.2. However, there is an even more obvious problem: no electron flux to and from the metal can in fact be *unlimited*. This fact was first noted by Bethe (1942) who applied non-diffusion concepts to the analysis of contact rectification phenomena. (See also Torrey and Whitmer (1948.)

By virtue of the kinetic gas laws, only a limited number of electrons in the metal can make impact with the $x = 0$ boundary in unit time. In no circumstances can any barrier field remove (and transport into the semiconductor) a greater number from that interface. A reverse current saturation effect would therefore be unavoidable, were it not for (a) the tunnel effect, (b) image forces, and (c) thermal side effects (see Persky 1972). Diode theory concerns itself with this saturation, but goes beyond that task, and gives rise to a rectification model of its own. In its simplest form, this is based on the assumption that the mean free path of electrons in the semiconductor is greater than or, at any rate, comparable with the barrier width. All electrons coming from the metal would then make their first collision in the semiconductor at a depth from which return to the metal is unlikely; hence all would 'count' towards the current.

In practice, it is easiest, but not entirely correct, to consider an imaginary boundary at $x=\lambda_B$ rather than the metal boundary at $x=0$, because one then has only one material to deal with. According to kinetic gas theory and Boltzmann statistics, the number of electrons with velocities between v_x and (v_s+dv_x) in the x direction which pass unit cross-section in unit time is

$$dn_1 = n_e v_x (m/2\pi kT)^{1/2} \exp(-mv_x^2/2kT)\, dv_x \tag{2.2.32}$$

where m is the standard electron mass. The minimum velocity required for a transition from the semiconductor into the metal is, of course, given by

$$E_{min} = \tfrac{1}{2}mv_{x0}^2 = e(V_D + V_B) \tag{2.2.33}$$

Since each electron carries a charge e, and since the current arises from all the electrons with energies greater than E_{min}, we have

$$\begin{aligned} j_+ &= n_e e\left(\frac{m}{2\pi kT}\right)^{1/2} \int_{v_{x0}}^{-\infty} v_x \exp(-mv_x^2/2kT)\, dx \\ &= n_e e\left(\frac{kT}{2\pi m}\right)^{1/2} \exp\left(-\frac{mv_{x0}^2}{2kT}\right) \\ &= n_e e\left(\frac{kT}{2\pi m}\right)^{1/2} \exp\{-e(V_D+V_B)/kT\} \end{aligned} \tag{2.2.34}$$

In accordance with convention, we have used Boltzmann equilibrium statistics here, which is intended to apply only at the edge of the barrier and in the bulk beyond, not within the barrier itself. Within the barrier, in the absence of collisions, there is no mechanism for 'thermalizing' the electron gas. Within the semiconductor, collisions provide that mechanism, but need space in which to achieve it. That need is ignored here (see also below).

It is, of course, necessary to envisage a corresponding flux in the opposite direction, i.e. a flux of electrons into the semiconductor, and this must equal the value of j when $V_B = 0$, even though its fundamental constitution may be very different. Thus:

$$j_- = n_e e\left(\frac{kT}{2\pi m}\right)^{1/2} \exp(-eV_D/kT) \tag{2.2.35}$$

giving a net current density

$$j = j_+ - j_- = n_e e\left(\frac{kT}{2\pi m}\right)^{1/2} \exp\left(-\frac{eV_D}{kT}\right)\left\{\exp\left(-\frac{eV_B}{kT}\right) - 1\right\}. \tag{2.2.36}$$

when the temperature dependence of n_e is taken into account, via $n_e = \mathcal{N}_c \exp(-\phi_n/kT)$, this yields the well-known

$$j = A^* T^2 \exp(-\phi_{ns}/kT)\left\{\exp\left(-\frac{eV_B}{kT}\right) - 1\right\} \tag{2.2.37}$$

with $A^* = 4\pi e m_n k^2/h^3$ (Richardson's constant) where an isotropic effective mass m_n has been used in place of m, as a (small) concession to band structure relationships. For free electrons ($m_n = m$), we would have $A^* = A = 120$ amperes $cm^{-2} K^{-1}$ (compare Sze 1969). For n-type Si, $A^* = 112$ amperes $cm^{-2} K^{-1}$, and for p-type Si, $A^* = 32$ amperes $cm^{-2} K^{-1}$ (Andrews and Lepselter 1970). Of course, A^* does in fact depend on the anisotropy of the effective carrier mass (Crowell 1965, 1969a), and thus indirectly also on the boundary field, as Andrews and Lepselter have pointed out. For semiconductors with ellipsoidal energy surfaces,

$$A^* = \frac{4\pi e k^2}{h^3}(l_x^2 m_y m_z + l_y^2 m_z m_x + l_z^2 m_x m_y) \tag{2.2.38}$$

where l_x, l_y, l_z are the direction cosines of the incoming electrons, relative to the crystal symmetry, and m_x, m_y, m_z are, as usual, the components of the effective mass tensor. In practice, and all these arguments notwithstanding, A^* is usually treated as a constant, in comparison with the voltage-dependence (Section 2.2.8). However, its absolute magnitude remains an 'adjustable parameter' of disturbing proportions, all the more since even the above expressions neglect all interface interactions between semiconductor and metal. It is, indeed, unlikely, considering the assumptions of the model, that the situation can be properly described by reference to the effective mass of carriers in the semiconductor alone. After all, one of the unbalanced fluxes comes from the metal and is supposed to penetrate into the semiconductor over a substantial distance without collisions. In the metal the electrons have a perfectly well-defined effective mass, but what their mass might be, in the semiconductor before

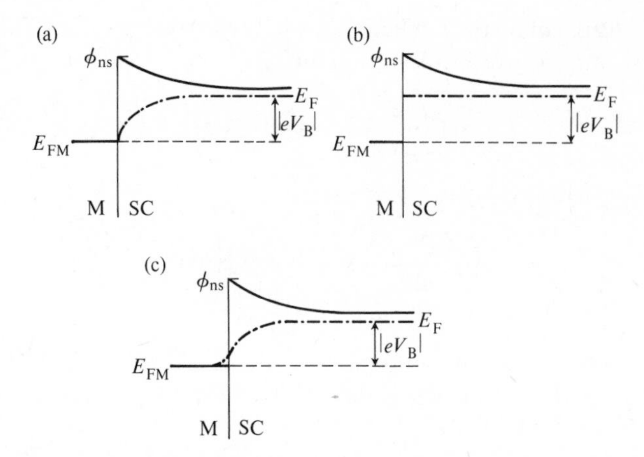

FIG. 2.19. Drift-diffusion and diode theories; schematic barrier contours with forward bias V_B. (a) Drift-diffusion; Fermi level here shown without any field penetration into the metal. (b) Diode theory; implied Fermi level here shown with step discontinuity at the boundary. (c) Compromise situation which is expected to prevail in practice.

thermalization, is not so clear. The longer the mean-free-path, the wider will be the region in which thermalization after transit will be achieved, and that region is also neglected by simplest diode models (though not altogether in Section 2.3.2).

There are, indeed, other problems. The conventional drift-diffusion theory assumes that the Fermi levels in the metal and the semiconductor coincide at $x = 0$, whereas the diode theory, as formulated above, assumes Boltzmann equilibrium within the semiconductor, which means a horizontal (or, in view of the current, almost horizontal) Fermi level within the barrier and a discontinuity at the boundary (Fig. 2.19). Such a discontinuity, implying a sharp boundary between adjacent electron distributions, is obviously implausible. In Section 2.3.2, the boundary condition will be somewhat relaxed, but this particular problem is not thereby solved. In discussions of the relationship between diode and drift-diffusion models, these differences are often overlooked.

On all these grounds, drift-diffusion models are here preferred, even though they are afflicted with a few problems of their own. Thus, for instance, Baccarani (1976) has shown that while the Boltzmann transport equation is valid for bulk material, it is always suspect when applied to the region near a boundary. Nevertheless, drift-diffusion models (modified by current-controlled boundary conditions as in Section 2.3.2, and occasionally by other corrections) have been surprisingly successful. This is so because it is frequently only the coarse predictions of a model that can be experimentally tested, and several models may have those in common.

In accordance with the conventions used above, eqn (2.2.12) may be written in a form identical with eqn (2.2.27):

$$J = J_s\{1 - \exp(-eV_B/kT)\}$$

where

$$J_s = (n_e e\bar{v})\exp(-eV_D/kT) \qquad (2.2.39a)$$

and

$$\bar{v} = (kT/2\pi m_n)^{1/2} \qquad (2.2.39b)$$

as an effective velocity. The use of m_n is again purely schematic. When electrons enter the semiconductor from the metal and reach bulk material without having made a collision within the barrier, they come to be increasingly far from the band edge. In as much as they can be regarded as belonging to the semiconductor, they should thus have an x-dependent effective mass before reaching their first energy-destroying encounter; but to what extent this is actually true remains unclear. Sze (1969) gives $2.2 \times 10^4\,\mathrm{m\,s^{-1}}$ as the room-temperature value of v for electrons in silicon, and $1.7 \times 10^4\,\mathrm{m\,s^{-1}}$ for holes; such differences of velocity are expected from the differences of effective mass. To ensure flux balance at the boundary, there has to be an adjustment in terms of carrier concentrations, which must then depart from the classical Boltzmann relationship, *even at zero current.* The thermionic emission model is therefore only superficially simple. In any event, the present version is approximate, and covers only the extreme case of very long mean-free-paths, just as the classical Schottky model represents an extreme case, namely that of very short mean-free-paths. Realistic cases must fall between these limits, depending on circumstances. Hybrid models are discussed in Section 2.3.

An interesting speculation is to contemplate the properties of a high barrier on a semiconductor with a narrow conduction band. Highly mobile electrons coming in from the metal might then find themselves in the upper half of the band. As far as is known, the consequences have never been analysed, and it remains uncertain whether such a system could be realized in practice. All the semiconductors which are characterized by long mean-free-paths have wide, often overlapping bands, and those (e.g. organic crystals) that are characterized by narrow bands have very short mean-free-paths.

Equation (2.2.37) denotes total saturation in the reverse direction. If, other things being equal, any model predicts a reverse current higher than that allowed here, it must be in error, at any rate as long as image-force and tunnelling considerations are unimportant. Equation (2.2.37) thus represents the *maximum degree of rectification* that any contact barrier can be expected to exhibit, as long as the donor content N_d is uniform.

(See Section 2.3.7 for a discussion of non-uniform doping.) Note again that the parameters of the metal enter into these matters only in as much as they influence V_D (via ϕ_{ns}). If a model is found to be in error, it may be that some of its restrictive assumptions have to be relaxed, e.g. as in Sections 2.2.7 and 2.2.8, or else that a hybrid model (Section 2.3) would be more appropriate.

2.2.7 *'Hot' carrier effects*

Local field variations play no role in the above analysis, nor does the drift velocity, since diode theory is concerned only with the relationships of thermionic emission. Incoming electrons 'see' the potential drop within the barrier, of course, and it is precisely this drop that causes the velocity distribution at $x = \lambda_B$ to be different from that at $x = 0$. In principle, the field could be arbitrarily high, but in the context of drift-diffusion models, the very magnitude of the barrier field leads to problems, as described below.

A typical barrier thickness might be 5×10^{-6} cm, and a typical barrier height 0.5 eV, giving a field of 10^5 V cm^{-1}. Under such conditions, Ohm's law ceases to be valid in materials of high carrier mobility, e.g. Ge and Si, as demonstrated by the classic experiments of Ryder and Shockley (1951), Shockley (1951), and Ryder (1953). Instead of maintaining proportionality to the field, the drift velocity saturates. This happens when the average energy of the electron cloud is increased significantly (by the field) above the thermal energy of $\frac{3}{2}kT$. The analysis of Sections 2.2.3 and 2.2.4 fails to take this into account, and thus tends to yield excessively high currents by pretending that the carrier mobility μ_n is constant. Drift velocities calculated on this basis can be many times higher than thermal velocities (see, for instance, Socha and Eastman (1982)) and, within the assumptions of the drift-diffusion model, such an outcome would be absurd. In fact, the mobility decreases as the field increases. On these grounds, and in order to arrive at an upper limit for the correction, Burgess (1953) made the opposite assumption, and took the drift velocity v_{dn} to be totally saturated, and thus constant within the barrier. A diminished mobility was expected to make the system more Boltzmann-like, but the constant (field-independent) v_{dn} was something new.

In order to develop this line of approach, another important assumption had to be made, namely one concerning Einstein's Relation in the presence of a high field. With μ_n field-dependent, the relation was taken to be

$$\mu_{Fn} kT/e = D_{Fn}, \qquad (2.2.40)$$

asserting the continued validity of a relationship in the accepted form. We do know that this relationship must be obeyed at zero current, because field and diffusion must then be in balance, even though the quiescent resident field may be very high. We therefore expect problems only when applied voltages are large enough to modify the internal field by truly substantial factors. Of course, the barrier field diminishes with increasing distance x, but Burgess assumed, for simplicity, that the drift velocity has its saturation value v_{sdn} everywhere within the barrier, and thus wrote:

$$j = -n(x)ev_{\mathrm{sdn}} + kT\mu_{\mathrm{Fn}}(\mathrm{d}n/\mathrm{d}x) \tag{2.2.41}$$

The principal interest is in the reverse direction, with positive values of V_{B}. Eliminating μ_{Fn} by considering

$$v_{\mathrm{sdn}} = -\mu_{\mathrm{Fn}}F = \text{constant} \tag{2.2.42}$$

leads to

$$\frac{\mathrm{d}n}{\mathrm{d}V} - \frac{en}{kT} = \frac{j}{kTv_{\mathrm{sdn}}} \tag{2.2.43}$$

which integrates as

$$n(x) = n_{\mathrm{c}} \exp(eV(x)/kT) - j/ev_{\mathrm{sdn}} \tag{2.2.44}$$

where n_{c} is a constant as far as the integration is concerned, but is dependent on j, and is so named because $n_{\mathrm{c}} = \mathcal{N}_{\mathrm{c}}$ at zero current. The concentration n_{c} is then determined by reference to the boundary conditions, of course. At $x = \lambda_{\mathrm{B}}$, we have

$$n(\lambda_{\mathrm{B}}) = n_{\mathrm{e}} = n_{\mathrm{c}}\exp\left(\frac{-\phi_{\mathrm{n}} + eV_{\mathrm{B}}}{kT}\right) - \frac{j}{ev_{\mathrm{sdn}}} \tag{2.2.45}$$

and at $x = 0$,

$$n(0) = n_{\mathrm{e}}\exp\left(\frac{-eV_{\mathrm{D}}}{kT}\right) = n_{\mathrm{c}}\exp\left(-\frac{\phi_{\mathrm{ns}}}{kT}\right) - \frac{j}{ev_{\mathrm{sdn}}}. \tag{2.2.46}$$

Together, the last two equations give, with $J = -j$,

$$J = n_{\mathrm{e}}ev_{\mathrm{sdn}}\exp\left(\frac{-eV_{\mathrm{D}}}{kT}\right)\frac{\{1 - \exp(-eV_{\mathrm{B}}/kT)\}}{[1 - \exp\{-e(V_{\mathrm{D}} + V_{\mathrm{B}})/kT\}]} \tag{2.2.47}$$

which can again be approximated to

$$J = J_{\mathrm{s}}\{1 - \exp(-eV_{\mathrm{B}}/kT)\} \tag{2.2.48}$$

and is thus identical in form to the expression yielded by diode theory, eqn (2.2.37), differing from it only in the detailed structure of J_{s}. Landsberg (1954) has shown how the analysis proceeds when the above simplifying assumptions are relaxed; the equation must then be solved numerically. As far as is known, no modern computer-work in this field has been published, but a more detailed discussion of the space-charge effects associated with current flow under high field conditions has been given by Krömer (1953). These effects arise from the fact that $n(x)$ versus x yields a different concentration profile when v_{dn} is saturated than when it is not.

Without assuming *complete* saturation, Krömer showed that the local extra carrier concentration $\Delta n(x)$ is proportional to jF, and that the constant of proportionality can be calculated from the lattice constant, the Debye temperature, and the width (in this case) of the conduction band. This excess $\Delta n(x)$ must then be reintroduced into Poisson's equation. The complete integration would again have to be performed by numerical methods. Analytic procedures yield only an implicit relationship between j, V_B, and $F(0)$. Because Burgess (1953) did assume complete saturation, these complications do not arise in the summary of his work given above. Once the drift velocity has reached its saturated limit, the precise (higher) value of F ceases to be important. Landsberg (1955) has given general arguments in support of the view that essential aspects of the diffusion theory do indeed remain unchanged, as long as μ and D change less than the field F does within the barrier as a function of x.

For more detailed models of hot-carrier behaviour within a Schottky barrier, see Blötekjaer (1966), Goldberg (1969), and Stokoe and Parrott (1974). It is necessary to consider the conditions under which charge transport and energy transport are both continuous. Blötekjaer and Goldberg have for this purpose used a modified form of the diffusion equation (2.2.7), namely

$$\begin{aligned} j &= en(x)\mu_n F(x) + \mu_n kT(\mathrm{d}n/\mathrm{d}x), \\ &= en(x)\mu_n F(x) + \tfrac{2}{3}\mu_n \frac{\mathrm{d}(n\bar{E}}{\mathrm{d}x} \end{aligned} \tag{2.2.49}$$

where $\bar{E}$ is the *average* energy of electrons, which is $\frac{3}{2}kT$ at low fields, but $\frac{3}{2}kT_e$ at high fields, T_e being the electron temperature. In addition, the power dissipated is evidently used in two ways. A part is passed to the lattice (ohmic loss), and a part is used to sustain the energy transfer by electrons from hot to less-hot regions. That depends on the electronic distribution function. With these notions, and the 'emission-controlled boundary conditions' discussed in Section 2.3.2, Stokoe and Parrott (1974) calculate a new voltage–current relationship in terms of T_e which is a function of x. At the edge of the barrier, at $x = \lambda_B$, $T_e = T$ is assumed though, as mentioned above, thermalization cannot actually be instant. At $x = 0$, T_e may differ from the lattice temperature T, but is not expected to do so to any appreciable extent. In between, $T_e \neq T$, and may be separately calculated on the basis of various approximations. As one would expect, reverse currents are always smaller than they would be if carriers were not hot. However, Stokoe and Parrott have also shown that, whereas the application of reverse voltages (leading to barrier fields greater than the quiescent fields) gives rise to carrier heating, the application of forward voltages (leading to barrier fields smaller than the quiescent fields) gives rise to carrier *cooling*, as long as $|V_B| < V_D$. Of course, when $V_B = -V_D$, T_e must again have its equilibrium value, which therefore goes through a minimum for a particular forward voltage applied; in Stokoe and Parrott's example: T_e (min.) $= 0.96\,T$ at $V_B = -0.9\,V_D$. The cooling effect is thus quite small and, as far as is known, there is no simple way of identifying it by experiment. The cooling concept implies that the average electron energy will be less than $\frac{3}{2}kT$, but this 'compression' of the electron gas into a narrower energy interval close to the bottom of the conduction band cannot lead to any substantial change of the effective mass, which means that the transport consequences of electron cooling are expected to be very small. The calculated results actually show that the current is reduced by cooling, just as it is by heating (though less), contrary to intuitive forecasts.

An entirely different (and more classical) approach to the problem is represented by 'ballistic' models of carrier transport, which assume that carriers of long mean-free-path (e.g. at low temperatures) can behave very much like electrons in vacuum, without 'seeing' the crystal field in the semiconductor. In that case, effective mass changes could be ignored, and this would make it possible for electrons to reach streaming velocities greatly in excess of the above saturation velocity v_{sdn} (Shur, 1981). It has been shown that such a pattern of assumptions can lead to agreement with many observations, e.g. on submicron devices at low temperatures, but it is also true that these notions involve conceptual difficulties of their own. (See also Shur and Eastman (1981).)

2.2.8 High field enhancement of reverse currents

In Section 2.1.5 we discussed high field corrections to the static barrier height, as necessitated by two processes: the Schottky effect and tunnelling. Since both effects depend on the local field close to the interface (approximately on $F(0)$), which in turn depends on the externally applied voltage, the high field direct corrections have a great impact on the voltage–current relationship primarily, of course, in the reverse direction. Since the effective barrier height is always lowered, the reverse currents are always enhanced. The proper way of introducing a correction into the model is to substitute the modified $V(x)$ contour into the integrating factor of eqn (2.2.13), but solutions could then be obtained only by numerical methods, e.g. as shown by Landsberg (1949, 1951a). However, in view of the many uncertainties involved, a more schematic method is considered permissible, namely the use of a simple correction factor $\exp(\Delta E/kT)$, where ΔE is the barrier lowering at $x=0$ from any cause. The procedure is somewhat crude, but is justified by the fact that image force and tunnel effect influence electron behaviour only at small distances x. For a Schottky barrier, for instance, the barrier-lowering under the influence of an applied voltage V_B can be calculated to a fair approximation by a modified eqn (2.1.31), namely

$$\Delta E = \Delta E_s = \left\{\frac{e^7 N_d(V_D + V_B)}{8\pi^2 \epsilon_\infty^2 \epsilon}\right\}^{1/4}. \tag{2.2.50}$$

A similar expression can be derived for a Mott barrier, of course. More accurate forms of analysis are frequently presented, but are rarely (if ever) worth while, in view of our general ignorance of the detailed structural parameters which prevail in any particular case. Indeed, a great deal of energy can be wasted on interpretive attempts which take simplistic models literally.

Image-force and tunnel-effect corrections lead to a pronounced 'softness' of the reverse characteristic, as shown in Fig. 2.20. For a barrier thickness $\lambda_M = 10^{-5}$ cm, the insert in Fig. 2.8 suggests that the barrier is effectively lowered by about $1.2kT$ when $V_B = 0.6$ volts (i.e. $24kT/e$ at room temperature), and this should increase the current by a factor of $\exp(1.2) = 3.3$. A point of inflection appears in the curve, because the correction is proportionately smaller at smaller voltages. Similarly, a greater thickness leads to a lower barrier field, and hence a smaller correction. By a suitable adjustment of the parameters, many other characteristics can be produced, but because these have no highly significant structure, they do not lend themselves to interpretive curve-fitting at

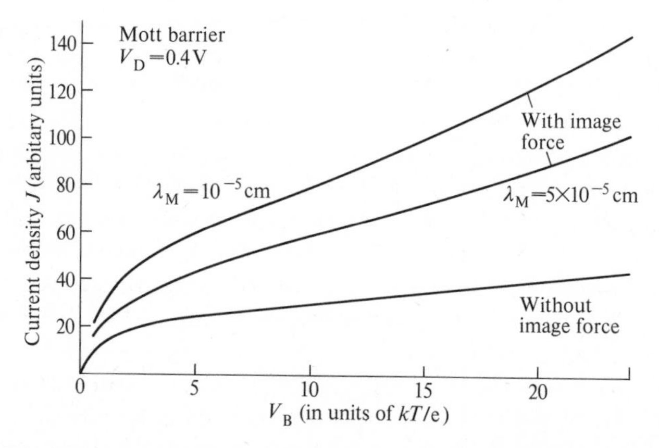

FIG. 2.20. Schematic image-force correction for the current through a Mott barrier on a semiconductor of high dielectric constant. $\epsilon = \epsilon_\infty = 16\epsilon_0$ assumed (here) for simplicity.

all convincingly. More than one type of correction can generally be 'fitted' to a given set of observations, and though such exercises are not always useless, their outcome should be treated with caution, and should always be tested against evidence derived by independent methods.

Among the many problems which can arise in practice are various kinds of edge effects and complications due to interface contamination. Ideally, experiments (intended for fundamental interpretation) call for an epitaxial metal layer on a single crystal substrate, but such a simple situation does not often present itself. In a rare case, Andrews and Lepselter (1970) managed to approach it by the use of silicide contacts (see Section 2.1.6), prepared by solid–solid chemical reaction with a silicon substrate. The resulting interfaces were believed to be structurally continuous and free from contaminants, including oxides. This made it possible to compare not only the *shapes* of calculated and observed voltage–current characteristics, but also absolute magnitudes. In this case, the conclusion was that the standard corrections for image force and tunnelling could *not* by themselves account for the observed softness of the reverse characteristic. The wave-function overlap effect mentioned in Section 1.3.2 was therefore invoked ('dipolar' correction), at any rate in schematic form, namely

$$\Delta\phi_{ns}/e = -\beta_C F(0) + \text{higher terms} \ldots \tag{2.2.51}$$

where β_C was regarded as an empirically adjustable constant, which measures the degree of overlap. For PtSi–n-Si contacts, typical values are $10 \leqslant \beta_C \leqslant 30$ Å, in good agreement with more fundamental calculations by Parker and co-workers (1968). Of course, such an additional constant is always a powerful instrument for bringing theory and experiment into harmony, and its full significance remains to be assessed. Meanwhile, good agreement was achieved for $ZrSi_2$–n-Si, RhSi–n-Si, RhSi–p-Si, and PtSi–n-Si interfaces over a substantial range of voltages (five orders-of-magnitude), currents (eleven orders-of-magnitude), and temperatures (276–344 K). As far as is known, pulse methods were not used for these measurements, but proportionality was established between current and contact area, indicating that self-heating played no significant role. All the present considerations apply to single-carrier systems and, in the course of comparisons between those theories and experiment, it is therefore important to ensure the absence of minority-carrier participation. Andrews and Lepselter (1970) were able to document this matter for their systems.

The tunnel effect leads (in principle) independently to an increase of current beyond the values calculated on the basis of drift-diffusion models, detailed analysis of tunnelling corrections has been provided by Rideout and Crowell (1970). The descriptive equations were ultimately solved by numerical methods. Calculated currents are enormously sensitive to the tunnelling distances envisaged, but about those we usually have no more than coarse information.

2.2.9 *Back-to-back barrier systems; field distribution and voltage–current characteristics*

We now proceed to a very different kind of problem, one related to macroscopic rather than microscopic transport. The fact that we often have to deal with *two* barriers has already been expressed in the context of Figs 1.1 and 1.4. Accordingly, current-carrying back-to-back barrier systems have been analysed at various levels of sophistication and generality. The ultimate need is for numerical solutions of the transport equations, but various analytic approximations are sometimes useful and instructive. Thus, Macdonald (1962) has shown that the diffusion equation can be integrated once, even when the free carrier space charge is taken into account. In the present terms, this yields

$$j = \frac{D_n e n_e}{L}\left[\exp\left(-\frac{eV_D}{kT}\right) - 1 + \frac{e(V_D + V_B)}{kT} - \frac{e\lambda_D}{2kT}\{F(L)^2 - F(0)^2\}\right] \tag{2.2.52}$$

where λ_D is as usual the Debye length, and where the unity term takes account implicitly of the voltage drop across the semiconductor bulk. $F(0)$ and $F(L)$ are the two prevailing boundary fields at $x = 0$ and $x = L$, where L is the thickness of the specimen, which may be arbitrarily larger than or smaller than the two barrier widths. This is a very general expression, which can be used when $F(0)$ and $F(L)$ both denote barrier fields, or else when one of them is zero, corresponding to an idealized low-resistance contact at one end. However, $F(0)$ and $F(L)$ are not explicitly known, which is why Macdonald proceeded by numerical methods from that point on. In many systems which are of practical interest, the voltage drop across semiconductor regions outside (between) the barriers is likely to be very small, and there is therefore something to be said for ignoring the bulk, as explained by Sze and co-workers (1971). When L is large compared with the barrier widths, the two contact resistances are effectively in simple series connection but, in the more interesting cases, L is small enough for the two-barrier regions to interact with one another.

Figure 2.21 shows the energy profiles and field distributions for various applied voltages. As long as the applied voltage is very small, the two barriers will be separate, one biased in the forward and one in the reverse direction. At some stage, as the applied voltage increases, the reverse-biased barrier will extend far enough to be continuous with the forward-biased barrier, and this is shown in Fig. 2.21(b). A further increase of

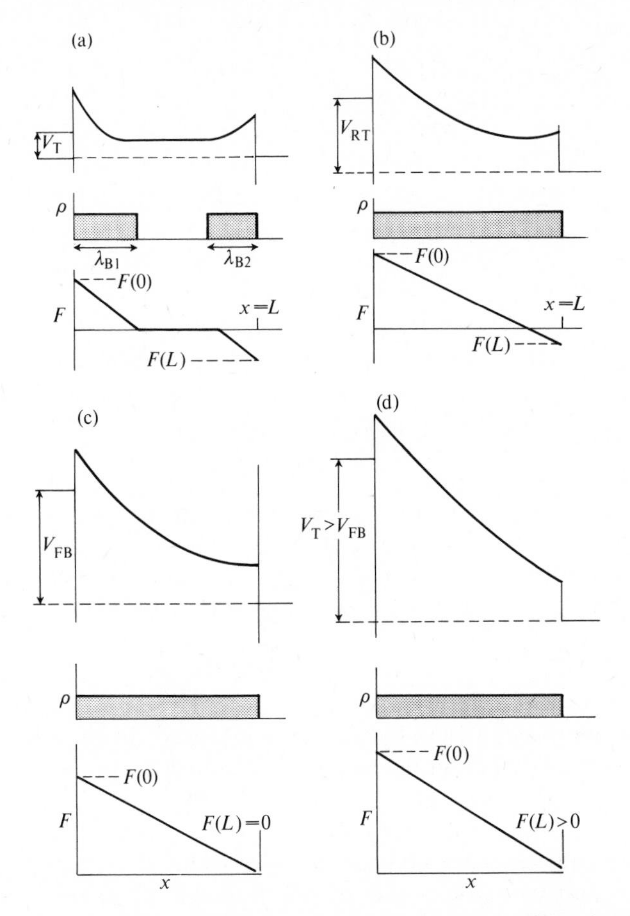

FIG. 2.21. Energy profile of a double-barrier system, following considerations by Sze and co-workers (1971). (a) Small voltage V_T applied to system. (b) Reach-through voltage V_{RT} applied. (c) Flat-band voltage V_{FB} applied. (d) $V_T > V_{FB}$. Shaded regions, ionic space charges. Free-carrier space-charges neglected.

applied voltage must eventually lead to a 'flat-band' condition for the previously forward-biased contact (Fig. 2.21(c)), and ultimately to a unidirectional field distribution, as in Fig. 2.21(d). It is convenient to look at these voltage ranges separately.

(a) Low voltages applied The situation of Fig. 2.21(a) is shown again in Fig. 2.22, which explains the nomenclature. The Fermi level of the undisturbed bulk portion is here regarded as the stable zero, from which the two applied barrier voltages V_{B1} and V_{B2} depart in either direction. Clearly, the *total* applied voltage is given by

$$V_T = V_{B1} - V_{B2} \tag{2.2.53}$$

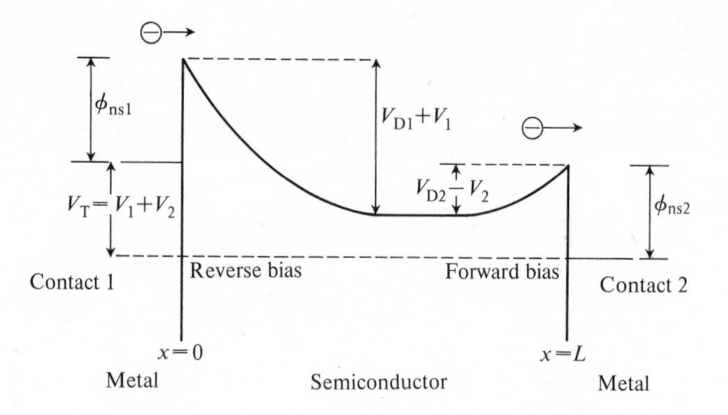

FIG. 2.22. Energy profile of a double-barrier system; detailed review of Fig. 2.21(a). $V_T < V_{RT}$.

where V_{BI} is positive and V_{B2} negative, in accordance with the terminology used above. However, it is more convenient here to deal with inherently positive quantities, and to take account of polarities explicitly. We shall therefore put $V_{B1} = V_1$ and $V_{B2} = -V_2$, and replace eqn (2.2.53) with

$$V_T = V_1 + V_2 \tag{2.2.54}$$

where V_T is as usual the total voltage across the semiconductor system. Neglecting minority carriers (as we do everywhere in this chapter), the reverse electron current on the left must be equal to the reverse forward current on the right, so that

$$j_{n1} = j_{n2} \tag{2.2.55}$$

where j_{n1} can be calculated conventionally in terms of V_1, and j_{n2} in terms of V_2. Sze and co-workers did so on the basis of diode theory (eqn 2.2.37), corrected for high field effects.

However, it is simple enough to predict the result without detailed calculations. The resistance of contact barrier **2** will be negligible compared with that of contact **1** for all but the smallest applied voltages. Therefore, the net characteristic of the assembly as a whole will be practically identical with that of the reverse characteristic associated (in this example) with contact **1**.

(b) *'Reach-through' and 'flat-band' voltages* A more interesting situation arises when the applied voltage is high enough for the barriers to touch (Fig. 2.21(b)). The applied voltage is then called the *reach-through voltage*, $V = V_{RT}$. According to eqn (2.2.1b), we then have two barrier thicknesses λ_{B1} and λ_{B2}, given by

$$\lambda_{B1} = \left\{\frac{2\epsilon}{eN_d}(V_{D1} + V_1)\right\}^{1/2} \tag{2.256a}$$

and

$$\lambda_{B2} = \left\{ \frac{2\epsilon}{eN_d} (V_{D2} - V_2) \right\}^{1/2} \tag{2.2.56b}$$

together with

$$\lambda_{B1} + \lambda_{B2} = L. \tag{2.2.57}$$

From these equations, and (2.2.54), V_{RT} can be calculated as:

$$V = V_{RT} = \frac{eN_d}{2\epsilon} L^2 - (V_{D1} - V_{D2}) - L \left\{ \frac{2eN_d}{\epsilon} (V_{D2} - V_2) \right\}^{1/2} \tag{2.2.58}$$

This still contains V_2 which can, however, be evaluated as shown below. Meanwhile, we note that a slightly higher applied voltage, $V = V_{FB}$ leads to the flat-band condition, corresponding to Fig. 2.21(c). Thus, eqn (2.2.58) can be adapted to yield

$$V = V_{FB} = \frac{eN_d}{2\epsilon} L^2 - (V_{D1} - V_{D2}) = \frac{eN_d}{2\epsilon} L^2 - \Delta V_D \tag{2.2.59}$$

where $\Delta V_D = V_{D1} - V_{D2}$. For a symmetrical system we have, of course, $\Delta V_D = 0$ which makes this equation particularly simple. Under all conditions V_{FB} may be calculated from the parameters of the system.

In order to proceed, we make use of a property that is characteristic of the Schottky barrier. From eqn (2.2.1a) we note that $[dE(x)/dx]_0 = N_d e^2 \lambda_B / \epsilon$ which is the slope of the energy profile at $x = 0$. The *average* slope derived from eqn (2.2.1b) is only

$$(V_D + V_B)/\lambda_B = \tfrac{1}{2}[dE(x)/dx]_0. \tag{2.2.60}$$

The actual slope at the origin, and thus the field, is therefore twice as great as the average slope. For the flat-band condition ($V_T = V_{FB}$), this is straightforward, since L is equal to the width of barrier **1**. We thus have

$$F(0)_{FB} = \frac{V_{FB} + \Delta V_D}{L/2} = \frac{2V_{FB} + 2\Delta V_D}{L}. \tag{2.2.61}$$

When $V_T < V_{FB}$, the situation is a little more complicated, but eqn (2.2.61) must obviously be approached as $V_T \to V_{FB}$, and the relationship must be linear, since the entire slab is electron-depleted and the space charge is uniform and constant. Thus,

$$F(0) = (V_T + V_{FB} + 2\Delta V_D)/L \tag{2.2.62}$$

and on similar grounds

$$F(L) = -(V_{FB} - V_T)/L$$

(This does not contain ΔV_D because of the way V_{FB} has been defined.) Together with $\lambda_{B1} = \epsilon F(0)/eN_d$ and eqn (2.2.59), these equations give

$$V_{D1} + V_1 = \frac{F(0)\lambda_{B1}}{2} = \frac{(V_T + V_{FB} + 2\Delta V_D)^2}{4(V_{FB} + \Delta V_D)} \tag{2.2.64}$$

and

$$V_{D2} - V_2 = \frac{F(L)\lambda_{B2}}{2} = \frac{(V_{FB} - V_T)^2}{4(V_{FB} + \Delta V_D)} \tag{2.2.65}$$

which are the results originally derived by Sze and coworkers (1971). The fact that $V_T = V_1 + V_2$ can easily be checked by subtracting eqn (2.2.65) from (2.2.64). Thus, V_2 can be calculated for any applied voltage V_T, such that $V_{RT} \leqslant V_T$, taking V_{FB} from eqn (2.2.59).

(c) *High voltages applied* Equation (2.2.62) is not limited to the $V_{RT} < V_T < V_{FB}$ range; it should also hold for $V_{FB} < V_T < V_{BD}$ where V_{BD} is the breakdown voltage. This is governed by $F(0)_{BD} (= F_{BD})$, which comes from eqn (2.2.62) when $V_T = V_{BD}$. Accordingly,

$$V_{BD} = F_{BD}L - V_{FB} - 2\Delta V_D. \tag{2.2.66}$$

F_{BD}, the breakdown field, is of course a characteristic of the material, depends on the Zener effect or else on impact ionization, whichever occurs earlier. At $V_T = V_{BD}$ a sudden increase of current is expected. The complete voltage–current relation would thus be something like the schematic contour shown in Fig. 2.23, and this is, indeed, what Sze and co-workers observed on thin silicon wafers with platinum electrodes.

The above approximations are possible only as long as the free-carrier space charge can be neglected. Section 2.4.3 will concern itself with cases in which that space charge has to be taken into account. Numerical

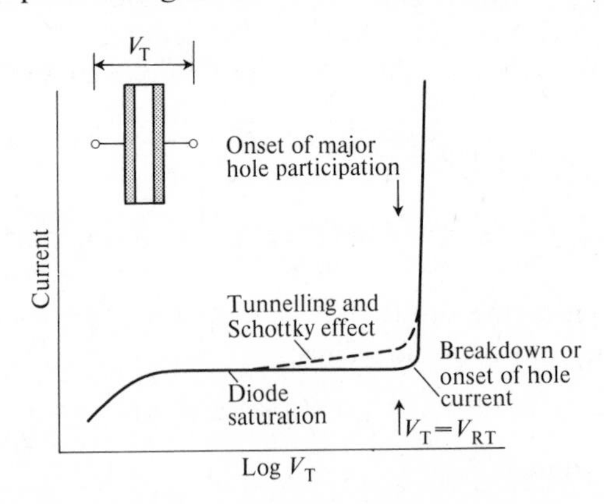

FIG. 2.23. Schematic voltage–current relationship of a double-barrier n-type system.

methods must then be used. The above treatment also assumes that the band gap is sufficiently wide to rule out all minority-carrier participation. Matters are much more complicated when minority carriers are present, and a complete analysis is not yet available.

Franceschetti and Macdonald (1979) have provided a two-carrier analysis, but only for a system with two totally blocking electrodes. Numerical methods had to be used. The capacitive behaviour of back-to-back contact systems (of the single-carrier type) is discussed in Section 4.1.5. A back-to-back barrier system with low-resistance contacts (at room temperature) has been analysed by Tantraporn (1970) on the basis of diode theory and the conditions needed to ensure current continuity. In passing, it is worthing noting that contact resistances which are low at room temperature may well be high at low temperatures; for some applications this is important.

Further reading

On calculations of the Richardson constant: Fonash 1972, Padovani 1969, Stratton 1962b.

On measurements of the Richardson constant: Borrego, Guttmann, and Ashok 1977, Srivastava *et al.* 1981.

On tunnelling through a variety of barrier structures: Mehbod *et al.* 1975.

On transport over a two-dimensional barrier: Buturla and Cottrell 1980.

On guard-ring methods of evaluating voltage–current characteristics: Saltich and Clark 1970.

On image-force barrier lowering: Rideout 1978.

On a generalization of diffusion relationships: Landsberg and Hope 1977, Simmons and Taylor 1983.

On barriers containing double-level donors: Meirsschaut 1976.

On interface and contact properties: Ludeke and Rose 1983, Sinkkonen 1983.

2.3 Current-carrying Schottky barriers; analysis of complex and hybrid models

2.3.1 General considerations; mechanisms of current flow

Attempts have often been made to combine 'the best of all models' into one, the idea being, of course, that any practical situation might be better described by a hybrid than by any of the idealized models discussed above. The magic term in this context is 'unified theory', and many claims made on behalf of such constructs are likely to be correct, though not always for the reasons given. The fact is that hybrid models are expected to give better agreement, not necessarily because they have more intrinsic merit, but because they have more adjustable constants. In most instances, we do not have the structural information to verify the physical correctness of the assumptions made. On the other hand, there are cases in which hybrid models hold undisputed sway, e.g. in connection with thin barriers on highly doped material (Chapter 6).

One of the earliest hybrid proposals was that of Billig and Landsberg (1950) whose composite model provides for a smooth change-over from a Schottky- to a Mott-barrier contour, governed by adjustable (but in practice unknown) parameters of a non-homogeneous donor distribution. In other contexts, hybrid models often arise from a more detailed analysis of boundary conditions. Thus, for instance, the basic models described in Section 2.2 assume that the concentration $n(0)$ of free electrons at the metal boundary is a constant, independent of current density, and this is one of the restrictive parameters that can be relaxed in a relatively straightforward way (see Section 2.3.2). Other hybrids arise from tunnelling considerations, as described below. In one way or another, a great many of the refinements introduced in the last twenty years have been concerned with hybrid models. Some add to our insight into physical mechanisms, others add only 'empty sophistication', in as much as the underlying assumptions are often difficult (and frequently impossible) to verify by independent methods. This difficulty arises from the fact that *almost* any experimental curve can be brought into some kind of harmony with one or other of the equations developed in Section 2.2, i.e. developed without the refinements which will now be introduced. In any event, the literature of the field is enormous, and only a few distinctive attempts at model hybrid model making can here be discussed. See Fig. 2.24.

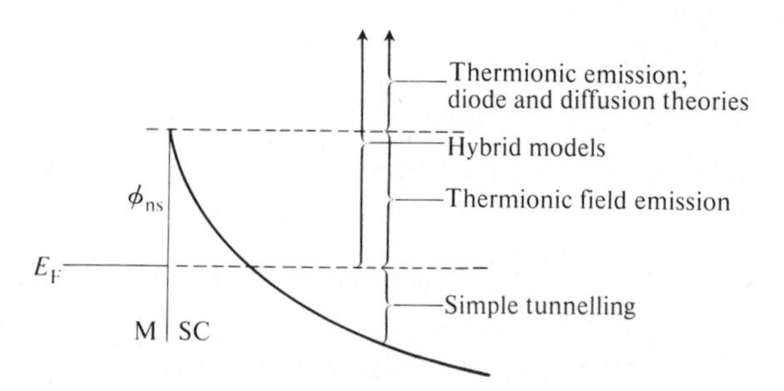

FIG. 2.24. Mechanisms of electron transmission over and through a barrier; classification of models.

The models analysed in Section 2.2 are mostly 'emission models', in as much as electrons must overcome the prevailing barrier height if they are to participate in charge transport. They do this within the framework of diode theory when the mean free path is very long, and within the framework of diffusion theory when it is very short (e.g. compared with the barrier width). When the mean free path is intermediate, as it is in many practical cases, neither classification may fit the circumstances, and a hybrid model is then needed. There is, of course, also a possibility that electrons may tunnel *through* the barrier, either at its base, or higher up, depending on the barrier profile. Such a tunnelling current may add merely a small correction to the emission current otherwise calculated, or else it may be the dominating component; much depends on circumstances. The terms *thermionic field emission* and *thermionically assisted tunnelling* have come to be used for the tunnelling contribution of thermally activated carriers, a process already encountered (albeit in highly schematic form) in connection with the high field effects discussed in Section 2.2.8. In general, we must distinguish between theories which concern themselves *only* with tunnelling at the Fermi level, e.g. as developed by Conley and Mahan (1967) and by Conley *et al.* (1966), those which envisage *only* thermionic field emission, e.g. as developed by Wilson (1932), and by Padovani and Stratton (1966b), and those hybrids which envisage thermionically activated tunnelling in addition to electronic emission *over* the barrier, e.g. as developed by Crowell and Rideout (1969b), and by Chang and Sze (1970). In the present context, the latter are the most relevant, with tunnelling processes representing a correction, to be introduced into models which depend primarily on diffusion-diode mechanisms for their main current-carrying capacity. Pure tunnelling theories are of interest only in connection with contacts of lowest resistance (Chapter 6), and need not be concerned with hybrid situations at all. Of course tunnelling from the Fermi level of a semiconductor can be envisaged only in degenerate material. A very useful survey and classification of theories has been provided by Rideout (1978), and in as much as it includes image force corrections, tunnel transmission through the barrier, as well as the effect of additional electron reflections above the barrier, the model discussed by Rideout and Crowell (1970) is one of the most comprehensive of its kind. Even then, some sacrifices had to be made, e.g. the assumption of long mean-free-paths. Accordingly, the model is a modified diode theory. Compared with tunnelling models which do *not* take account of image forces, the Rideout and Crowell theory produces higher reverse currents, as one would expect, but the image force also has some effect on the characteristics in the forward direction (see Section 2.3.3).

2.3.2 *Emission-controlled boundary concentrations; diffusion-diode theory*

The explicit assumptions which are characteristically associated with the classic Schottky model are given by eqns (2.2.11), in particular, the assumption of a constant carrier concentration at the semiconductor-metal interface ($x=0$). The question is where, if anywhere, does this constancy prevail when current is flowing. Implicit in Schottky's theory was the assumption of a vanishingly small mean free path, and on such terms, the notion of equilibrium at $x=0$ tends to be acceptable. A contrasting situation is described in Section 2.2.6, where the opposite assumption is made, namely that the mean free path is large compared with the barrier width. Nothing resembling thermionic electronic equilibrium can then exist anywhere in the barrier, because there are no collisions capable of establishing it. In practice, all plausible cases must fall somewhere between these assumptions.

A first-level correction is simple enough. (The present formulation is based on suggestions by A. Many; personal communication.) Let us suppose that the mean free path l_s (scattering length) corresponds to a small but non-zero fraction of the barrier width. For electrons going into the semiconductor, each free path would originate somewhere within the metal or at the metal surface and terminate in the semiconductor within the barrier. Conversely, for electrons going into the metal, each free path would originate within the barrier and end by some kind of collision within the metal. Because the probability of scattering within the metal is very high, the paths will terminate very close to the metal surface. It is convenient to consider the situation which prevails across a region of width l_s at the interface, without specifying precisely where this region begins and ends. On the metal side, there will be a concentration n_m of electrons capable of overcoming the barrier, and they will give rise to a thermionic emission current $-e\bar{v}_m n_m$, where $\bar{v}_m$ is given by eqn (2.2.39b). The symbol is here used with a subscript to identify the metal. On the semiconductor side of the region under consideration, there will be a corresponding electron concentration n_{sc} above the barrier height. Accordingly, a current will flow, given by

$$e(\bar{v}_{sc}n_{sc}-\bar{v}_m n_m)=j \tag{2.3.1}$$

where $\bar{v}_{sc}$ and $\bar{v}_m$ are taken as inherently positive. If n_m is taken as constant, which is reasonable, this equation reflects the dependence on n_{sc} on j, namely

$$n_{sc}=\frac{j+e\bar{v}_m n_m}{e\bar{v}_{sc}}. \tag{2.3.2}$$

This changes the situation depicted (before correction) by Fig. 2.15; the

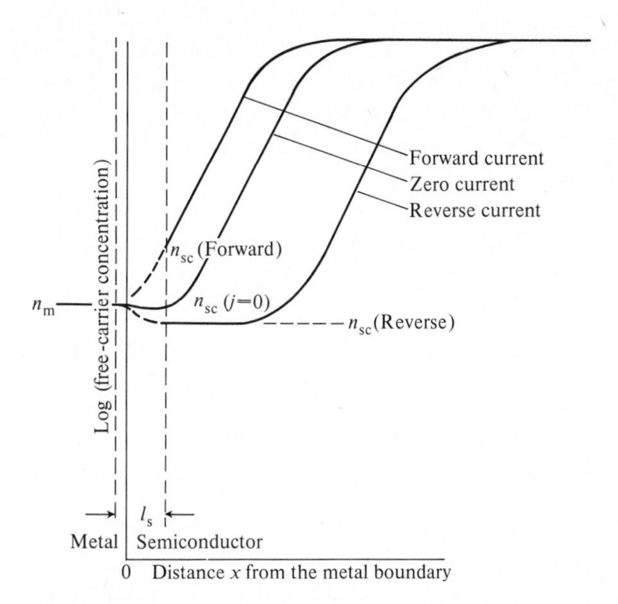

FIG. 2.25. Electron concentration contours within a Schottky barrier; schematic representation, after correction for emission-controlled boundary condition. n_m = concentration of electrons in the metal close to $x = 0$, with energies above ϕ_{ns}. (Compare Fig. 2.15)

contours become more like those shown in Fig. 2.25, though their exact shape has never been established. A small concentration gradient would exist across l_s even at zero current, associated with the ratio $\bar{v}_m/\bar{v}_{sc}$. This ratio depends on effective masses which are, of course, different in the two materials, but no simple generalization can be made, and no further sophistication is worthwhile within the crude framework established here. Figure 2.25 assumes $\bar{v}_{sc}/\bar{v}_m$ *slightly* greater than 1.

With this modification, the voltage–current relationship previously calculated can be reassessed. It is simplest to assume that no collisions at all take place within the l_s region, and that this region therefore contributes nothing to the resistance. For $x > l_s$ normal diffusion theory would prevail, with n_{sc} as the new boundary value. In special situations, it is possible to take this correction somewhat farther, but because of the many detailed assumptions involved, it is generally unsafe to do so. One plausible elaboration concerns the high reverse direction for which diffusion currents become unimportant. This permits the current density to be written as

$$j \approx n_{sc} e \mu_n F(0) = -n_{sc} e v_{dn} \tag{2.3.3}$$

where $F(0)$, and therefore j, are negative. With eqn (2.3.1) this gives

$$n_{sc} = \frac{(\bar{v}_m/\bar{v}_{sc}) n_m}{1 + (v_{dn}/\bar{v}_{sc})} \tag{2.3.4}$$

where $v_{dn} = -\mu_n F(0)$ is the drift velocity in the prevailing boundary field. In principle, this could be a useful boundary condition (replacing 2.2.11) for numerical calculations. In practice, its scope is limited by the fact that, in the high reverse direction, Schottky effect and tunnelling become important concerns and, compared with those, the present correction is expected to be small. In any event, such a numerical computation has not yet been made. On the other hand, its general character is clear enough. According to eqn (2.2.25), the current density at ($j = -J$) high reverse voltages would be given by

$$j = -\sigma\{(V_D + V_B)2N_d e/\epsilon\}^{1/2} \exp(-eV_D/kT) = \sigma_R F(0) \qquad (2.3.5)$$

and it is precisely the value of σ_R that would be modified by the present correction. The resulting relationship would be

$$j = \sigma_R \frac{F(0)}{1 - \mu_n/\bar{v}_{sc}\{F(0)\}} \approx -\sigma_R \frac{z_1(V_B)^{1/2}}{1 + z_2(V_B)^{1/2}} \qquad (2.3.6)$$

where z_1 and z_2 are constants. With $F(0)$ negative (V_B positive) this implies a *lowering* of the reverse current, though only by a small amount, since the drift velocity must always be small compared with $\bar{v}_{sc}$ within the framework of the assumptions here made. Even the shape distortion is not likely to be noticed in the course of graph-fitting attempts.

For didactic purposes there is another useful way of looking at the problem. Equation (2.3.4) can alternatively be written as

$$n_{sc} = \frac{n_m \bar{v}_m}{\bar{v}_{sc} + v_{dn}} = \frac{n_m \bar{v}}{\bar{v} + v_{dn}} \qquad (2.3.7)$$

if the difference between $\bar{v}_{sc}$ and $\bar{v}_m$ were taken as negligibly small, for simplicity. The current density at high reverse voltages (only) would then become, by eqn (2.3.3):

$$j = \frac{n_m \bar{v}}{\bar{v} + v_{dn}} e\mu_n F(0) = \frac{n_m \bar{v} e}{\{\bar{v}/\mu_n F(0)\} - 1}. \qquad (2.3.8)$$

Let us now call x_T the distance within which the barrier energy drops by kT. Then $x_T = -kT/eF(0)$ and $\bar{v}/\mu_n F(0) = -\bar{v} e x_T/\mu_n kT$ which can be further simplified to

$$\frac{\bar{v}}{\mu_n F(0)} = -\sqrt{\left(\frac{3}{2\pi}\right)} \frac{x_T}{l_s} \qquad (2.3.9)$$

bearing in mind that $\mu_n = el_s/m_n v_T$ and $v_T = (3kT/m_n)^{1/2}$. For the high reverse current density, this gives

$$j = -n_m \bar{v} e \Big/ \left\{1 + \sqrt{\left(\frac{3}{2\pi}\right)} \frac{x_T}{l_s}\right\} \qquad (2.3.10)$$

which illustrates that what really matters is not so much how l_s compares with the barrier width, but more immediately how it compares with x_T (though, of course, x_T and λ_B are related). Physically, the significance of x_T arises from the fact that electrons, entering from the metal and penetrating to $x > x_T$, tend to remain in the semiconductor, because any collision they may then make is unlikely to send them back over the barrier into the metal. If they make such a collision at a point $x < x_T$, an appreciable amount of backstreaming must be expected. In other words, incoming electrons do not really belong to the semiconductor until they are thermalized within it. The factor $(3/2\pi)^{1/2}$ is of the order of unity, and the distinction between collisions at $x > x_T$ and $x < x_T$ is unsharp, of course. Crowell and Sze (1966b) have given a sophisticated discussion of this aspect.

For large x_T/l_s ratios (relatively shallow barrier, small mean free path), we have from eqn (2.3.8) $j \rightarrow n_m e\mu_n F(0)$, as the Schottky diffusion theory predicts. The larger the mean free path, the greater is the necessary correction. If we had $l_s \gg x_T$, then the current density would tend to $j \rightarrow -n_m \bar{v} e$, which corresponds to the findings of pure diode theory. Thus, for high reverse voltages applied, a bridge has been achieved between the two theories, which are treated in Section 2.2. as limiting and mutually exclusive models.

No corresponding simplification (based on eqn (2.3.3)) is possible for *low* reverse currents or in the forward direction, a fact which is often overlooked. However, it is easy enough to see that the net effect of the present correction is the same as in the reverse direction, namely a lowering of the current density. In the high reverse direction this is due to the diminished value of n_{sc}, resulting in a lower field current. In the forward direction, it is due to the increased value of n_{sc}, resulting in a smaller concentration gradient, and thus a smaller diffusion current. In a sense, the fact that the mean free path is substantial makes the system less responsive to externally controlled fields, precisely because it is then more under the control of thermal kinetic processes.

If a complete synthesis between the diffusion and diode theories cannot be obtained analytically, it is still of interest to examine whether the diffusion theory can be *amended* by the present considerations *within* its own terms of reference, i.e. for very small (but non-zero!) mean free paths. For this purpose, eqn (2.3.3) cannot be used but under low reverse of forward conditions, it is possible to obtain a simple expression for the current via eqn (2.3.2) *without* neglecting diffusion. To do this, we reconsider the integration of eqn (2.2.9) of the Schottky model, now with a new boundary condition, namely n_{sc} in place of $n(0)$.

Strictly speaking, we would now have to integrate not from from zero to λ_B, but from l_s to λ_B, but the old limits will in fact be allowed to stand, taking l_s as small. This yields

$$j\int_0^{\lambda_B} \exp(-eV/kT)\,\mathrm{d}x = eD_n n_e \exp\{(\phi_n - eV_B)/kT\} - eD_n n_{sc} \exp(\phi_{ns}/kT). \tag{2.3.11}$$

Though the shape of the barrier is slightly altered by the new circumstances, the integral on the left is not expected to be sensitively affected and, for present purposes, we shall assume it to remain unchanged. This gives, by analogy with eqn (2.2.24), and bearing in mind that $J = -j$:

$$j = e\mu_n\{(V_D + V_B)2N_d e/\epsilon\}^{1/2}\{n(0)\exp(-eV_B/kT) - n_{sc}\}. \tag{2.3.12}$$

Substituting from eqn (2.3.2), and lumping constants into B:

$$j \approx B(V_D + V_B)^{1/2}\{\exp(-eV_B/kT) - (1 + j/n(0)e\bar{v})\} \tag{2.3.13}$$

which finally yields

$$j \approx \frac{B(V_D + V_B)^{1/2}\{\exp(-eV_B/kT) - 1\}}{1 + B(V_D + V_B)^{1/2}/n(0)e\bar{v}}. \tag{2.3.14}$$

It will be seen that this does indeed reduce to eqn (2.3.6) for large positive values of V_B. The constant B 'contains' the mean free path, which thus governs the magnitude of the correction in the denominator.

By a somewhat different argument (from which, however, the inherent limitation to high reverse voltages emerges much less clearly), Crowell and Sze (1966*b*) arrived at essentially similar equations and, in particular, at an equivalent of eqn (2.3.6), but for arbitrary mean free paths. They also showed that the current can be expressed not only in terms of $F(0)$, as here, but in terms of the applied voltage V_B, as long as an *effective* drift velocity $\bar{v}_{dn}$ is used in place of $\mu_n F(0)$. Taking account of the image force, they calculated

$$\bar{v}_{dn} = \mu_n kT\left[\int_{x_m}^{\lambda_B} \exp\left\{-\frac{E(x_m) - E'_{(x)}}{kT}\right\}\mathrm{d}x\right]^{-1} \tag{2.3.15}$$

where $E'(x)$ describes the barrier contour *before* the image force correction. As a result, but *only* as long as $V_B > 0$ and V_D sufficiently high, eqn (2.3.14) can be written in the alternative form

$$j = \frac{e\bar{v}_{dn}\mathcal{N}_c \exp(-\phi_{ns}/kT))}{1 + \bar{v}_{dn}/\bar{v}_{sc}}\{\exp(-eV_B/kT) - 1\} \tag{2.3.16}$$

which leads to the conclusions already discussed above for the limiting cases of $\bar{v}_{dn}/\bar{v}_{sc} \gg 0$ and $\bar{v}_{dn}/\bar{v}_{sc} \ll 0$. Crowell and Sze also interpreted $\bar{v}_{sc}$ in terms of Bethe's diode theory, which gave $\bar{v}_{sc} = A^*T^2/e\mathcal{N}_c$, A^* being Richardson's emission constant, as defined quasi-classically by eqn

(2.2.37). (See also Section 2.3.4) The effective drift velocity $\bar{v}_{dn}$ is always smaller than it would be without image-force correction (even while hot-electron effects are neglected). Electron tunnelling takes over for very high values of V_B, for which eqn (2.3.16) is not intended to apply. See also Schultz (1954) for earlier calculations without image-force corrections.

Equation (2.3.14) itself is not limited to the reverse direction, but when the implied *forward* characteristics are plotted semi-logarithmically, they appear to be linear, just as they do for the uncorrected equation, as long as we are within a plausible range of V_B values. It is not, therefore, easy to assess the merit of the correction by reference to measured results and graphic methods. The corrected version is conceptually more satisfying, but it is not actually much more useful than the simple Schottky model in the interpretation of practical situations.

2.3.3 *The tunnelling contribution; parametric description of rectifying characteristics*

In Section 2.1.6, tunnelling considerations were applied to barriers in the quiescent state. A full calculation of the additional tunnelling currents which flow over a wide range of applied voltages is, of course, more complicated, but the essential concepts were introduced into the field at a very early stage by Wilson (1932), whose model envisaged unactivated tunnelling as the *only* current-supporting process. To do this, he calculated the number of electrons at any particular energy that are eligible for tunnelling, as well as the wave-mechanical transparency of the barrier to electrons of that energy, as discussed in Section 2.1.5. Finally, it was necessary to integrate over all the relevant energies, i.e. from the bottom of the conduction band to the top of the barrier. In order to calculate the transparency factor, some assumptions had to be made about the shape of the barrier. Mott and Schottky barriers had not yet been thought of, and Wilson, by way of a working hypothesis, assumed his barrier to be rectangular. The result of the integration, based on earlier work by Frenkel (1930), yielded a current density which, with the present conventions, may be written as

$$j/j_0 = \exp\{(\beta - 1)eV_B/kT\} - \exp(\beta eV_B/kT), \qquad (2.3.17)$$

where β is a constant between $\frac{1}{2}$ and 1, independent of V_B, but dependent on the barrier height and thickness. It can be shown that β approaches unity for barriers of very low height. Under such conditions, we would have

$$j/j_0 = 1 - \exp(eV_B/kT) \qquad (2.3.18)$$

which resembles the 'ideal' rectification characteristic discussed in Section

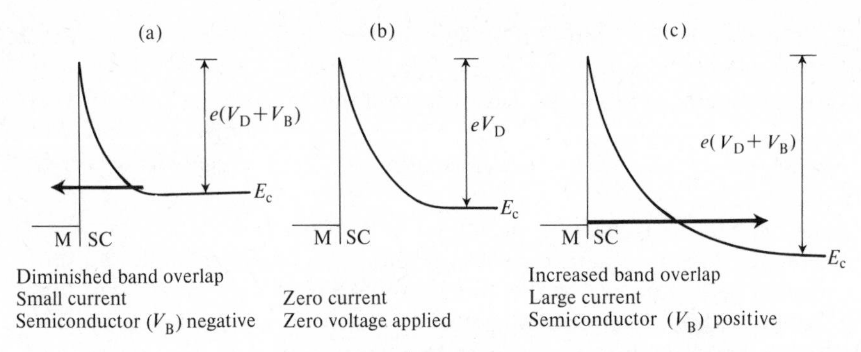

FIG. 2.26. Principle of rectification by wave-mechanical tunnel effect; Wilson 1932. Horizontal arrow denotes tunnelling current. Forward current would flow when V_B is positive (contact on n-type material).

2.2.6. This makes it (slightly!) tempting to designate β as another form of 'non-ideality factor). Note, however, that in this case, and contrary to all the cases discussed in Sections 2.1 and 2.2, the current density is now large, instead of small, when voltages $V_B>0$ are applied in what is ordinarily the reverse direction. In other words, this equation predicts rectification in the direction *opposite* to that predicted by the diffusion-diode theory, and thus opposite to that ordinarily observed in practice. It is easy to see by reference to Fig. 2.26 why it should be so, on the basis of band overlap, even though overlap is not the only relevant consideration. If the operating mechanism were as described, then making the semiconductor positive would enormously increase the scope for tunnelling (as it does in a modern Esaki diode), and would thus lead to the forward direction of current flow. Conversely, making the semiconductor negative diminishes the overlap, and leads only to a very small *reverse* current in n-type material. Compare Fig. 1.6. This conflict was not noticed for a time, because the semiconductor rectifiers which Wilson had in mind were those based on cuprous oxide and selenium. These materials have long since been firmly established as p-type conductors, but in Wilson's time their status was still misunderstood. Had they really been n-type, as was assumed, Wilson's polarity prediction would have been correct. In due course, the error was discovered and, indeed, the now-familiar models by Mott and Schottky were originally devised at least in part as responses to Wilson's crass conflict with experimental observations.

Equation (2.3.17), with its parametric non-ideality factor β, represents a variety of rectifying characteristics. Within the framework of tunnelling theory, the range of β is, of course, restricted, but it is of interest to consider what happens when β takes on values outside this range. Thus, for $\beta=0$, we would have

$$j/j_0=\exp(-eV_B/kT)-1 \qquad (2.3.19)$$

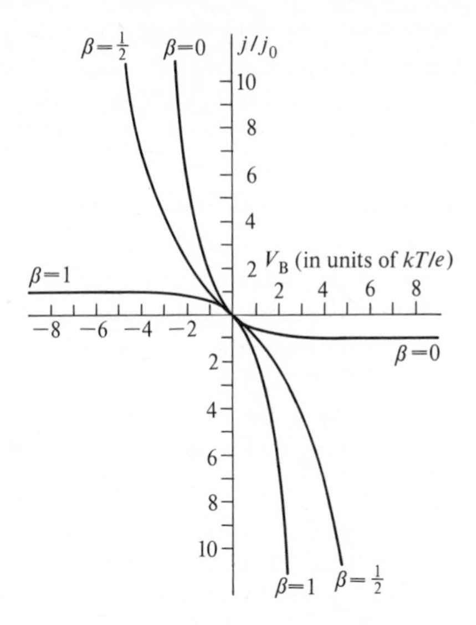

FIG. 2.27. Parametric representation of contact characteristics, in accordance with eqn (2.3.17), with β taking values between 0 and 1.

which is the 'ideal' rectifying characteristic of Section 2.2.6. We see that values of β between zero and unity encompass rectification characteristics of both polarities, from one form of ideality to another. For $\beta=\frac{1}{2}$, eqn (2.3.17) becomes symmetrical; Fig. 2.27 shows these relationships. Whenever tunnelling currents represent *additional* components of a system primarily controlled by diffusion-diode considerations, they must, of course, exhibit the same basic properties that they show when dominant. Accordingly, their contribution is greatest in what is ordinarily the reverse, and smallest in the forward direction. Indeed, when sufficient tunnelling is superimposed on a diffusion-diode model, the 'reverse' direction can become the direction of easy current flow (Crowell, 1969b). Note that the reverse direction is here shown as fully saturated in both curves, but this is only a consequence of the simplifying assumptions made. We know that the diffusion theory does not lead to saturation, and it has been shown by Conley and co-workers (1966) that the same is true for thermionically assisted tunnelling, when the distortion of the barrier by image forces is taken into account. Both types of curves therefore show a maximum of differential resistance ($\delta V_B/\delta j$) at some small voltage in the direction of smaller current flow.

Although the introduction of $\beta<\frac{1}{2}$ is here purely schematic, it does reflect the reality of practical rectifier performance in one respect: ideality is rarely fulfilled. Small departures from it, arising from the nature of Mott and Schottky barriers, have already been discussed in Sections 2.2.3

and 2.2.4. Larger departures can arise from a variety of other causes, e.g. image-force lowering of the barrier height, edge effects, inversion layers, artificial barrier layers, and minority-carrier injection and extraction. The image-force effects are particularly easy to see. At high forward voltages they are negligible, which means that currents conform to 'ideal' values. Towards zero voltage, image-force barrier-lowering increases the currents, meaning that the $\log|j|$ versus V_B line will have a smaller-than-ideal slope, from which a $\beta > 0$ might in principle be evaluated. However, in such complex circumstances, strict linearity of the $\log|j|$ versus V_B relationship cannot really be expected; β ceases to be constant. Sze and co-workers (1964), and also Rideout and Crowell (1970), have used the 'local' value of the slope to evaluate ΔE_s, the amount of barrier-lowering, as a function of V_B, and have even calculated the high frequency permittivity ϵ_∞ from eqn (2.2.50). Such a procedure would appear to be risky,† but through its use Rideout and Crowell did succeed in obtaining good and self-consistent agreement between calculated and measured results on W–GaAs and Au–GaAs contacts, as far as the shape of the forward characteristic was concerned. The magnitude of reverse currents was also accurately assessed. The evaluation of contact parameters from the observed characteristics was achieved by computer techniques.

In the context of discussions which are mainly concerned with diffusion-diode theory, it is sometimes convenient to express eqn (2.3.17) in a different form, by putting $1-\beta = 1/\eta$. With this constant, familiar from Section 2.2.5, we have

$$j/j_0 = \exp\left(-\frac{eV_B}{\eta kT}\right) - \exp\left\{\left(1-\frac{1}{\eta}\right)\frac{eV_B}{kT}\right\} \qquad (2.3.20)$$

which, when β is close to zero (and, of course, only then), can be approximated to

$$j/j_0 = \exp(-eV_B/\eta kT) - 1. \qquad (2.3.21)$$

Obviously $0 < \beta < 1$ implies $1 < \eta < \infty$. Empirical values for β (or η) have been determined for a variety of practical situations; see Rideout (1978) for a summary.

† The problem arises in many places under different disguises. A certain relationship is expected to be (say) linear, on the basis of a particular model. It is then tested by reference to observations. If the relationship in question is indeed found to be linear, a unique coefficient is deduced from its slope, and the model is declared sound. When the observations fail to yield a straight line, the model should be declared 'proven false', but investigators are not always willing to take such a drastic step. The analysis is often continued *as if* the model had been successful, evaluating local slopes at each point instead of a general one for the line as a whole. In this way, at any rate superficially, the model can be saved, but only at the price of turning one of its constants into a variable, and forgetting that *every* curve is a straight line if consideration is limited to a sufficiently small segment of it.

In other contexts, it is sometimes convenient to express 'non-ideality' by defining a pseudo-temperature T_0 in the form

$$j/j_0 = \exp\{-eV_B/k(T+T_0)\} - 1 \qquad (2.3.22)$$

and Levine (1971) has made effective use of this mode. However, although β (and η or T_0) can be used to 'codify' various voltage–current characteristics empirically, departures from the ideal values $\beta = 0$ and $\beta = 1$ are not readily interpretable in unique terms related to the charge flow mechanism. See Section 2.2.5 and Crowell (1977).

To summarize, the situations in which tunnelling processes play the principal current-carrying role are those in which the barriers are very thin, and the corresponding barrier resistances very low. Those will be further discussed in Chapter 6. In other situations, tunnelling is responsible for incremental corrections to currents caused primarily by diffusion-diode processes, and are therefore most important in the reverse direction. As a consequence, the reverse characteristics are even less saturating than they would otherwise be, but it is rarely possible to isolate and identify tunnelling as sole cause of non-saturation, because other mechanisms tend to be at work. Some of those arise from lateral inhomogeneities of the barrier. If such inhomogeneities involve the existence of locally thin barrier regions, then tunnelling currents will flow in parallel with thermionic emission currents. The total effect on the forward characteristic will be small but, in the reverse direction, tunnelling regions ('hot spots') may appear, equivalent to local short circuits. The total voltage–current characteristic would then be more like a sum of currents represented by eqns (2.3.18) and (2.3.19), and would thus tend towards low resistance in both directions.

2.3.4 Analytic and numerical approaches to the tunnelling problem

One of the most detailed models of tunnelling through a Schottky barrier is that due to Chang and Sze (1970), which goes beyond previously discussed models, in as much as it includes consideration of the image force, and makes use of transmission coefficients derived from numerical solutions of the Schroedinger equation. The use of numerical methods was necessary, because the region near the top of the barrier is one in which the potential changes very rapidly with distance, which makes the WKB approximation inapplicable. The values derived by Chang and Sze are thus substantial improvements upon the purely schematic results presented in Fig. 2.8. Chang and Sze's procedure is also more general, in as much as it is formulated in terms of Fermi–Dirac rather than Boltzmann statistics. On the other hand, free-carrier space charges are neglected (as they usually are), which limits the treatment to well-conducting materials. Under such conditions it was possible to show, amongst other things, that the reverse saturation current goes through a minimum as the doping increases from very low to very high levels. On intrinsic material, the only barrier that can be envisaged is one of the kind discussed in Section 3.1.3, and that is bound to be very thin, implying high currents. On highly doped materials, barriers are likewise thin. The appearance of a resistance maximum is therefore entirely expected. By way of an example, a barrier of $\phi_{ns} = 0.85$ eV on Si has its maximum reverse resistance for a donor content of 10^{16} cm^{-3}. The results also demonstrate that thermionic electron transitions *over* and tunnelling transitions *through* the barrier are not simply additive, since the very presence of the barrier causes some electrons of energy $>\phi_{ns}$ to be reflected. Only those *not* reflected constitute the thermionic emission current. For electrons of energy ϕ_{ns}, the transmission coefficient is not only below unity, but field dependent.

Along similar lines, Crowell and Sze (1966c) were able to calculate the Richardson emission constant (then denoted by A^{**}, to distinguish it from the semi-classical value A^*). Among other key results are revised quantum-mechanical transmission coefficients, obtained without recourse to the WKB approximation (Crowell and Sze 1966b). They are shown in Fig. 2.28(a), which applies specifically to a gold contact on GaAs. Electron energy increments ΔE are measured relative to the top of the barrier. For the lowest field shown (10^3 V cm^{-1}) the barrier is almost non-transparent, and the effect of quantum-mechanical reflection disappears quickly with increasing energy. As the fields increase, the transparency increases rapidly, and high-energy reflection becomes more important. Figure 2.28(b) gives (schematically) the ratio of the tunnel current, to the thermionic current, as a function of $F(0)$. At first, the ratio varies only slowly for rising applied fields, and even diminishes slightly as a result of weighting caused by the electron velocity distribution. At a later

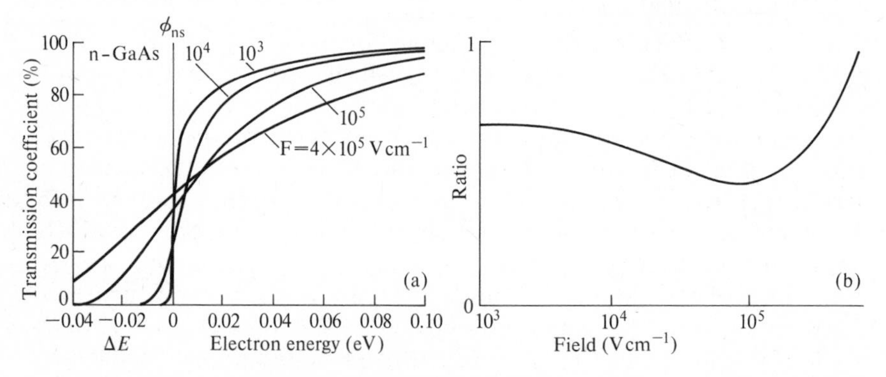

FIG. 2.28. Electron tunnelling through a barrier. After Crowell and Sze (1966b). (a) Transmission coefficients calculated without the WKB approximations. (b) Ratio of tunnel current to thermionic current as a function of field.

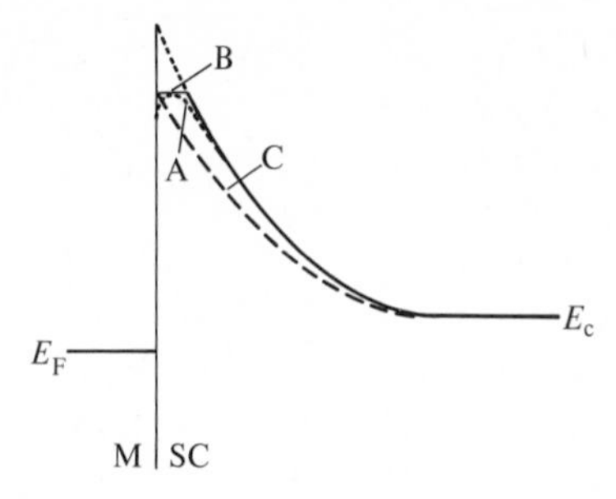

FIG. 2.29. Schottky barrier approximations. After Vilms and Wandinger (1969). Curve A: Schottky barrier with image force. Curve B: Schematic representation of A. Curve C: Schottky barrier of the same height as B.

stage, it increases sharply. Of course, all such curves depend on the nature of the semiconductor (number of minima, effective mass tensors), and on temperature.

On the basis of analytic procedures developed by Padovani and Stratton (1966a,b), Murphy and Good (1956), and Stratton (1964), Vilms and Wandinger (1969) calculated total tunnel currents for the barrier models B and C shown in Fig. 2.29. Of these three contours, curve A represents the realistic profile of the Schottky barrier, corrected for the effect of image force. The full curve B is a schematic representation thereof, and C a normal Schottky barrier of the same total height. Calculations based on C obviously overestimate the tunnel current; those based on B underestimate it, but a consideration of both leads to an acceptable approximation. Fair agreement with experimental results on Si and GaAs was obtained. The matter is of special interest in connection with low resistance contacts. See also Chapter 6. Similarly, Shannon (1977b) obtained excellent agreement for Ni–n-Si contacts with non-uniform doping profiles (achieved by ion implantation) which produced virtually triangular barrier profiles. This served to simplify the comparison between theory and experiment. Tunnelling distances between 25 Å and 30 Å at 300 K were calculated from the results.

2.3.5 *Current-dependent barrier heights*

In all the discussions so far, the basic height ϕ_{ns} of the contact barrier 'as seen from the metal' has been regarded as a current-independent constant, but there are several reasons for believing that this might not (or not always) be justified. Any non-constancy of ϕ_{ns} would, of course, affect the voltage-current characteristic in a most sensitive way. Conversely, if any observed characteristic were to depart from expectations formulated on the basis of ϕ_{ns} being constant, such departures could always be interpreted as consequences of a variable barrier height, whether they in fact arise from this cause or not. For this very reason, this all-powerful interpretational tool should be employed only with the greatest caution. Caution must take the form of internal consistency tests; in each case it is necessary to ask whether there are other experiments and criteria (besides the voltage–current characteristics themselves) which are in harmony with the variable barrier height hypothesis. On the other hand, there is nothing esoteric about the notion of a voltage-dependent barrier height ϕ_{ns}. We have already seen (Section 1.3.3) that the interface states are really extended states of the metal, and their occupancy may vary with applied voltage for two reasons:

(a) When current flows, $n(0)$ changes, and this calls for an adjustment of the charge trapped in the communicating in interface states. In the forward direction ($V_B<0$), $n(0)$ increases (see Fig. 2.15), and the negative charge in interface states should increase likewise. Accordingly, the Fermi level at the interface should creep up (in relation to the band edges) which means that the barrier should be *lowered* (see Fig. 1.15) if the interface states are not tightly bunched. Correspondingly, ϕ_{ns} should increase in the reverse direction.
(b) When an external voltage is applied to a barrier, the total charge within the barrier changes, and a corresponding (but opposite) charge is capacitively induced in the metal, to be accommodated 'at' the metal side of the interface. The occupancy of interface states will therefore change, and so will the amount of band bending within the metal (Fig. 1.14). When $V_B<0$ (forward direction), the extra charge on the metal will be positive (diminished band-bending), as will the extra charge in interface states. Both should lead to a *raising* of the barrier; indeed the two mechanisms are not easily distinguishable from one another. Their effect is summarized by Fig. 2.30, and is clearly opposed to effect (a) above. It should come into play even if the interface states were tightly bunched.

Variable barrier heights have been invoked to account for the two principal departures of observed voltage–current relationships from expectations based on (say) simple diode theory: failure of the log $|j|$ versus

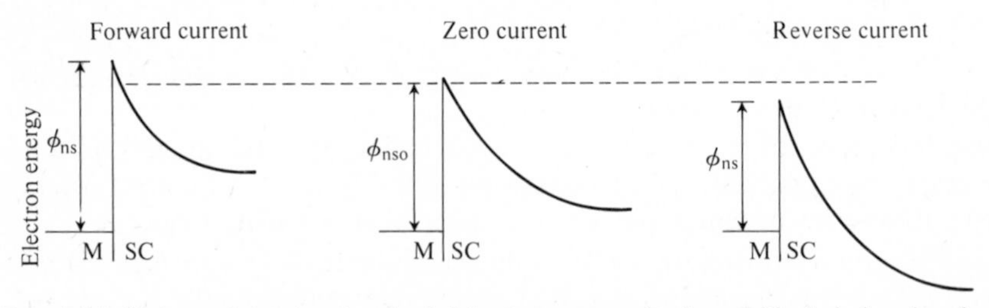

FIG. 2.30. Voltage-dependent barrier height ϕ_{ns}, as a result of capacitively induced surface charges on the metal and its associated interface states.

V_B relationship in the forward direction to exhibit a slope as large as e/kT, and failure of the reverse characteristic to saturate. In each case, alternative corrective mechanisms are, in fact, available, but Levine (1971) has successfully reconciled variable barrier heights with observations on materials of such highly diverse surface structures as Si, GaAs, ZnS and $SrTiO_3$. To do this, Levine adopted essentially model (b) above, but ascribing *all* effects to a changing interface occupancy, rather than band bending in the metal. With such assumptions, it becomes a simple matter to interpret changes of barrier height in terms of the density of interface states. To illustrate this, let us consider a system of constant surface state density, as shown on Fig. 1.15(a), and a barrier of constant static capacitance C_{BS}, such as would be implied by a Mott structure. For any applied voltage V_B, the extra charge in interface states would be

$$\Delta Q_{ss} = -C_{BS} V_B \tag{2.3.23}$$

and that would be related to the barrier height via

$$\phi_{ns} = \phi_{ns0} + \alpha_s \Delta Q_{ss} \tag{2.3.24}$$

where α_s is some constant, inversely proportional to the density of interface states per unit energy. The diode equation (2.2.37) for V_B large and negative would then become

$$j = A^* T^2 \exp(-\phi_{ns0}/kT) \exp\left\{-\left(\frac{e-\alpha_s C_{BS}}{kT}\right) V_B\right\} \tag{2.3.25}$$

giving a $|d(\log|j|)/dV_B|$ of $(e-\alpha_s C_{BS})/kT$ instead of e/kT or else a slope of $e/k(T+T_0)$ where $T_0 = \alpha_s C_{BS} T/(e-\alpha_s C_{BS})$. Thus, determination of the slope (if constant, as here implied) would lead to information on α_s. The procedure could be adapted to more complicated models, but is is inherently tied to the idea that *some* basic equation (in the example given here, the simplest diode equation, uncorrected for image forces) must be accepted as *correct* for a constant ϕ_{ns} on *a priori* grounds. Variable barrier height is then invoked to account for departures from expectations, under conditions which prevent the soundness of the expectations themselves from being verified.

One of several possible checks on the self-consistency of the conclusions can be derived from observations of the temperature dependence of the system at constant current, as described in detail by Levine (1971). Alternatively, Crowell and Sze (1966b) have inferred α_s not only from an analysis of current–voltage characteristics, but also from capacitance measurements as a function of voltage. The latter yield a barrier height uninfluenced by image force and, in that sense, a simpler type of result; see Section 4.1. For a material like Si, with densely bunched interface states, $\phi_{ns} - \phi_{ns0}$ can never be very great; it is only the exponential relationship between current and barrier height that gives the matter real importance. Conversely, one might envisage a material with a low and quasi-uniform density of interface states, in which case the present effects would have a dominating importance. All manner of intermediate situations could, of course, arise, but no definite evidence has come to light. Unfortunately, there is no known way of disentangling the contributing effects due to mechanisms (a) and (b) above, nor is it really safe to neglect band bending in the metal in the context of (b).

2.3.6 *Laterally non-homogeneous barriers*

The problem of barriers with random lateral variations has already been referred to in Section 2.1.3. It presents itself in various ways and is to some extent amenable to analysis. Thus, suppose that we have a 'patchy' barrier in which the barrier height is subject to local variations, slight enough to maintain the essentially one-dimensional flow character of the system. There would be a mean barrier height $\bar{\phi}_{ns}$ and a local barrier height $\phi_{ns}(y, z)$, with a standard deviation $\Delta\phi$. Then

$$\mathrm{d}\mathcal{A} = \frac{\mathcal{A}}{\sqrt{2\pi}(\Delta\phi)} \exp\left\{-\frac{(\phi_{ns} - \bar{\phi}_{ns})^2}{2(\Delta\phi)^2}\right\} \mathrm{d}\phi_{ns} \tag{2.3.26}$$

would be the small area $\mathrm{d}\mathcal{A}$ in which the barrier height is expected to be tetween ϕ_{ns} and $\phi_{ns} + \mathrm{d}\phi_{ns}$. (Check $\int_0^\infty \mathrm{d}\mathcal{A} = \mathcal{A} =$ total area.) For illustrative purposes, we shall assume here that the current for a homogeneous system is given by the uncorrected idode formula, eqn (2.2.37). Through every small area $\mathrm{d}\mathcal{A}$, there would be a saturation current $j_s(y, z)\,\mathrm{d}\mathcal{A}$ in the high reverse direction, whereby $j_s(y, z)$ is given by

$$j_s(y, z) = A^*T^2 \exp\{-\phi_{ns}(y, z)/kT\}. \tag{2.3.27}$$

The total current density will be

$$j = \frac{1}{\mathcal{A}} \int_{-\infty}^{\infty} j_s(y, z)\,\mathrm{d}\mathcal{A},$$

which can be integrated numerically. (The lower integration limit has no physical meaning, but is otherwise harmless.) Figure 2.31 shows the relationships, and demonstrates *inter alia* a result which can also be derived analytically, namely that a standard deviation of (say) kT implies an effective lowering of the barrier by $kT/2$.

The same kind of analysis can be used for parameters other than the barrier height, and has been so applied by Chang and Sze (1970) to explore the effect of randomly variable doping on the reverse saturation current. As long as streamline current flow is maintained, the *shape* of the voltage–current relationship is not affected by lateral non-homogeneity. Of course, large values of $\Delta\phi$ would quickly invalidate this assumption.

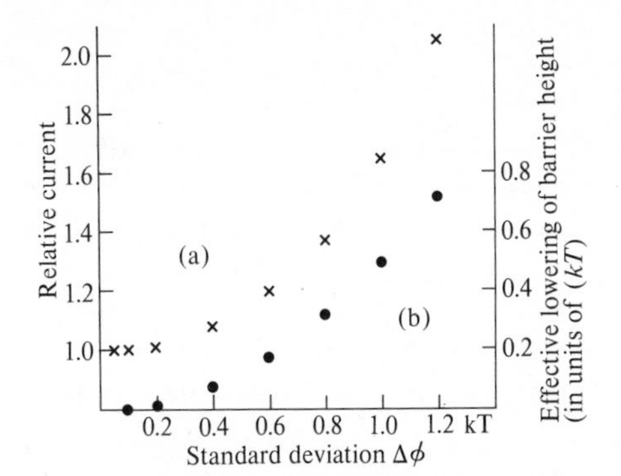

FIG. 2.31. Lowering of the effective barrier height by lateral non-homogeneity. (a) Relative current as a function of standard derivation $\Delta\phi$. (b) Equivalent effective barrier lowering.

The problem of gross heterogeneity should also be addressed but is, of course, much less tractable. In the last analysis, it must be recognized, in this field as in all others, that results obtained from specimens of highly non-uniform and uncontrolled structure are unsuitable for interpretation in fundamental terms. On the other hand, minor fluctuations of barrier height, donor density, etc, tend to be found even in the most carefully made systems.

Gross variations of barrier height would certainly lead to 'hot spots' which may, indeed, come to carry most of the current. Each spot would be associated with its own spreading resistance (see Section 1.1.3), but if the spots were very close together, the spreading regions would overlap, and the current flow would again tend to be one-dimensional. Figure 2.32

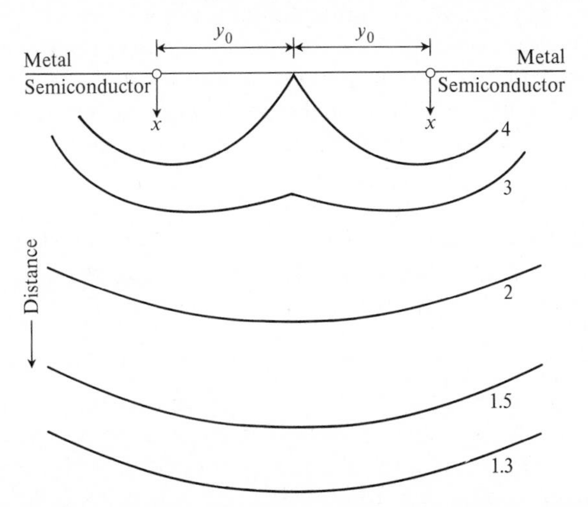

FIG. 2.32. Bulk current flow pattern in the vicinity of two hot-spots at a distance $2y_0$ apart. Equipotential contours for four different potentials (in arbitrary units).

shows two point sources, a certain distance $2y_0$ apart. Close to the interface, the two sources are 'seen' as distinct, by reference to the drawn equipotentials, but by the time x if is of the same order as y_0, the current flow is virtually one-dimensional. With sources randomly distributed over the y–z plane, this averaging-out effect would occur even nearer to the interface. There remain the problems associated with the region $x<y_0$, but no analysis of their importance is available.

There is an interesting set of experiments by Wronski *et al.* (1974), which yielded results in apparent conflict with the expectations raised here. By means of co-sputtering, it is possible to make metal–insulator mixtures in which the 'insulator' can occupy volume fractions between 0 and 90 per cent. Such materials are known to become metallically conductive when the metal content is about 50 per cent. A test was made to see whether there is a simultaneous change of contact properties when such mixtures are used to form Schottky barriers on Si, GaAs, and CdS single crystals. Astonishingly, there was not; mixtures containing as little as 10 per cent of metal behaved as 'metallic' contacts. In the composition range 60–90 per cent insulator (SiO_2, Al_2O_3) in the metal (Ni, Au) the barrier heights measured were independent of composition. At this stage we must suppose that the sputtering process leads to a compositional profile which corresponds to a continuous metal layer on the semiconductor surface, but this remains to be confirmed. Moreover, other observations are on record which yield quite different conclusions. Thus, Lim and Leaver (1980) found that Cermet–Si contacts are very much dependent on the metal content of the Cermet. The matter is in need of further clarification. Meanwhile, we have an account of interesting experiments by DiStefano (1971), who examined lateral inhomogeneities at a Si–SiO_2 interface by means of scanning internal photoemission. He found substantially lower barrier heights in regions of sodium contamination, and was able to display three-dimensional contour maps of the local current density under a voltage bias, very much along the lines described by the purely schematic Fig. 2.32.

All other aspects apart, electrode edge effects are, of course, a form of lateral heterogeneity that can never be totally eliminated. Diode breakdown processes tend to be initiated at contact edges, as one would expect, because the field is augmented there (Jäger, 1969).

2.3.7 *Barriers with non-uniform doping profiles*

In all the cases so far discussed, the donor concentration N_d was considered a constant; in practice it may be a function of x. This can arise in various ways, e.g. through the diffusion of donors into the semiconductor from the metal contact, or from some other source before the contact is actually established. For an arbitrary doping profile, Poisson's equation has to be solved numerically, and little exploratory work has been done along those lines. There are, however, profiles (i.e. N_d as a function of x) which can be analytically solved and, of these, one is of special interest, namely that described by

$$N_d(x) = N_d + \{N_d(0) - N_d\}\exp(-x/x_0) \tag{2.3.28}$$

which provides for an asymptotic approach to uniform bulk conditions. When this is used in Poisson's equation (2.1.2b) under the condition $dE/dx = 0$ when $x = \lambda_0$, and $E(0) = \phi_{ns}$, we obtain

$$\frac{dE}{dx} = \frac{e^2 N_d}{\epsilon}(x - \lambda_0) - \frac{x_0 e^2}{\epsilon}\{N_d(0) - N_d\}\left\{\exp\left(-\frac{x}{x_0}\right) - \exp\left(-\frac{\lambda_0}{x_0}\right)\right\} \tag{2.3.29}$$

and

$$E(x) = \phi_{ns} + \frac{e^2 N_d}{2\epsilon}\{x^2 - 2x\lambda_0\} + \frac{x_0 e^2}{\epsilon}\{N_d(0) - N_d\} \times \left\{x_0 \exp\left(-\frac{x}{x_0}\right) + x\exp\left(-\frac{\lambda_0}{x_0}\right) - x_0\right\}. \tag{2.3.30}$$

Further, since $E(x) = \phi_n$ where $x = \lambda_0$, we have

$$\phi_{ns} - \phi_n = eV_D = \frac{e^2 N_d \lambda_0^2}{2\epsilon} - \frac{x_0 e^2}{\epsilon}\{N(0) - N_d\} \times \left\{x_0 \exp\left(\frac{\lambda_0}{x_0}\right) + \lambda_0 \exp\left(-\frac{\lambda_0}{x_0}\right) - x_0\right\}. \tag{2.3.31}$$

If the bulk concentration N_d were uniformly extended to $x = 0$, we would have simply

$$eV_D = e^2 N_d \lambda_{o0}^2 / 2\epsilon \tag{2.3.32}$$

which serves to identify an 'ordinary' barrier width λ_{o0} in the absence of applied voltages. Accordingly, eqn (2.3.32) can be written as

$$\left(\frac{\lambda_{o0}}{\lambda_0}\right) = 1 - 2\left(\frac{x_0}{\lambda_0}\right)^2\left\{\frac{N_d(0) - N_d}{N_d}\right\}\left\{\exp\left(-\frac{\lambda_0}{x_0}\right) + \left(\frac{\lambda_0}{x_0}\right)\exp\left(-\frac{\lambda_0}{x_0}\right) - 1\right\}. \tag{2.3.33}$$

When λ_0/x_0 is very large (rapid decay of the incremental donor concentration), we find

$$\left(\frac{\lambda_{o0}}{\lambda_0}\right)^2 \approx 1 + 2\left(\frac{x_0}{\lambda_0}\right)^2\left\{\frac{N_d(0) - N_d}{N_d}\right\} \tag{2.3.34}$$

which means $\lambda_0 < \lambda_{o0}$. When λ_0/x_0 tends to zero (incremental donor concentration spreading deeply into the semiconductor), the term in curly brackets tends to $-(\lambda_0/x_0)^2$, which means that

$$\left(\frac{\lambda_{o0}}{\lambda_0}\right)^2 \to 1+\left\{\frac{N_d(0)-N_d}{N_d}\right\}=\frac{N_d(0)}{N_d} \tag{2.3.35}$$

as, of course, it must.

From these equations, with an appropriate change of integration limits, one could also find out how such a barrier is deformed when external voltages are applied. As far as is known, the corresponding voltage–current relationships have never been calculated.

Of course, there are other doping profiles for which Poisson's equation can be solved, particularly those which can be represented as polynomials, which means virtually all plausible profiles. Thus, the reader may like to verify that a donor profile given by

$$N_d(x)=N_d(1+a_1x+a_2x^2+\ldots) \tag{2.3.36}$$

corresponds to a quiescent barrier width λ_0 given by

$$eV_D=\frac{e^2N_d}{\epsilon}\left(\frac{\lambda_0^2}{2}+\frac{a_1\lambda_0^3}{6}+\frac{a_2\lambda_0^4}{12}+\ldots\right). \tag{2.3.37}$$

Lubberts and Burkey (1975) have given an example of a case solved by numerical methods, but their use of equilibrium statistics limits its validity to zero current conditions.

The question arises whether any type of barrier doping profile is capable of enhancing rectification and this is indeed found to be so. For instance, consider a system doped in the manner shown in Fig. 2.33. Forward voltages applied would reduce the barrier width at a rate controlled by N_{d1}, reverse voltages would increase it at a rate controlled by N_{d2}. In principle, the barrier could be made to widen enormously in the reverse direction, and could thus be associated with higher-than-normal resistances, even though image-force and tunnel effects would tend to minimize the consequences. Nevertheless, parameters could in principle be chosen which would go beyond the rectification limit previously calculated. Since the barrier height ϕ_{ns} is unaffected, this particular

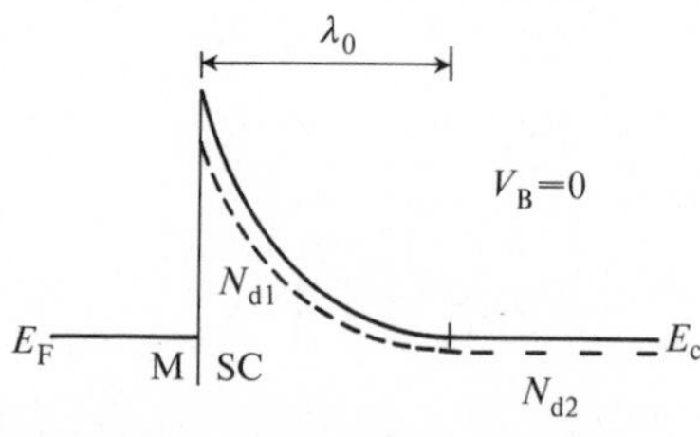

FIG. 2.33. Non-uniformly doped barrier system, $N_{d1} \gg N_{d2}$, for improved rectification.

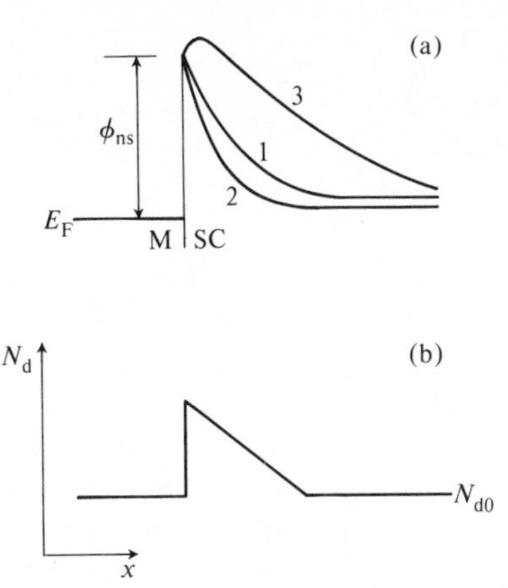

FIG. 2.34. Types of 'camel' diode. (a) After Shannon (1976). (1) Normal, uniform doping. (2) Donor concentration higher than for (1). (3) Donor concentration mostly as in (1), but with greatly increased n-type doping close the the metal-semiconductor interface. (b) After Allyn *et al.* (1980). Sawtooth-shaped composition profile (created by molecular beam epitaxy), as the origin of an asymmetric barrier.

scheme would work only in systems dominated by carrier diffusion. There are, however, methods of controlling the *effective* barrier height as such, by doing a semiconductor region in the immediate vicinity of the metal interface p-type. Of course, this would give rise to a narrow p–n junction which could not be realistically analysed in terms of one carrier type alone, but, if the p-type layer were highly doped (and of very short carrier lifetime) it could play the role otherwise played by the metal. Shannon (1976) has shown how this comes about, and a simple representation is given in Fig. 2.34(a). High values of donor concentration N_d lower the *effective* barrier height because of image force and tunnelling. Conversely, high N_a values in the boundary region lower the barrier field and (assuming ϕ_{ns} approximately constant) lead to an energy maximum in excess of ϕ_{ns}. Even if the carriers had long mean-free-paths, they would be affected by the new barrier height. It should be noted that all pure *diode* systems rectify to the same degree. For very long mean-free-paths the barrier resistance would be affected by the new doping profile, but not the voltage–current asymmetry.

A complete picture of how such structures behave is not yet available, but Shannon's experiments (actually using p-Si substrates with Sb implantation) yielded a very interesting level of agreement with expectations. Similar results on silicide contacts have been reported by Studer (1980). The corresponding structures are often referred to as 'camel diodes', imaginatively (if confusingly) named after the energy hump. Since, in such structures the highly doped semiconductor region plays the role of the metal, the actual presence of a metal is no longer required. Rectifying devices can in fact be made by creating non-uniformly doped regions entirely within a semiconductor. Thus, Allyn and co-workers (1980) have demonstrated rectification with sawtooth-like doping profiles of the kind shown in Fig. 2.34(b).

Rectifying systems sometimes exhibit 'forming' processes, also referred to as 'current creep', i.e. instabilities which arise from the movement of ionized donors or acceptors (and/or traps) in the high field of the barrier. Such electrolytic processes modify the barrier profile, leading sometimes to the improvement of rectification, sometimes to its destruction. The problem used to be serious in the days of the selenium rectifier (e.g. see Rose and Schmidt (1947) and Henisch and Ewels (1950)) but has been largely eliminated by modern technology, clean preparational methods, and the use of stable semiconductors in single-crystal form. This holds for commercial devices, but in the 'less formal' contacts encountered in the laboratory, forming and current creep still play a notable role. Ionized donors tend toward the metal, and thereby establish structures approximating to that shown in Fig. 2.33. Such structures, up to a point, imply improved rectification, but in as much as they are associated with increased boundary fields, they could also lead to lower effective barrier heights. Changes of this kind may be partly reversible. (Fig. 2.35.) Donor centres may also separate out at the interface, forming a new phase there with the metal, which is likely to reduce the current ('negative creep). There is a good deal of evidence that this is what happens at metal contacts with selenium, but it is impossible to make general predictions. As one would expect, higher temperatures favour creep processes. In principle, this opens the possibility of establishing a specially desirable distribution of active centres at a high temperature, and then 'freezing it in' for normal use. If ionized donors were completely mobile, did not separate out at the interface, and were freely replaced from the bulk, the

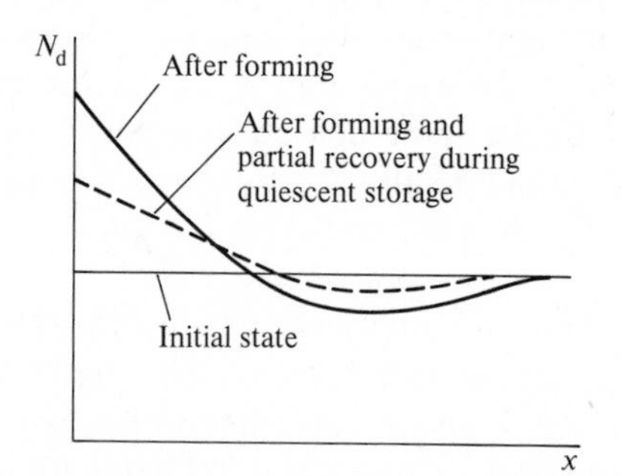

FIG. 2.35. Redistribution of donor centres during 'forming' and recovery; schematic contours. Forming can take place under the influence of applied voltages, or else under the built-in, quiescent barrier field ('self-forming').

donor distribution would in the course of creep approach a Boltzmann profile, for which field and diffusion are in balance. Such a distribution would be characteristically associated with every applied (forming) voltage.

In the ordinary way, 'forming' is now regarded as an undesirable disturbance, but there were times when it held some promise for device development in its own right. Thus, Landsberg (1951b) examined the low frequency characteristics of a barrier on the assumption that its profile is established by the diffusion of ionized centres under the influence of the applied signal, reaching a new steady state during each half-cycle in the manner described above. See also Jaffé (1952).

Corresponding to current creep, there is of course a simultaneous capacitance creep, likewise resulting from the modified distribution; e.g. see Schmidt (1941). Since the dry electrolytic processes discussed here are associated with time-constants, they affect AC properties, conductive as well as reactive, in a complicated way. Moreover, it is not always easy to distinguish them from other effects, e.g. those associated with the presence of deep traps.

2.3.8 *Voltage–current relationships in theory and experiment*

References to comparisons between theoretical predictions and experimental results will be found scattered throughout this book. Hundreds of such attempts have in fact been made and, in the nature of things, all confirm at least some of the experimenter's expectations. None confirm *all* the expectations that one might conceivably entertain, because every measurement tests *two* matters which are inextricably intertwined:

(a) whether the logical consequences of original model assumptions have been correctly formulated, and
(b) whether the experimental system in hand really has the intended structural characteristics.

Problem (b) calls for careful analytic and descriptive procedures for which time is hardly ever found, either in the laboratory or in publications. As a result, much is usually taken for granted.

If an experimenter has problems in knowing his own system, so has any later evaluator of his results, which is why a systematic overview of experimental verification attempts would not be meaningful. Selective examples serve the purpose much better. Here we shall be concerned specifically with evidence for the validity of the above models, as derived from voltage–current relationships. (Tests involving capacitance relationships will be referred to in Chapter 4.) One of the classic pioneering efforts in that direction is due to Kahng (1963), and involves a particularly careful analysis of gold contacts on n-type silicon of varying resistivity and meticulously controlled surface treatment. By a variety of tests and calculations, it was shown that those particular contacts were free from minority-carrier injection (Chapter 3), which made the assessments relatively simple. This is certainly *not* an assumption that one would want to make more generally without extensive supporting evidence; the fact is that contact properties can vary widely from structure to structure. The results discussed below thus refer only to a particular type of contact on a particular material, but serve beyond that to outline a plausible investigatonal procedure. The obvious questions are:

(i) do voltage current characteristics have any of the predicted shapes,
(ii) do they depend on temperature in the predicted way, and
(iii) do they depend on structural parameters (as far as these can be ascertained) in accordance with expectations.

Kahng's (1963) own analysis begins with (i), and specifically with the forward characteristic; see Section 2.2.5. For his Au–n-Si diodes, the forward characteristics did indeed behave in accordance with eqn (2.2.31) for forward voltages greater than about $2kT$, with a non-ideality factor η close to unity ($\eta = 1.09$), while bulk resistivities varied between 0.12 and 51.0 ohm-cm. Typically, such a relationship might be obeyed over four decades of current, and this is also true for aluminium contacts on n-Si (Gutknecht and Strutt 1972). Occasionally, as for gold contacts on SiC, conformity over many more decades is reported (Wu and Campbell 1974). For PtSi–n-Si Lepselter and Sze (1968) observed linearity over a current range covering almost 8 powers (of 10). At high currents there is always a deviation which has been clearly shown to depend on bulk resistivity, as one would expect. The lower the series resistance (taken as constant, in the absence of injection), the greater is the range over which $\log|J|$ versus V_B does in fact yield a straight line. However, at lowest bulk resistivities, when the high-current departures are negligible, a new low-current departure makes its appearance. (Figure 2.36.) Its origin is still comewhat of a mystery.

An η factor can be established from the higher current range, and eqn (2.2.31) even after making allowance for the series resistance R_b. We then have, for forward voltages in excess of kT/e

$$J \approx -J_s \exp\left\{-\frac{e(V_B - JR_b)}{\eta kT}\right\} \tag{2.3.38}$$

(compare Section 2.2.5.) V_B and J are both negative. Accordingly,

$$\log\left\{\frac{-J}{\exp(-eV_B/\eta kT)}\right\} = \log J_s + \frac{JR_b e}{\eta kT} \tag{2.3.39}$$

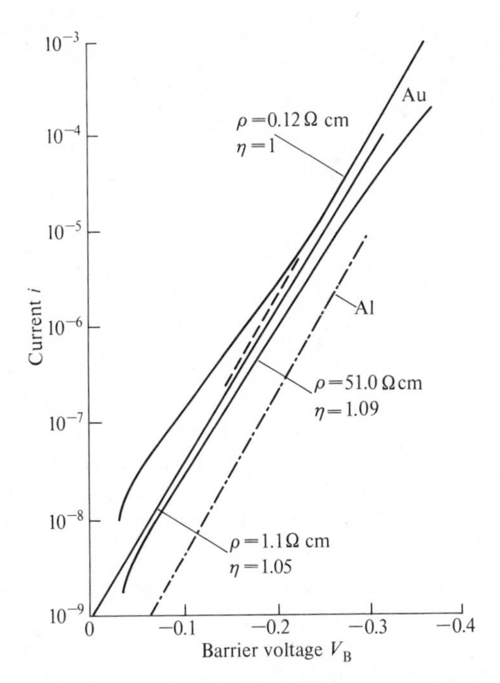

FIG. 2.36. Experimental tests of voltage–current characteristics; in the forward direction. Summary of results obtained on Al and Au–n–Si contacts. Full and broken lines: Au contacts; Kahng (1963); contact area 1.8×10^{-3} cm^2; $\phi_{ns}=0.79\pm0.02$ eV. Dotted line: Al contacts on a (100)-Si surface; Gutknecht and Strutt (1972); contact area 4.87×10^{-4} cm^2.

and that plot is capable of yielding J_s as well as R_b for further comparisons. In particular, Kahng used eqn (2.3.39) to explore the temperature-dependence of J_s, taking J_s to be given by eqns (2.2.37) and (2.2.39). It was then possible to evaluate Richardson's constant A^*, as well as the barrier height ϕ_{ns}. The barrier height so found diminished somewhat with increased doping content, qualitatively and quantitatively as one would expect from the modified barrier shape due to the operation of image forces (Section 2.2.8). On the other hand, A^*, which should have been a constant, showed major variations, signifying that the simplest diode notions could not be maintained in all respects. Basic to the expectation of a particular and constant value of A^* is the simplifying assumption of electronic equilibrium at the metal–semiconductor interface, and it is entirely plausible that this should be unfulfilled.

The reverse characteristic can be subjected to similar scrutiny, voltage and temperature dependence being again the principal criteria. According to *pure* diode theory (eqn 2.2.39a), J should saturate completely; according to *pure* diffusion theory (eqn (2.2.25) and (2.2.27), J should depend on $(V_D+V_B)^{1/2}$, and both could be subject to the influence of image forces (eqn (2.2.50)). It is thus plausible to test whether any power law of the form $(V_D + V_B)^z$ is obeyed. The answer is, only moderately well, with $z \approx \frac{1}{2}$ only at 100 °C, and z as high as 3 at −40 °C. There are indications that z is actually smaller than $\frac{1}{2}$ above 100 °C. Some form of voltage-dependent barrier lowering must therefore be at work. Edge effects may also be involved.

The temperature dependence of reverse currents yields similarly complex results, but it is reassuring to find that at high temperatures, at any rate, the activation energy evaluated has the same value as the barrier height obtained from the forward characteristic, namely 0.79 ± 0.02 eV. The crystal orientation was not specified. Gutknecht and Strutt (1972) have reported that, at any rate for Al–n-Si contacts, the barrier height is 0.72 eV on (111)- and 0.81 eV on (100)-surfaces, values similarly confirmed by the forward, reverse, and capacitance characteristics. No orientational effects were observed for PtSi–n-Si contacts, for which an explanation has been suggested in terms of surface state coupling. However, there can be no certainty at this stage. (In passing, the highest barriers ever reported (~0.9 eV) on p-type silicon appear to be associated with hafnium contacts (Saxena 1971). As one would expect, such contacts are of low resistance (quasi-ohmic) when applied to n-type material.) At low temperatures (which makes diode currents difficult) the reverse characteristics yield a *lower* activation energy, suggesting that the currents are then controlled by minority-carrier generation in the bulk material. (Extraction currents; see Section 3.4.2.)

At sufficiently high temperatures (e.g. above room temperature) when diode currents are thought to prevail, one can test for image-force effects as an alternative for testing the $(V_D + V_B)^a$ power law, i.e. by plotting $\log J$ against $(V_D + V_B)^{1/4}$, as suggested by eqn (2.2.50), or something equivalent (e.g. $F(0)^{1/2}$). In this way Kahng obtained excellent agreement with the predictions of barrier lowering based on simple image-force theory, making recourse to more complex mechanisms unnecessary. In contrast, a series of results on silicon reported by Shannon (1977a) demanded not only image force but tunneling corrections. Shannon was able to extend the analysis to systems of varying donor content, the argument being as follows. For any constant donor content, the shape of the barrier can be simply calculated, neglecting free-carrier charges. Accordingly, the tunnelling probability can be assessed as a function of energy, yielding a certain reverse current. If the reverse current actually found were to differ from the predicted value, the difference could be ascribed to a divergence of the barrier shape from the assumed contour, and thus to a variation of $N_d(x)$. From measurements of J and dJ/dV_B, Shannon was able to obtain the $N_d(x)$ profile, and compare it with similar results obtained from capacitance measurements (see Chapter 4). The agreement (at 77 K) was remarkably good, considering the general complexity of such systems.

The conclusion is that whereas practical situations are inevitably more complicated than idealized models, it is reasonable to believe that our picture of barrier transport processes is essentially correct. The above examples represent relatively simple cases. As systems become more complicated it becomes increasingly difficult to interpret their characteristics *uniquely* in terms of structure. This is well illustrated by another comprehensive experimental series (Pike and Sweet 1975), performed on Si–Ge alloys, with interfacial layers of varying thickness. For a corresponding analysis of silicide–Si contacts, see Pellegrini (1975a).

For another illustration of the difficulties inherent in these problem, it is instructive to consider results obtained by Wilson and Allen (1975) on ZnSe Schottky diodes. For substantial voltages, these diodes had reverse characteristics of the form

$$J = J_0 \exp(V_B/V_0) \tag{2.3.40}$$

with V_0 almost temperature independent, and therefore not interpretable as in any sense equivalent to kT/e. Typically, diodes conformed to this empirical relationship over 5 decades of current. Even more surprisingly, a correlation was found to exist between the V_0 and J_0 values measured on different diodes, in the sense that $\log(J_0)$ diminished linearly with $1/V_0$ over a sensational 30 decades of J_0. There is no immediate prospect of arriving at a simple model for this form of behavior. Wilson and Allen point out that there is at least one other such example in semiconductor physics, namely the Meyer–Neldel Rule (Meyer and Neldel 1937), which established the correlation between σ_0 and E_{geff} in

$$\sigma = \sigma_0 \exp(-E_{geff}/2kT) \tag{2.3.41}$$

where E_{geff} is an effective band gap, experimentally determined.

Further reading

On boundary value corrections: Baccarani 1976, Viktorovitch and Kamarinos 1976.

On a general overview of Schottky barrier characteristics and models: Padovani 1971.

On electronic states at $Si–SiO_2$ interfaces: Cheng 1977.

On rectification at ion-bombarded surfaces: Allen and Farnsworth 1956.

On current-controlled interface state occupancy: Card 1975b, Rhoderick 1975.

On camel diodes: Roy and Daw 1980, Shannon 1979, Wu 1981.

On current creep: Hoffmann *et al.* 1950, Lehovec 1951.

2.4 Current-carrying barriers: numerical solutions

2.4.1 *Semi-infinite Schottky systems*

For single carrier situations, the starting point is, as always, eqn (2.2.7), but without the simplifying assumptions made in Section 2.2, analytic solutions are not feasible. Instead, we use numerical methods. (To be sure, numerical computations on single-carrier systems can be subsumed under computations on two-carrier systems, but this involves complications which are often unnecessary and didactically non-transparent.) There are several equivalent ways in which the equation can be prepared for this kind of treatment. A convenient one is to express the current density in terms of F_∞, the bulk field: $j = n_e e\mu_n F_\infty$, and to introduce new normalized variables in the form

$$\begin{aligned} \boldsymbol{X} &= x/\lambda_D & \boldsymbol{N}(x) &= n(x)/n(e) \\ \boldsymbol{E} &= E/kT & \boldsymbol{V}(x) &= eV(x)/kT \\ \boldsymbol{F}_\infty &= (\mathrm{d}\boldsymbol{E}/\mathrm{d}\boldsymbol{X})_\infty & \boldsymbol{S}(x) &= (\mathrm{d}\boldsymbol{N}/\mathrm{d}\boldsymbol{X})_x \end{aligned} \tag{2.4.1}$$

where λ_D is the Debye length. $\boldsymbol{N} = 1$ in the bulk material.

Of course, quite different forms of normalization can be used (compare Sections 2.4.2 and 3.2.2) depending on convenience. With these changes, and in conjunction with Poisson's equation, eqn (2.2.7) can be transformed into the dimensionless equation

$$\frac{\mathrm{d}^2\boldsymbol{N}}{\mathrm{d}\boldsymbol{X}^2} = \boldsymbol{N}(\boldsymbol{N}-1) - \frac{\boldsymbol{S}}{\boldsymbol{N}}(\boldsymbol{F}_\infty - \boldsymbol{S}) \tag{2.4.2}$$

for complete donor ionization. In this form, the equation is very suitable for step-by-step computation, e.g. by Runge–Kutta procedures. Such calculations yield $\boldsymbol{N}$ as a function of $\boldsymbol{X}$, with $\boldsymbol{F}_\infty$ as a fixed parameter, proportional to the current density. For the model itself one assumes a certain value of $E(0)$, which defines the barrier height. One further assumes an arbitrary initial value of $\boldsymbol{S}(0)$, and computes its consequences. If the $\boldsymbol{S}(0)$ value chosen is too high, $\boldsymbol{N}-1$ tends to be positive as $\boldsymbol{X}$ becomes very large. Conversely, if $\boldsymbol{S}(0)$ is too small, $\boldsymbol{N}-1$ tends to become negative. The correct choice of $\boldsymbol{S}(0)$, whether made by inspection, or else automatically by the computer, makes $\boldsymbol{N}-1 \to 0$ as $\boldsymbol{X} \to \infty$, and is recognized by this criterion. This constitutes the well-known

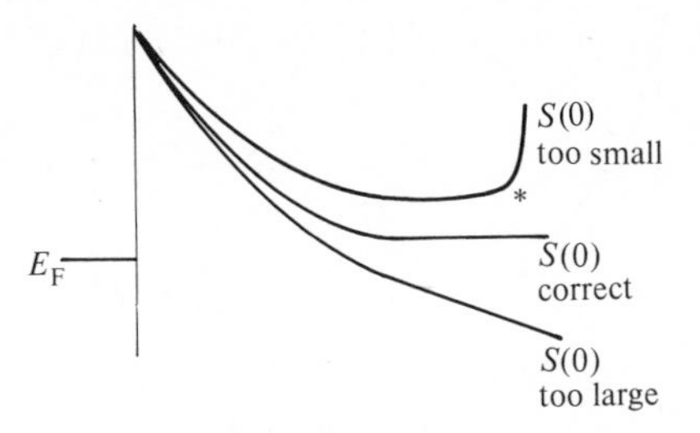

FIG. 2.37. Numerical calculation of barrier characteristics. The 'shooting method' applied to a semi-infinite barrier system. Note instability at *.

'shooting method', which has a tempting simplicity but also severe limitations (see below).

The procedure is even more clearly illustrated in terms of energy contours, and it is simple enough to show that

$$\mathrm{d}\boldsymbol{E} = \frac{\boldsymbol{F}_{\infty} - \boldsymbol{S}}{\boldsymbol{N}}\,\mathrm{d}\boldsymbol{X} \tag{2.4.3}$$

where $\boldsymbol{S}$ and $\boldsymbol{N}$ depend on $\boldsymbol{X}$. Since $\boldsymbol{N}(0)$ is fixed, an initial choice of $\boldsymbol{S}(0)$ amounts to an initial choice of $(\mathrm{d}E/\mathrm{d}x)_0$, and corresponds to a certain boundary field. Figure 2.37 illustrates the situation schematically. The energy contour is then derived from the initial value $E(0)$ and the step-by-step application of eqn (2.4.3).

In principle, the method would appear to be flexible; in practice $\boldsymbol{S}(0)$ must be fixed with extraordinary precision (e.g. to 16 places of decimals!) to achieve solutions which are stable over long distances. Of course, none of the contact parameters are actually defined to anything like that accuracy in nature; nevertheless, the reasons for the onset of instability during step-by-step computation can be readily understood. We know from analytic work with *linearized* equations that the expression for $\Delta\boldsymbol{N}(=\boldsymbol{N}-1)$ contains terms which vary with $\exp(\boldsymbol{X})$. (See eqn (3.2.25).) Since all the solutions in semi-infinite systems must tend to zero with increasing x, the coefficients of $\exp(\boldsymbol{X})$ terms must identically vanish. In analytic work, they are accordingly *set* equal to zero, and this is indeed one of the *results* of the method. In numerical computation, there is no direct equivalent to the zero-setting process, though the terms must in fact be zero when the ideal solution has been reached. Before it is reached, these terms may be small but are not zero, and if a non-zero term is in effect multiplied by $\exp(\boldsymbol{X})$, there must be some value of $\boldsymbol{X}$ for which the product becomes really large and disturbing.

This is an inherent difficulty, which would exist even if $\boldsymbol{S}(0)$ could be fixed with *total* accuracy, since truncation errors are always present, and are cumulative in the procedure used. The shooting method is therefore

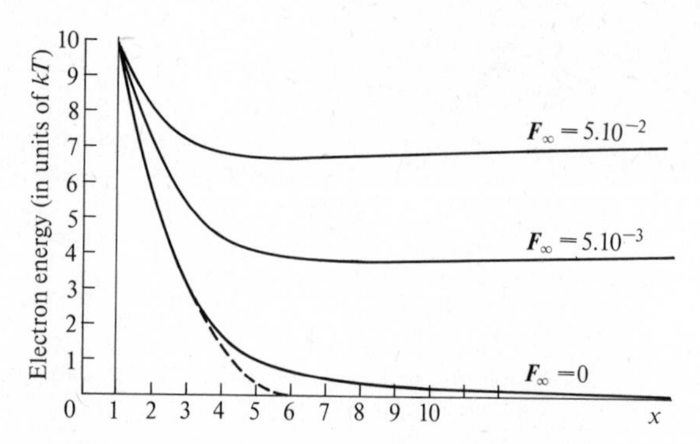

FIG. 2.38. Numerically calculated energy contours for a Schottky barrier on relatively-well-conducting material. Forward currents. Fixed barrier height $eV_D = 10kT$; variable bulk field F_∞ (normalized). Broken line: static characteristic ($F_\infty = 0$), with electronic space-charge neglected.

restricted to calculations over small distances, e.g. of the order of 10–12 Debye lengths, and for some purposes this is not enough. In any event, it is highly desirable to adjust the step size of the computation, making it small where parameters change rapidly, and larger where they do not. Additional problems arise in two-carrier systems (Section 3.5), as one would expect.

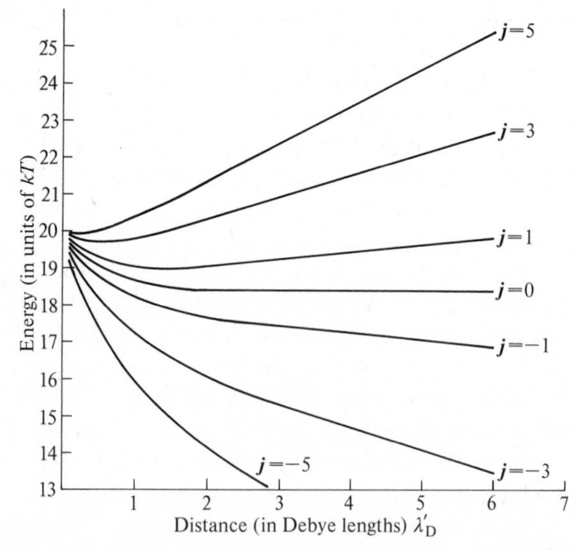

FIG. 2.39. Numerically calculated energy contours for a low Schottky barrier on high resistivity material. After Park and Henisch (unpublished). $n_e = 5\ n(0)$, leading to $eV_D = 1.61kT$. Variable current densities in the forward and reverse directions, here normalized in the form $\mathbf{j} = j\lambda'_D/eD_n n(0)$ where j = actual current density and $n(0)$ the electron concentration at the metal–semiconductor interface. ϕ_{ns} is shown as $20kT$, but this is arbitrary; the absolute barrier height does not enter into the calculations. λ'_D is here defined in terms of $n(0)$.

Figure 2.38 shows sample contours calculated by this method for a comparatively low barrier height of 10 kT. For $\boldsymbol{F}_\infty = 0$ (zero current) the barrier has in principle an infinite width. By the time a forward bulk field as small as $\boldsymbol{F}_\infty = 5 \times 10^{-3}$ has been reached, the barrier has already narrowed from infinity to about $8.5\lambda_D$ (in this example). Figure 2.38 also shows how the bulk field compares with the barrier field. Even so, these results apply only to a conventional case of high bulk conductivity. If the bulk conductivity were low, the bulk and barrier fields could become comparable, as shown in Fig. 2.39. The normalized current has been used as a variable, and it will be seen that $\boldsymbol{j} = 5$ leads in this case to the flat-band condition. The field at the contact interface is then zero, and the bulk field very high.

Results of this kind raise the question of how voltage–current relationships could and should be specified. When the bulk field is negligible, there is no problem; when the bulk field is substantial, forward characteristics (at any rate) could be specified in terms of V_a (see insert in Fig. 2.40), defined at the energy minimum. No such minimum occurs in the

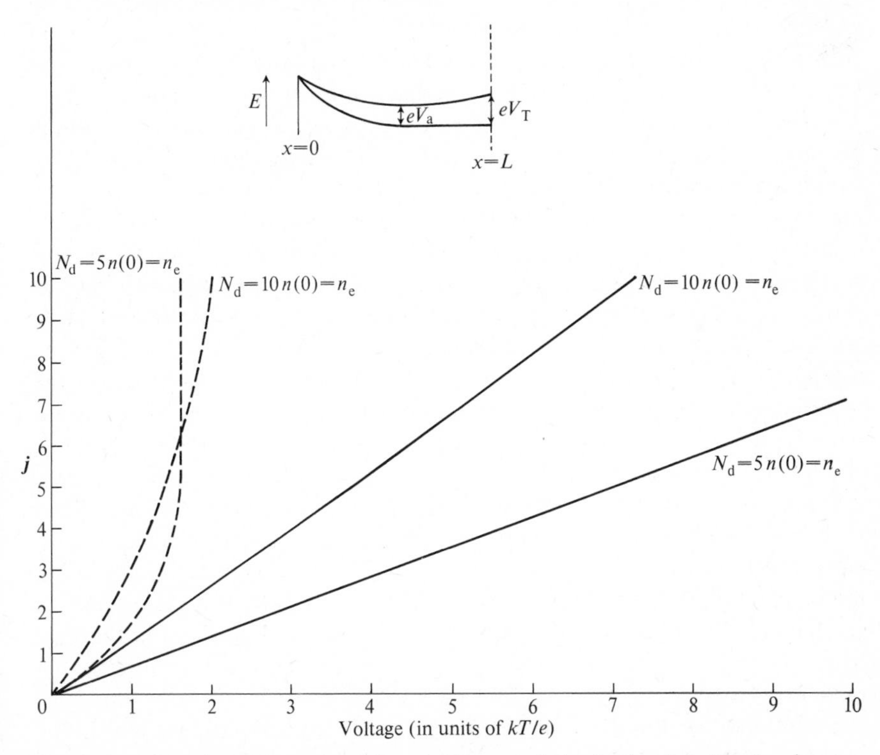

FIG. 2.40. Comparison of computed forward characteristics for low Schottky barriers on high-resistivity material. Broken lines: $\boldsymbol{j}$ versus $V_a/(kT/e)$. Full lines: $\boldsymbol{j}$ versus $V_T/(kT/e)$. Both for two values of n_e: $n_e = 5\ n(0)$, $eV_D = 1.61kT$; $n_e = 10\ n(0)$, $eV_D = 2.31kT$. Current density $\boldsymbol{j}$ normalized as in Fig. 2.39. System width here taken as $L = 6\lambda_D'$.

reverse direction, and the problem remains. Alternatively, one could specify the relationship in terms of the total voltage V_T for any particular specimen thickness L. By adding a linear ballast, this tends to suppress the rectification effect. Indeed for a contact on high resistivity material all signs of rectification quickly disappear, as Fig. 2.40 shows. The system then behaves, quite misleadingly, as if no barrier were involved. In the examples shown, this fact is particularly emphasized because the diffusion potentials are so low, but it is precisely for such barriers (in which the free-carrier charge is never negligible) that numerical methods are required. For higher barriers, the differences between V_T and V_a would, of course, be smaller. It should be remembered that the j–V_a characteristic is purely notional, since V_a is never directly accessible to measurement.

The limitations of the 'shooting method' can be overcome through the use of matrix procedures, as described in the following Section.

2.4.2 *Finite Schottky systems*

Though semi-infinite models have a certain mathematical elegance and appeal, they do not, of course, correspond to anything that exists in practice. Finite models have computational advantages but involve an additional problem, in as much as a second set of boundary conditions has to be fixed. The case which is of immediate interest here is that in which a low resistance contact (ideally, a zero resistance contact) appears at a distance L from the rectifying boundary. In calculations, this could be simulated by making the charge density zero, i.e. in the present terms, by making $n(L) = n_e$, as long as all the donors are considered ionized. Since, in any event, neutrality is closely approached at substantial distances, this is a tempting procedure. On the other hand, the above discussion has shown how enormously sensitive the computed concentrations at large distances are to the precise (assumed) value of $F(0)$, and it might be thought that the arbitrary imposition of the condition $n(L) = n_e$ on the system could have equally drastic consequences. However, Moreau, Manifacier, and Henisch (1982) have shown that it is not so. This was done by comparing numerical results for given boundary conditions and two specimens of different thicknesses, e.g. $L = 1000\lambda_D$ and $L = 25\lambda_D$. After making appropriate allowance for the ohmic bulk component, the results were the same to a high degree of accuracy, proving that (in the cases under review) even a thickness of $L = 25\lambda_D$ was sufficient to simulate a semi-infinite specimen. This is, indeed, what practical workers in the field have always assumed, and for single-carrier systems the assumption is evidently justified.

Finite models are best handled by matrix techniques, in which the region under discussion is divided into a number of fixed (and preferably unequal) steps, each of which is governed by a difference equation of its own. The problem is then one of solving (say) fifty to one hundred equations simultaneously, which also means that the method is free from most of the cumulative errors mentioned above in connection with Runge–Kutta. The algorithm† used for this purpose is, of course, different from that described above (Section 2.4.1), but we begin again with the diffusion equation (2.2.7). Normalizing carrier concentrations and distances as usual, this time in the form

$$\boldsymbol{F}=\frac{F(x)\lambda_{\mathrm{D}}}{kT/e}; \qquad \frac{\boldsymbol{E}}{\boldsymbol{X}}=-\frac{eV}{kT}$$
$$j=\frac{j\lambda_{\mathrm{D}}}{eD_{\mathrm{n}}n_{\mathrm{e}}} \qquad \boldsymbol{X}=\frac{x}{\lambda_{\mathrm{D}}}. \tag{2.4.4}$$

The normalized current can then be expressed as

$$\boldsymbol{j}=\boldsymbol{NF}+\mathrm{d}\boldsymbol{N}/\mathrm{d}\boldsymbol{X} \tag{2.4.5}$$

λ_{D} is defined here in the normal way, by reference to the bulk carrier concentration n_{e}. $\boldsymbol{j}$ is constant, in the sense that it is independent of $\boldsymbol{X}$. The equation may then be integrated for $\boldsymbol{N}$, and this may be done by means of the integrating factor $\exp(\boldsymbol{E})$ without making any distinction between general and particular solutions, i.e. in the form:

$$\boldsymbol{N}(x)=\exp(-E)\left\{\int_0^{\boldsymbol{X}}\exp(\boldsymbol{E})\cdot\boldsymbol{j}\cdot\mathrm{d}\boldsymbol{X}+C\right\} \tag{2.4.6}$$

where C is non-constant and remains to be determined by reference to the boundary conditions. In a finite system, the boundary voltages $\boldsymbol{E}(0)$ and $\boldsymbol{E}(L)$ are fixed, and the current has to be calculated for each pair of such values.

For $\boldsymbol{X}=0$, we have $\boldsymbol{N}(0)=\exp\{-\boldsymbol{E}(0)\}+C$, from which C is readily obtained, for substitution into eqn (2.4.6). At the other end of the specimen, we must know $\boldsymbol{V}(L)$ and $\boldsymbol{N}(L)$ and with these values eqn (2.4.6) gives $\boldsymbol{j}$ in the form

$$\boldsymbol{j}=\frac{\boldsymbol{N}(L)\exp(\boldsymbol{E}(L)-\boldsymbol{N}(0)\exp(\boldsymbol{E}(0))}{\int_0^{L/\lambda_{\mathrm{D}}}\exp(\boldsymbol{E})\,\mathrm{d}\boldsymbol{X}} \tag{2.4.7}$$

(Compare eqn (2.2.14).) The integration in the denominator is performed numerically, in the first instance on the basis of an assumed potential contour, e.g. a linear contour. This contour is then corrected and updated in stages. Thus, an assumed $\boldsymbol{E}$ as a function of $\boldsymbol{X}$ leads to a definite value

† The author is indebted to Dr. Yves Moreau of the Centre d'Electronique de Montpellier, Montpellier, France, for drawing his attention to this method.

of $\boldsymbol{j}$ via eqn (2.4.7), and that leads in turn to definite values of $\boldsymbol{N}$ as a function of $\boldsymbol{X}$ via eqn (2.4.6). Each of these values can then be substituted into Poisson's equation, in the normalized form

$$d^2\boldsymbol{E}/d\boldsymbol{X}^2 = 1 - \boldsymbol{N} \tag{2.4.8}$$

yielding a series of simultaneous equations which can be numerically solved for improved values of $\boldsymbol{E}$ as a function of $\boldsymbol{X}$. This new $\boldsymbol{E}$-contour is then substituted for the previous $\boldsymbol{E}$-contour, and so on. After a few iterations, the potential profile of the barrier reaches a stable shape even though the initially assumed shape (e.g. linear) was enormously oversimplified. Complicated profiles need smaller (and, accordingly, more) computational steps.

The first solutions for finite systems (albeit limited to $L = 10\lambda_D$) were given by Macdonald (1962). Amongst other things, they showed that the semi-logarithmic current–voltage plot in the forward direction was indeed linear over a wide range (as the conventional Schottky relationships suggest), but that its slope was in that instance $e/1.15kT$ instead of e/kT (Fig. 2.41). This is a further confirmation of the fact that the popularly assumed e/kT slope is an unrealistic expectation (see also Section 3.5.1). For a contact on high-resistivity material, the effect would be even more pronounced.

Special situations arise in the case of very thin systems, a thin system being here defined as one in which $L < \lambda_0$. The total thickness is thus less than the barrier width would be if the same contact were made on semi-infinite (or, at any rate, on sufficiently thick) bulk material. The only known analysis of this kind is one due to Moreau and Manifacier (1980). Figure 2.42(a) shows results for such a model ($L = 3\lambda_D$), with $n(L) = 0$ at all times. Accordingly, there has to be an electric field at this point, even

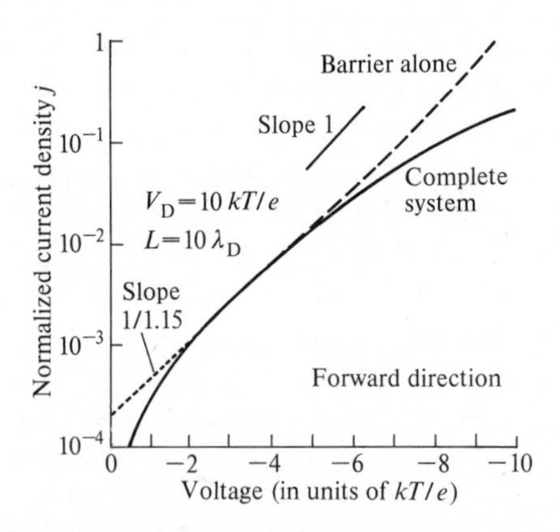

FIG. 2.41. Numerically calculated forward characteristics; single-carrier system. After Macdonald (1962). $\Delta n(L) = 0$ assumed. Broken line V_a; full line V_T.

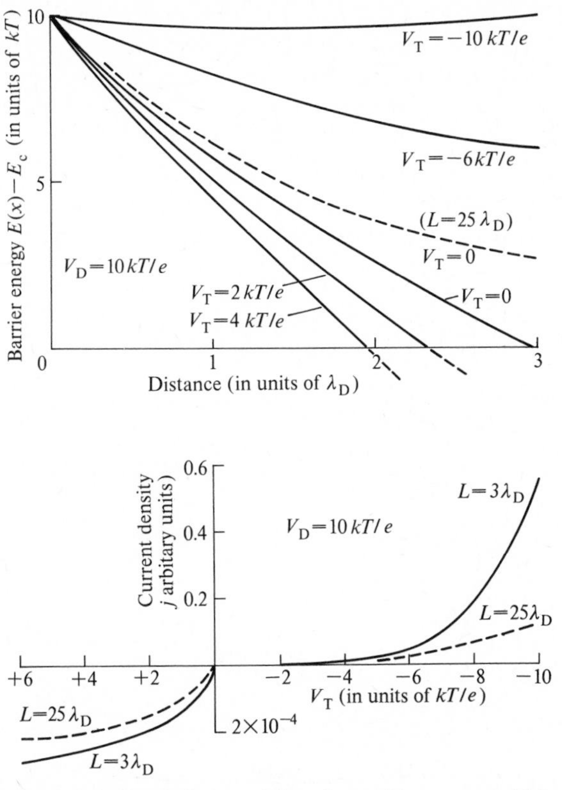

FIG. 2.42. Numerically calculated characteristics of very thin systems with barriers. Moreau and Manifacier (1980). Solid lines: $L=3\lambda_D'$. Broken lines: $L=25\lambda_D$. λ_D defined here in terms of the bulk value n_e. (a) Energy contours for various forward and reverse (total) voltages V_T applied. (b) Voltage current characteristics.

for zero current; the barrier 'is not allowed to end', though it comes close to ending as higher forward voltages are applied. Figure 2.42(b) gives the corresponding voltage–current characteristic, in comparison with that of the same barrier on a specimen of thickness $L=25\lambda_D$. It will be seen that the series resistance is all important in the high forward direction, as one would expect. In the reverse direction (positive values of V_T), the situation is controlled by the total electron exhaustion near the interface. Specimen thickness then has only a comparatively small influence on the current.

One matter which can be instructively pursued by numerical techniques concerns the problem of barrier thickness, already mentioned in Section 2.4.1. In the traditional Schottky model, the barrier (in the absence of current flow) ends at $x=\lambda_0$. In fact, when the free-charge-carriers are taken into account (see Section 2.1.2) it never ends, since dE/dx is zero

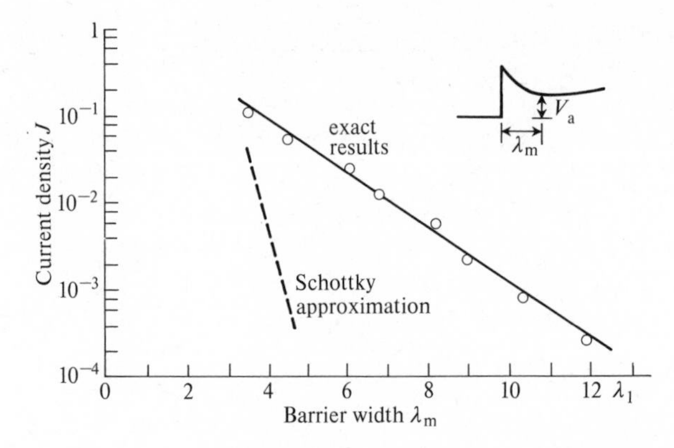

FIG. 2.43. Comparison of barrier widths as a function of current. Single carrier system; $N_d = 10^{17}\,\text{cm}^{-3}$. Complete ionization. $\boldsymbol{j} = j\lambda_D/eD_n n_e$. Full line: barrier widths λ_m numerically computed, based on V_a (see insert). Broken line: barrier widths λ_B calculated for a classical Schottky model, based on $V_B + V_D$. After Moreau, Manifacier, and Henisch (1982).

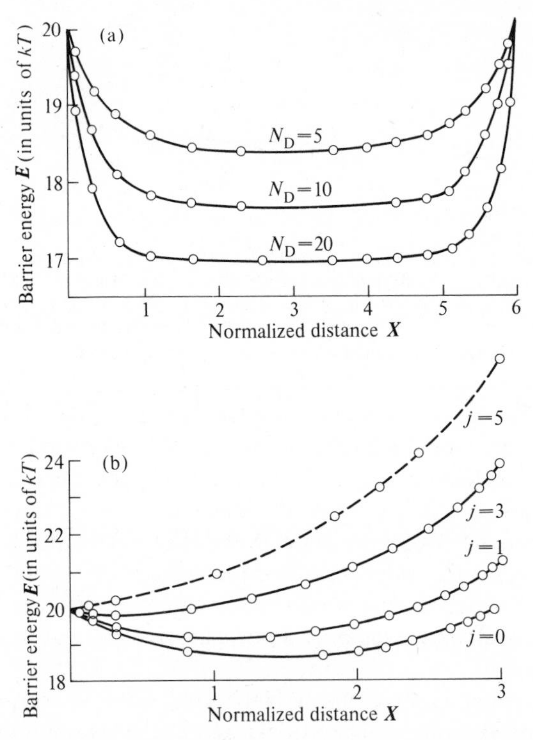

FIG. 2.44. Typical energy contours for systems with two interacting barriers. Specimen thickness L and distances $\boldsymbol{X}$ normalized to λ'_D (a) $L = 6\lambda'_D$, zero current, various donor contents normalized to $n(0)$. (b) $L = 3\lambda'_D$, various currents, normalized as in Fig. 2.39. After Park and Henisch (unpublished). λ'_D defined by reference to $n(0)$.

only at infinity. As increasing forward currents flow, the energy minimum (potential maximum) moves rapidly from $x=\infty$ to very small distances from the metal interface, and those distances actually provide a more meaningful concept of barrier thickness than the Schottky model. Figure 2.43 shows how the two quantities compare, as a function of forward current density, and the relative crudity of the Schottky approximation is thereby demonstrated. Of course, this discrepancy must also show itself in the calculated barrier capacitance.

All the above comments apply to a single barrier, but practical systems often involve *two* barriers, as envisaged in Section 2.2.9. However, the analytic treatment given there neglected free-carrier space-charges, and did so for the very purpose of achieving analytic solutions. Better solutions are possible by computer methods, using either matrix or shooting methods. Thus, the shooting method of Section 2.4.1 can easily be adapted to two barriers, by 'aiming' not at $\mathbf{N}-1\to 0$ as $\mathbf{X}\to\infty$, but at $\mathbf{N}=\mathbf{N}(0)=\mathbf{N}(L)$ at $\mathbf{X}=L$, assuming that the two barriers are identical at zero current. When the two barrier thicknesses are much smaller than the specimen thickness L, the barriers are in non-interacting series connection, and the problem is simple. When $L<2\lambda_0$, the two barriers interact in a complex manner which is totally inaccessible by analytic methods. Figure 2.44 shows typical results for a poorly conducting material.

2.4.3 *Thick composite barriers*

Barrier structures of the kind illustrated in Fig. 1.19 have also been the subject of analysis by numerical methods, but mostly in the context of thermionic field emission through the 'insulating' layer. When the (say) oxide barrier is at all high, tunnelling is in fact the only plausible transport mechanism, and for that to happen, the barrier must be thin, with λ_a less than 30 Å (and preferably much less). If it is thick, the contact will appear as totally insulating for practical purposes. This leaves open the possibility that we might have a barrier classified as 'thick' (i.e. too thick for tunnel penetration), but also as 'low' (i.e. not very much higher than the Schottky barrier within the semiconductor itself; $\phi_{mi}-\phi_{ns}<kT$). Such a structure constitutes an abrupt heterojunction between two semiconductors, and is more appropriately treated under that heading. A one-dimensional analysis has been made by Moreau (1983b), using drift-diffusion theory and the matrix method outlined in the previous section. The main point of interest is not so much the precise voltage–current characteristic (which must obviously resemble that of a Schottky barrier of equal total height) but the effect of the permittivity ratio ϵ_i/ϵ where ϵ_i is the permittivity of the insulating layer and ϵ that of the semiconductor. There is also the problem of interface states between the semiconductor and insulator. In their absence, the electric fields on either

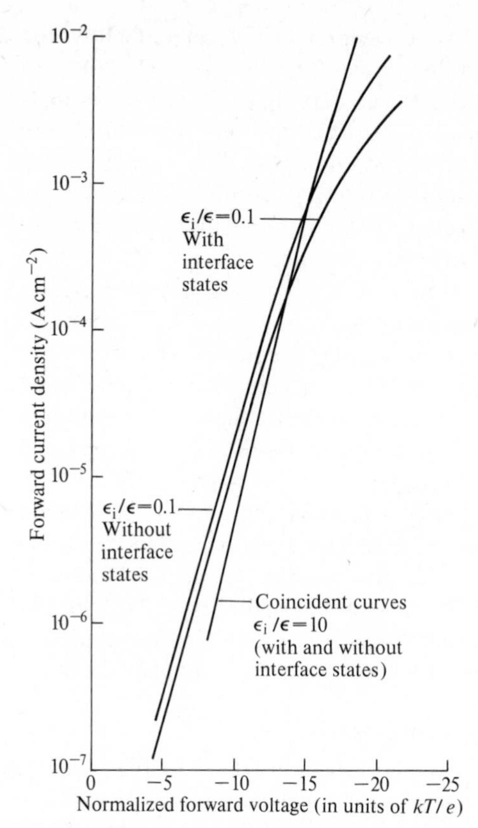

FIG. 2.45. Voltage–current characteristics of composite barriers; varying permittivity ratio ϵ_i/ϵ; with and without interface states. Idealized single-carrier systems. $\lambda_e = 5\lambda_D$; $L = 25\lambda_D$ Total barrier height $\phi_{mi} = 21kT$ Conductivity band step at the interface: $5kT$. 'Insulator' assumed to be intrinsic material. Semiconductor: $N_d = 10^4 \times n_i = n_e$. Interface state concentration (at Fermi level): equivalent to $4n_e$. Current normalized as in Fig. 2.43. After Moreau (1983b) and Henisch *et al.* (1984).

side of that boundary would have to be in the ratio ϵ/ϵ_i; in their presence, the field ratio can be very different, all the more because the interface charge may be current-dependent. Both factors govern the relative voltage drops across insulator and semiconductor. The model therefore has many descriptors, and supports a great variety of barrier contours.

Figure 2.45 shows some typical computed results, and provides in passing another contribution to the discussion of 'non-ideality' aspects. Though the systems evaluated differ greatly from one another, this shows itself little or not at all in the corresponding values of the non-ideality factor η. Even the enormous change from $\epsilon_i/\epsilon = 0.1$ to $\epsilon_i/\epsilon = 10$ has only an unspectacular effect on the η sector. This has also been shown to hold for two-carrier systems (associated with even more descriptors), even under illumination. We conclude again that whereas non-ideality factors can serve as useful empirical summaries of contact behaviour, they have no merit as structure-diagnostic tools (see also Sections 5.2 and 6.3.2).

Further reading

On charge trapping in oxide layers: Kolk and Heasell (1980a)

3

TWO-CARRIER SYSTEMS

3.1 Inversion layers

3.1.1 Inversion layer profiles; minority-carrier injection

In all the models discussed above, minority carriers were excluded from consideration. This procedure is justified in many practical cases, but certainly not always, as the world of the transistor so eloquently testifies. The simultaneous participation of two carrier types complicates phenomena, of course, and quantitative insights must then rely even more on numerical solutions. In some instances, these are available, in others they remain to be provided. In the models which follow, the bulk semiconductor is assumed to be non-degenerate. A discussion of space-charge effects in degenerate material with two types of charge carriers has been given by Seiwatz and Green (1958).

Minority carriers (holes in the examples discussed below) come into prominence, even in darkness, whenever barriers are high enough to bring the minority-carrier band-edge into the vicinity of the Fermi level. This is, of course, a matter of degree, culminating in a situation which, for a barrier on n-type material, makes p > n. An *inversion layer* is then said to exist, as shown on Fig. 3.1(a). It involves a distortion of the barrier shape, since the free holes constitute an x-dependent space charge, added to the (ordinarily fixed) space charge in ionized donors. As a result, the barrier profile is no longer parabolic; its slope increases sharply towards the metal. λ_I marks the distance from the metal over which $p > n$, the thickness of the inversion layer. Figure 3.2 shows how the carrier concentrations vary throughout the barrier. At zero current, the barrier field is in equilibrium with the concentration gradient of holes, as indicated. When a barrier voltage is applied, e.g. in the forward direction, as envisaged in Fig. 3.1(b), this equilibrium is not maintained. The result of its displacement is a net diffusion current of holes into the semiconductor at the energy minimum (where $F = 0$). This is, indeed, what we mean by *minority-carrier injection.* Once past the field-free minimum, the injected holes continue to diffuse, but also find themselves on the bulk field and drift with it, until they decay. While they do so, they constitute a positive space charge and, in a 'lifetime' semiconductor (only), this will be quickly neutralized by the free majority carriers. This provides for increments Δn and Δp, both positive. This is reflected by the contours of the (schematically drawn) quasi-Fermi levels in Fig. 3.1(b) and by the concentration

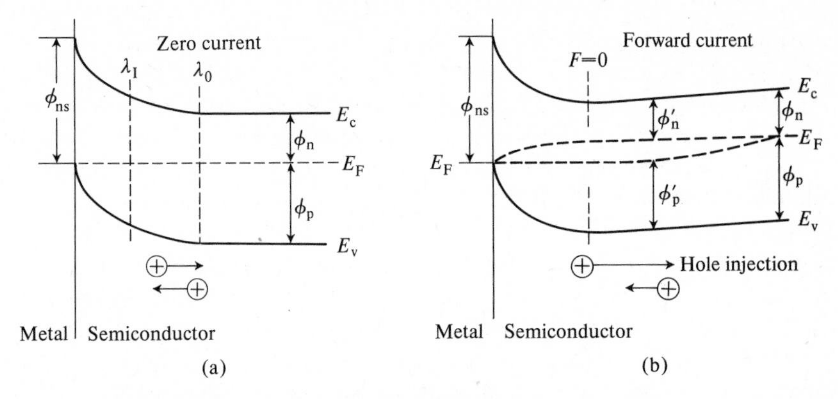

FIG. 3.1. Energy contour of a barrier on n-type material in the presence of an inversion layer; hole injection in the forward direction. Inversion layer thickness λ_I.

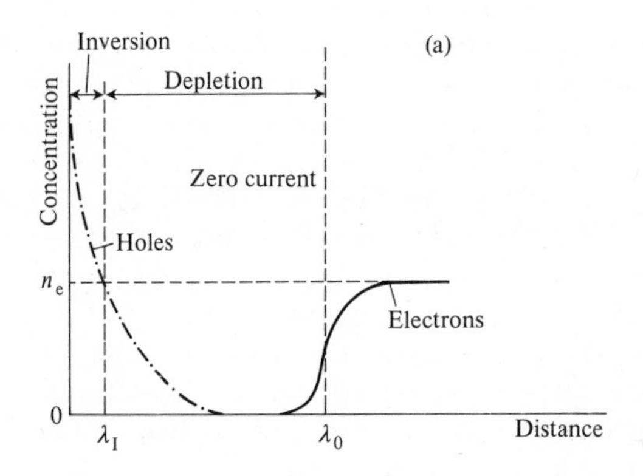

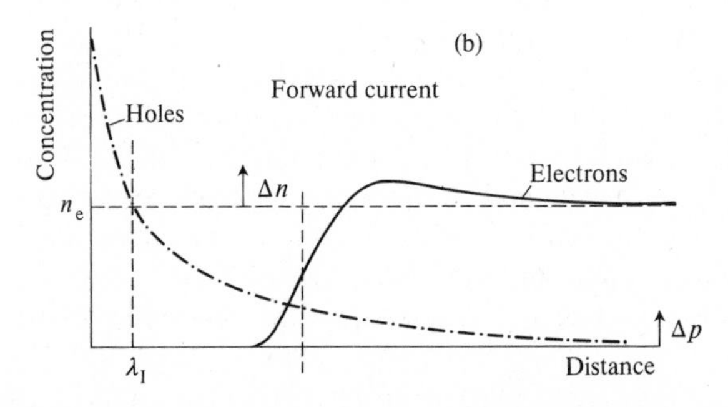

FIG. 3.2. Concentration contours through a barrier on n-type material in the presence of an inversion layer. Inversion layer thickness λ_I. Equilibrium bulk concentration of electrons n_e.

contours in Fig. 3.2(b). (Quasi-Fermi levels, defined by $n = \mathcal{N}_c \exp(-\phi_n'/kT)$ and $p = \mathcal{N}_v \exp(-\phi_p'/kT)$, are useful graphic devices for introducing non-equilibrium carrier concentrations into an energy band diagram, but because they are not themselves accessible to experimentation and measurement, their use cannot introduce new insights. They merely (albeit instructively) describe the situations established by other arguments.)

For 'lifetime' material, these relationships are qualitatively correct, whether we have an inversion layer or not, but they become important (in the sense of affecting the bulk material underneath) only when such a layer is present or, at any rate, when the formation of such a layer is approached by virtue of a sufficient barrier height. However, Scharfetter (1965) has shown that even contacts commonly regarded as 'non-injecting' may be significantly influenced by a small amount of minority-carrier participation in the total current. To neglect injection completely on *a priori* grounds can therefore be risky; it should never be done without carefully assembled supporting evidence. For high barriers, the need to consider minority-carrier participation is immediately obvious, and such cases are not in the least exotic. Quoting other sources, Green (1976) resports eV_D for vacuum cleaved n-Si as 0.8 eV for most metals, 0.81 eV for Au, 0.87 eV for PtSi, and 1.0 eV for Pt; and for p-Si: 0.90 eV for Hf and 0.82–1.0 eV for Cr (see also Section 2.3.8). There are even cases known in which the barrier height is greater than the semiconductor band gap. Walpole and Nill (1971) have reported this for Pb on degenerate n-PbTe, and for Au on p-InAs.

Since there is always a location at which $F = 0$ in the forward direction, the origin of minority-carrier injection as far as the bulk is concerned is *always* a diffusion process. The hole current then gradually changes its character, from diffusion to drift, throughout the decay region. The electron current does the same, but towards the left, being a field current in the bulk and a diffusion current at the energy minimum. *Within* the barrier there is again a field component, but now in the opposite direction (compare Fig. 2.13). The response of a *relaxation* semiconductor to injected minority carriers is, of course, entirely different, as discussed in Section 3.3. However, in both cases the nature of the current carriers (i.e. whether holes or electrons) *in the metal* is totally irrelevant, and to that extent the term 'injection' is linguistically unfortunate. The metal serves only for charge replacement at the interface, where the minority carriers are thermally generated in response to the externally displaced equilibrium.

The exact shape of barriers with inversion layers can be ascertained by numerical methods. In the presence of current flow, these calculations tend to be complex, but some qualitative features are immediately obvious. The presence of the inversion layer makes the region immediately adjoining the metal conductive, and thereby alters the field distribution. However, because of the high screening power of the dense hole concentration, the boundary field itself is not likely to be greatly influenced by the application of barrier voltages. Since it is the boundary field which controls tunnel effect and image-force corrections, the *effective height* of such barriers is also relatively stable, though considerably lower than it would be without the inversion layer.

Though the region immediately adjoining the metal is conductive by virtue of the high hole concentration, that conductivity is actually considerably lower than one might expect under bulk conditions. This is so, because the mobility of carriers whose movement is confined by a cusp-like potential contour is smaller than the bulk mobility. The original calculations of this effect are due to Schrieffer (1955), and comprehensive reviews of the situation have been provided by Many and co-workers (1965), and by Frankl (1967); (see also Section 6.1).

The extent to which an inversion layer influences contact behaviour depends, of course, also on the donor content. At high levels of doping, it will make itself felt only when the barrier is also high. At low levels of doping, even a low barrier may be materially influenced by the presence of minority carriers, even though no actual inversion layer is formed. Figure 3.3(a) shows this. A barrier height of $V_D = 0.1$ eV is low enough for the free-electron concentration to matter. When the donor content is

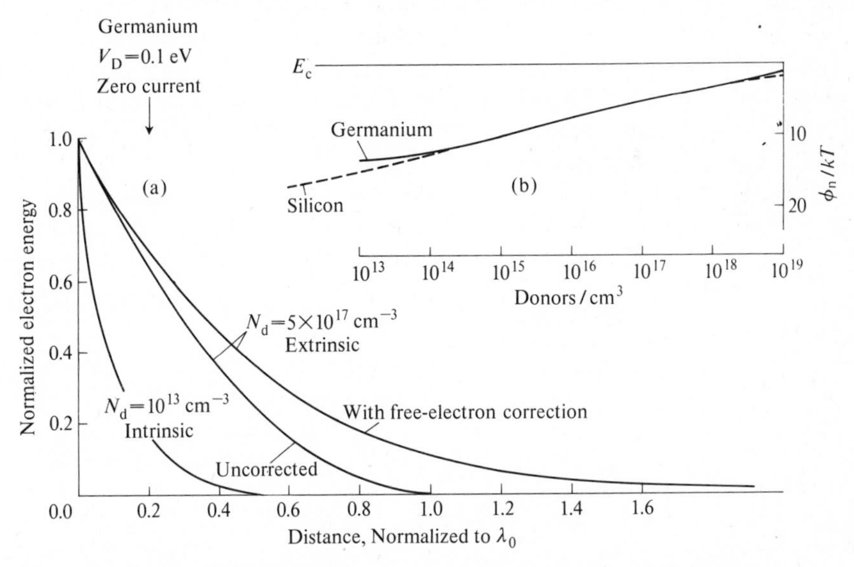

FIG. 3.3. Barriers in extrinsic and intrinsic germanium. (a) Energy contours. (b) Position of the Fermi level below the conduction band. Classical Schottky barrier thickness λ_0.

very low, then the minority-carrier space-charge is all important, and leads to an enormous increase in boundary field, as well as to a considerably smaller barrier width. Since the designations 'high barrier' and 'low barrier' are relative, Fig. 3.3(b) gives the position of the Fermi level (in terms of ϕ_n/kT) for Ge and Si for different donor content, numerically derived from the neutrality condition, with $E_d = 0.01$ eV for Ge and 0.044 eV for Si. There is overlap over a wide range, but at the higher doping concentrations the different impurity activation energies make themselves felt; at lowest concentrations the difference in band gaps comes into play.

In many ways, relationships involving electrons and holes are of symmetrical structure, but a barrier is an intrinsically asymmetrical entity, and it is natural enough that electrons and holes should interact with it in different ways. In Section 2.1.5 we discussed the image force to which electrons are subject when in the close vicinity of a conducting surface. Its effect was to lower the barrier height (Fig. 2.8). When we have holes in an inversion layer, as here envisaged, the situation is quite different. Like an electron, a hole is attracted to the metal surface but, unlike an electron, it has no barrier to overcome. Attraction to the metal causes its energy to be lowered, and since the energy of a hole is measured downward on electron energy diagrams, the outcome is a raising of the E_V contour close to the metal (Fig. 3.4). We are not dealing here with a genuine modification of the collective relationship which constitutes the band structure, but only with a variation of local potential. It would not be permissible to consider a hole and an electron simultaneously near the metal surface within the framework of the simple image-force argument. Moreover, unlike the electrons, holes find themselves in a region which represents a considerable charge carrier reservoir. Within such a region screening effects cannot be neglected.

It is one of the characteristics of injecting contacts on lifetime materials that they diminish any bulk resistance R_b which may be in series connection, by making additional charge-carriers available. The effect is greater

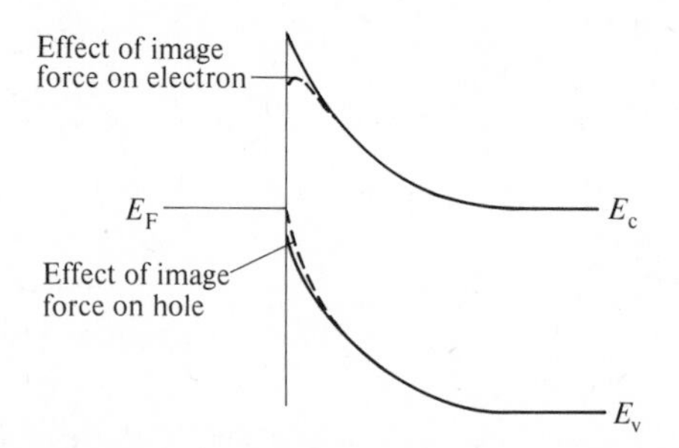

FIG. 3.4. Image force effects for electrons and holes.

for point contacts than for planar systems, because R_b then arises mostly from the region in the immediate vicinity of the point. A diminished R_b results in an improved rectification ratio. However, it should also be noted that a contact is not absolutely necessary for injection, if a barrier of sufficient height exists on the *free* surface of the semiconductor. Capacitively applied fields can achieve the same barrier distortion, and can transfer minority carriers from the inversion layer (which is then a fixed reservoir for them) into the bulk material. A suddenly applied voltage (capacitively coupled electrode positive) thus gives rise to a hole injection transient. That transient decays, and the steady state re-establishes the Boltzmann relationship.

3.1.2 *Inversion layer profiles*

For the case of extrinsic conduction, with all the donor centres ionized, the barrier profiles can be found analytically, as first shown by Banbury (1952). The charge density within the barrier is no longer eN_d (as it is without inversion layer) but

$$\rho(x) = eN_d + e\mathcal{N}_v \exp\left\{\frac{-\phi_p + E(x) - \phi_n}{kT}\right\} \tag{3.1.1}$$

and this value must now be used in Poisson's equation (2.1.2a). Accordingly, the first stage of integration yields:

$$\frac{1}{2}\left(\frac{dE}{dx}\right)^2 = \left(\frac{N_d e^2}{\epsilon}\right)\left[E(x) + kT\left(\frac{\mathcal{N}_v}{N_d}\right)\right.$$
$$\left.\times \exp\left\{\frac{-\phi_p + E(x) - \phi_n}{kT}\right\}\right] + \text{constant} \tag{3.1.2}$$

where the integration constant must be such as to make $dE/dx = 0$ when $E(x) = E(\lambda_B) = \phi_n$ Hence, taking the negative root,

$$\frac{dE}{dx} = -\left(\frac{2N_d e^2}{\epsilon}\right)^{1/2}\left[(E(x) - \phi_n) - kT\left(\frac{\mathcal{N}_v}{N_d}\right)\right.$$
$$\left.\times \exp\left(\frac{-\phi_p}{kT}\right)\left\{1 - \exp\left(\frac{E(x) - \phi_n}{kT}\right)\right\}\right]^{1/2} \tag{3.1.3}$$

In the present context, we are interested mainly in the region near the interface, for which $E(x) - \phi_n$ will differ from eV_D by only a few units of kT. Hence the first term in the square brackets can be regarded as constant in comparison with the rapidly varying exponential term. We may therefore put $E(x) - \phi_n = eV_D$, and though small improvements could obviously be made, they are not worth while. With $a_1 = (2N_d e^2/\epsilon)^{1/2}$ and $a_2 = kT(\mathcal{N}_v/N_d)\exp(-E_g/kT)$, and neglecting unity within the last curly brackets, eqn (3.1.3) can be written as

$$\frac{dE}{dx} = -a_1\{eV_D + a_2 \exp(E/kT)\}^{1/2} \tag{3.1.4}$$

from which

$$x = -\frac{1}{a_1}\int \frac{dE}{\{eV_D + a_2 \exp(E/kT)\}^{1/2}} \tag{3.1.5}$$

This integral can be evaluated (see Gradshteyn and Ryzhik 1965). An integration constant must be chosen, such that $x = 0$ when $E(x) = \phi_{ns}$. Under these conditions:

$$x = \frac{kT}{a_1(eV_D)^{1/2}} \times \log \frac{[\{eV_D + a_2 \exp(E/kT)\}^{1/2} + (eV_D)^{1/2}][\{eV_D + a_2 \exp(\phi_{ns}/kT)\}^{1/2} - (eV_D)^{1/2}]}{[\{eV_D + a_2 \exp(E/kT\}^{1/2} - (eV_D)^{1/2}][\{eV_D + a_2 \exp(\phi_{ns}/kT)\}^{1/2} + (eV_D)^{1/2}]}. \tag{3.1.6}$$

Figure 3.5 shows this relationship for contacts of different barrier heights on Ge containing 10^{17} donors/cm^3, and its various features are instructive:

(a) Note that the slope for high values of x is always the same, as it must be, because it depends only on N_d/ϵ, once the screening effect of the inversion layer is over. All the contacts considered have the same N_d/ϵ value, of course.

(b) The curvature of the contour increases as ϕ_{ns} increases. For $\phi_{ns}/kT = 18$, the inversion layer is still negligible, as one would expect, since ϕ_p/kT for this material is 21.5, and eV_D only 12.7. As soon as ϕ_{ns} becomes

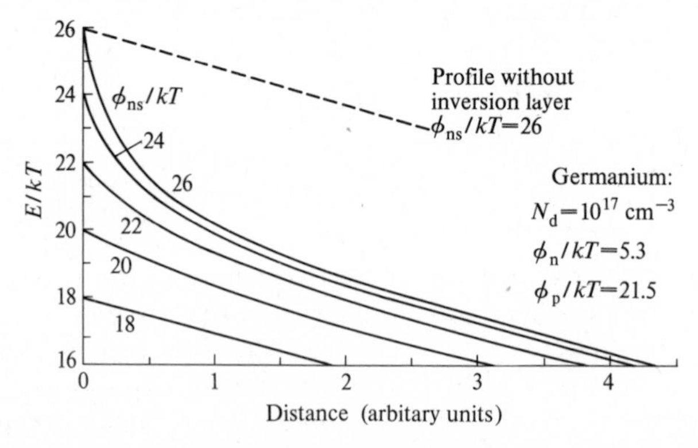

FIG. 3.5. Barrier contours for different barrier heights; formation of the inversion layer on extrinsic material. Equation (3.1.6); 0.8 distance unit $\approx 10^{-7}$ cm.

comparable with eV_D, the inversion layer (or, at any rate, the additional hole concentration) makes itself felt. For $\phi_{ns}/kT=26$, we have $eV_D=20.7$ which is just within the range that makes the tip of the barrier degenerate in minority carriers (a fact which does not significantly influence Fig. 3.5).

(c) Note that, as ϕ_{ns} increases, the contours come more and more closely together. This means that the barrier thickness becomes less and less dependent on the nominal barrier height, again due to the screening effect of the inversion layer. Because of image force and tunnelling, the effective height of the barrier increases much more slowly than the nominal barrier height.

3.1.3 Static barriers on intrinsic material

There is just one case in which the barrier problem on a two-carrier system can be solved explicitly, namely that of a barrier on intrinsic material. This has been shown by Dousmanis and Duncan (1958) and, in the form given below, by Manifacier (1979). One approximation is involved, namely $m_n=m_p$, which makes $\mathcal{N}_c=N_v=\mathcal{N}$: equal densities of states in the two bands. Non-degeneracy is assumed. Under such conditions, we have for values of x within the barrier

$$\frac{d^2E}{dx^2}=\frac{e^2\{p(x)-n(x)\}}{\epsilon}$$

with

$$n(x)=\mathcal{N}\exp\left\{\frac{-(\phi_n+\Delta E)}{kT}\right\}$$

and

$$p(x)=\mathcal{N}\exp\left\{\frac{-(\phi_p-\Delta E)}{kT}\right\} \tag{3.1.7}$$

where $\Delta E=E(x)-E_c$, and $\phi_n=\phi_p=E_g/2$. Thus,

$$\frac{d^2\,\Delta E}{dx^2}=\frac{2e^2\mathcal{N}}{\epsilon}\exp\left(-\frac{E_g}{2kT}\right)\sinh\left(\frac{\Delta E}{kT}\right) \tag{3.1.8}$$

which may be written for convenience as

$$\frac{d^2\,\Delta E}{dx^2}=\frac{S_c}{2kT}\sinh\left(\frac{\Delta E}{kT}\right)$$

where

$$S_c=\frac{4kT\mathcal{N}e^2}{\epsilon}\exp\left(-\frac{E_g}{2kT}\right). \tag{3.1.9}$$

The first stage of integration then gives

$$\left\{\frac{\mathrm{d}\Delta E}{\mathrm{d}x}\right\}^2 = S_c \cosh\left(\frac{\Delta E}{kT}\right) + \text{constant} \qquad (3.1.10)$$

and since $\mathrm{d}\Delta E/\mathrm{d}x$ must tend to zero as ΔE tends to zero, the constant must be -1. Thus

$$\frac{\mathrm{d}\Delta E}{\mathrm{d}x} = -S_c^{1/2}\left\{\cosh\left(\frac{\Delta E}{kT}\right) - 1\right\}^{1/2}. \qquad (3.1.11)$$

The negative root has to be taken, because ΔE diminishes as x increases. Since

$$\sinh^2(u/2) = \tfrac{1}{2}(\cosh u - 1)$$

eqn (3.1.11) can be written as

$$\frac{\mathrm{d}\Delta E}{\mathrm{d}x} = -S_c^{1/2}\left\{2\sinh^2\left(\frac{\Delta E}{2kT}\right)\right\}^{1/2} = -(2S_c)^{1/2}\sinh\left(\frac{\Delta E}{2kT}\right). \qquad (3.1.12)$$

By a series of algebraic manipulations, this can be made to yield ΔE as a function of x. To do this, we consider first that

$$\frac{\mathrm{d}}{\mathrm{d}x}\sinh\left(\frac{\Delta E}{2kT}\right) = \frac{1}{2kT}\cosh\left(\frac{\Delta E}{2kT}\right)\frac{\mathrm{d}\Delta E}{\mathrm{d}x}$$

$$= -\left(\frac{S_c}{2}\right)^{1/2}\frac{1}{kT}\cosh\left(\frac{\Delta E}{2kT}\right)\sinh\left(\frac{\Delta E}{2kT}\right). \qquad (3.1.13)$$

The factor $(S_c/2)^{1/2}(1/kT)$ turns out to be $1/\lambda_{\mathrm{Da}}$, where λ_{Da} is the *ambipolar Debye length, given by*

$$\lambda_{\mathrm{Da}} = \left\{\frac{\epsilon kT}{e^2(p_e + n_e)}\right\}^{1/2}. \qquad (3.1.14)$$

We can thus write

$$\int \frac{\mathrm{d}\sinh(\Delta E/2kT)}{\cosh(\Delta E/2kT)\sinh(\Delta E/2kT)} = -\frac{x}{\lambda_{\mathrm{Da}}} + \text{constant}$$

$$= \int \frac{\mathrm{d}\sinh(\Delta E/2kT)}{\sinh(\Delta E/2kT)\{1+\cosh^2(\Delta E/2kT)\}^{1/2}}. \qquad (3.1.15)$$

The value of $\int\{\mathrm{d}u/u(1+u^2)^{1/2}\}$ can be found in standard tables of integrals: $-\log\{1+(1+u^2)^{1/2}/u\}$. Equation (3.1.15) thereby becomes

$$\log\left[\frac{1+\{1+\sinh^2(\Delta E/2kT)\}^{\frac{1}{2}}}{\sinh(\Delta E/2kT)}\right] = \frac{x}{\lambda_{\mathrm{Da}}} + \text{constant} \qquad (3.1.16)$$

or

$$\frac{1+\cosh(\Delta E/2kT)}{\sinh(\Delta E/2kT)} = B_c \exp(x/\lambda_{Da}).$$

When $x = 0$, we have $\Delta E = eV_D$, which gives

$$\frac{1+\cosh(eV_D/2kT)}{\sinh(eV_D/2kT)} = B_c \tag{3.1.17a}$$

and thus

$$\frac{1+\cosh(\Delta E/2kT)}{\sinh(\Delta E/2kT)} = \frac{1+\cosh(eV_D/2kT)}{\sinh(eV_D/2kT)} \exp(x/\lambda_{DA}). \tag{3.1.17b}$$

Since $\tanh(u/2) = \sinh u/(1+\cosh u)$

$$\tanh\left(\frac{\Delta E}{4kT}\right) = \tanh\left(\frac{eV_D}{4kT}\right)\exp(-x/\lambda_{Da})$$

or (3.1.18)

$$\left(\frac{\Delta E}{kT}\right)_x = 4\tanh^{-1}\left\{\tanh\left(\frac{eV_D}{4kT}\right)\exp\left(-\frac{x}{\lambda_{Da}}\right)\right\}$$

which can also be written as

$$\left(\frac{\Delta E}{kT}\right)_x = 2\log\left\{\frac{1+\tanh(eV_D/4kT)\exp(-x/\lambda_{Da})}{1-\tanh(eV_D/4kT)\exp(-x/\lambda_{Da})}\right\} \tag{3.1.19}$$

as a descriptive expression of the barrier contour. Figure 3.6 gives two

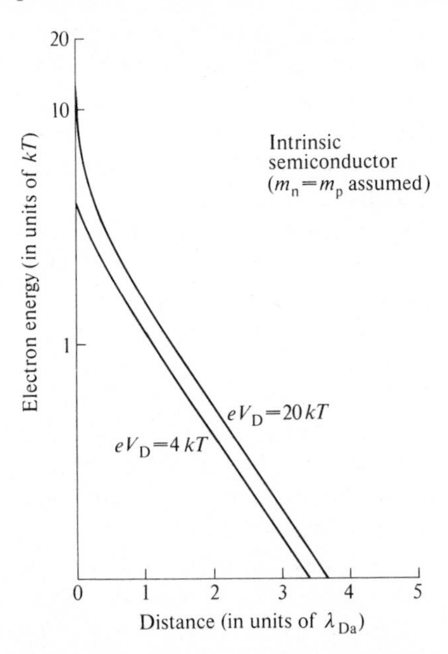

FIG. 3.6. Barrier contours near the metal-semiconductor interface for two different barrier heights on intrinsic material. $m_n = m_p$ assumed.

typical relationships calculated in this way, one for $eV_D = 4kT$ (barrier low compared with the band gap at room temperature) and one for $eV_D = 20kT$ (barrier height greater than half the band gap). The latter shows the enormous boundary field ordinarily associated with inversion layers (though on intrinsic material the hole-rich region cannot be so called); the former does not. For large values of x, both curves reflect the exponential decay of eqn (3.1.18). It must be so, since free carrier disturbances always disappear exponentially at a rate governed by the Debye length, which the two curves have here in common; hence the identical slopes on the semi-logarithmic plot. The shape of the contour is, of course, significantly different from that of a Schottky barrier, in which the space charge density is constant. Note also that the barrier width is not very sensitive to barrier height. The above calculations represent only the static situation; see Section 3.5.3 for comments on current-carrying contacts on intrinsic material.

3.1.4 *The injection ratio*

In Section 1.2 we were concerned with the current composition ratios γ_1 and γ_2, of which only γ_1 was relevant to the forward direction. It was defined as the fraction of the total current carried by minority carriers 'at' the contact, without then enquiring very closely about the precise location. In the present context, it is appropriate to call $\gamma_1 = \gamma$, *the injection ratio,* provided $\gamma_n < \gamma \leqslant 1$, and to define γ as j_p/j at $x = \lambda_B$. Indeed, the injection ratio *has* no simple definition at $x = 0$, because the composition of the current changes very rapidly at that point, from the transport mode characteristic of the metal to that characteristic of the semiconductor. (On the scale of Fig. 1.1, the points $x = 0$ and $x = \lambda_B$ are indistinguishable.)

We can crudely classify injection ratio as 'high' (e.g. $\gamma = 0.7$–1.0) and 'low' (e.g. say $0.7 > \gamma > \gamma_n$. $\gamma < \gamma_n$ would, of course, amount to minority-carrier *exclusion.* Early measurements on contacts of *low* injection ratios were reported by Many (1954, 1955) using an ingenious bridge method. γ was found to increase with forward current, and similar observations have been described by Scharfetter (1965) and by Yu and Snow (1969).

It will be clear that γ is by no means an independent variable of the system, being neither a characteristic surface property, nor a bulk parameter. It is, in principle, the outcome of the complete transport equations which govern the two-carrier system (Section 3.5), with their appropriate boundary conditions. On one side (the left, in all diagrams here), these boundary conditions depend on the nature of the surface and the height of the barrier. On the other side (the right, in all diagrams here), they are determined by the equilibrium properties of the bulk material, unless a second contact is envisaged *within* the disequilibrium range, in which case matters are more complicated. Since the transport equations cannot be analytically solved, precise calculations of γ have to rely on numerical methods, and are critically dependent on the assumptions made at $x=0$. To circumvent this need, several attempts have been made to obtain at least rough estimates by analytic methods which are, of course, likewise dependent on the specific assumptions made. The results should never be pressed into service outside the range set by these inherent limitations. With these cautions, the analysis can proceed as follows.

At $x=\lambda_B$, the current is a pure diffusion current, since the field is zero. Thus

$$j_p = -eD_p(\mathrm{d}p/\mathrm{d}x). \tag{3.1.20}$$

We also have the continuity relationship

$$\frac{\mathrm{d}p}{\mathrm{d}x} = \frac{p_e - p(x)}{\tau_p} - \frac{1}{e}\frac{\mathrm{d}j_p}{\mathrm{d}x} \tag{3.1.21}$$

and, together, these equations yield

$$p(x) - p_e = \{p(\lambda_B) - p_e\}\exp(-x/L_p) \tag{3.1.22}$$

where L_p is the diffusion length for holes, namely $(D_p\tau_p)^{1/2}$. This procedure actually assumes that the effects of the field are negligible *everywhere*, and not only at $x=\lambda_B$, which means that the considerations are more appropriate to low than to high current levels. Equation (3.1.22) also leans on the fact that λ_B is negligible compared with L_p, which means that it does not really matter whether the decay curve begins precisely at $x=0$ or $x=\lambda_B$. Moreover, further approximations are now needed, since some assumption must be made as regards the value of $p(\lambda_B)$. Thus, Gunn (1954) took $p(0)$ to be fixed, with $p(x)$ dependent on $p(0)$ and the local potential in the Boltzmann manner. This gives

$$p(\lambda_B) = p_e \exp(-eV_B/kT), \tag{3.1.23}$$

where V_B is the voltage applied to the barrier (negative here, because we are dealing with the forward direction). Use of the Boltzmann equation implies, strictly speaking, a limitation to zero current, but it will be shown in Section 3.5.3 that the departures from the Boltzmann concentrations are small, even in the presence of current flow. As one would expect, the approximation is also better for high lifetime than for low lifetime material. For present purposes, we shall accept eqn (3.1.23) as sufficiently accurate to support an injection ratio estimate. (Compare Fig. 2.15 for a *single*-carrier system, in which $n(0)$ and $n(\lambda_B)$ were *both* regarded as fixed (full line).) The last two equations give

$$p(x) - p_e = p_e\{\exp(-eV_B/kT) - 1\}\exp(-x/L_p) \tag{3.1.24}$$

and

$$j_p = p_e e(D_p/L_p)\{\exp(-eV_B/kT)-1\}\exp(-x/L_p) \qquad (3.1.25)$$

where again the difference between $x=0$ and $x=\lambda_B$ has been neglected. It will be seen that D_p/L_p has the character of a velocity.

To calculate γ we need the total current, and one might simply take this to be given by an equation *of the form* (2.2.39a) without specifying the detailed structure of the saturation current. This would make j_s an empirical constant, and give

$$\gamma = j_p/j = p_e e(D_p/L_p)/j_s \qquad (3.1.26)$$

which should be a fair approximation at low current densities (see also Swanson 1954), always assuming that j_s is well defined. However, true saturation in the reverse direction is rare, which means that a reliable value of j_s may not be available. In an attempt to proceed nevertheless, one might consider that

$$n(\lambda_B) = n_e + p(\lambda_B) \qquad (3.1.27)$$

as long as neutrality prevails (i.e. in a pronounced *lifetime* semiconductor). This would modify the diode current of majority carriers, and lead directly to

$$\gamma = \frac{p_e(D_p/L_p)}{p_e(D_p/L_p)+\{n_e+p(\lambda_B)\}\{kT/2\pi m_n\}^{1/2}\exp(-eV_D/kT)} \qquad (3.1.28)$$

at $x \approx \lambda_B$, from which a 'zero current' approximation can be derived by allowing $p(\lambda_B) \to p_e$. A somewhat greater degree of generality is achieved by substituting for $p(\lambda_B)$ from eqn (3.1.23), a schematic procedure which deserves something less than our total confidence, amongst other things because of the neglected field terms. This yields

$$\gamma = 1\bigg/\left[1+\left\{\frac{kT}{2\pi m_n}\right\}^{1/2}\left\{\frac{L_p n_e^2}{p_e D_p}\right\}\left\{1+\frac{n_i^2}{n_e^2}\exp(-eV_B/kT)\right\}\exp(-eV_D/kT)\right]. \qquad (3.1.29)$$

The zero current approximation follows from $V_B=0$. On this basis, γ would increase with increasing barrier height, which is entirely as expected, and which Yu and Snow (1969) have experimentally confirmed for a variety of metal contacts on silicon, with ϕ_{ns} between 0.65 and 0.85 eV. In this respect, the schematic model is entirely successful. It also predicts that γ should decrease with increasing forward current, though only for substantial forward voltages ($eV_B<0$). Eventually it *must* always do so, but the experiments by Many and Scharfetter quoted above recorded an initial increase, especially for initially small injection ratios. Scharfetter has accounted for this in terms of the field current, neglected in the above analysis. One of the consequences of field current at $x > \lambda_B$, is to increase the value of $(dp/dx)_{\lambda_B}$, and thus also the value of the injection ratio. One would expect this effect to dominate at low currents when γ is small, as it certainly did in Scharfetter's experiments. Yu and Snow (1969) have similarly reported an increase with current density, their γ values being of the order of 10^{-2} to 10^{-4} at lowest currents. Such small values are not in harmony with eqn (3.1.29). There are evidently circumstances in which the simplifying assumptions fail to give a valid picture of the situation (see Section 3.5.3).

Some of the conflict of expectations relating to the initial value and current-dependence of γ has been resolved by Clarke *et al.* (1974), who have shown (using numerical methods) that the *size* of the contact is an important parameter. In a large-area system, the decay of injected carriers with distance is relatively slow; in a radial system it is very rapid, and this corresponds to an enhancement of the prevailing minority-carrier diffusion current. Thus, other things being equal, point contacts are expected to have higher injection ratios than planar contacts. Clarke *et al.* also showed that these differences are not related to the different current densities as such, but to the different concentration contours. Figure 3.7

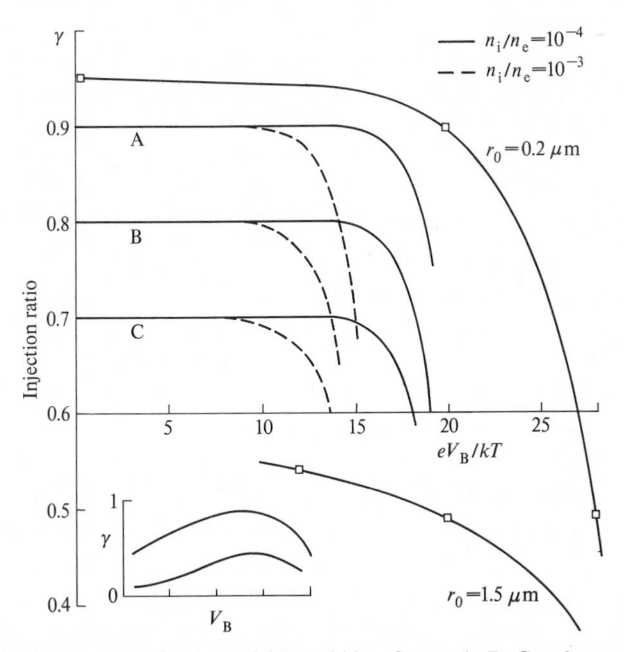

FIG. 3.7. Injection ratio as a function of forward bias. Curves A, B, C: values calculated in accordance with eqn (3.1.29) for different 'zero current' γ values. Square block-points: numerically computed values for the contact radii given (Clarke *et al.* 1974). Insert: schematically drawn values which reflect actual situations for large-area contacts; compare Green and Shewchun (1973).

shows some of the results of their computation (square block-points). The differences between small and large contacts are clear. For small contacts, there is excellent agreement with the results of Braun and Henisch (1966a) and with experimental observations by Banbury and Houghton (1954 and 1955); see also Section 3.5.1. For large contacts, the computed injection ratios are certainly smaller, but even then not as small as some of those mentioned above. and an initial rise with current is not reflected by the available results. However, somewhat different computations by Green and Shewchun (1973) do show such a rise for contacts associated with initially small γ values. It will be shown in Section 3.5.3 that this is also true for contacts on intrinsic material. Of course, when an injection ratio is initially very high, nothing can be expected except a decrease with increasing current, as $\Delta p(\lambda_B)$ becomes comparable with n_e.

All these models assume that there is a potentially unlimited minority-carrier generation rate at $x=0$, but other provisions could be made. The minority carrier bulk lifetime τ_p enters the above considerations via L_p. In practical cases, the specimen thickness may be small compared with the diffusion length, and this would make the results independent of τ_p. Yu and Snow (1969) have described such a system. It gives results very similar to those reviewed above, except that L_p in eqn (3.1.26) is replaced by the actual specimen thickness (see also Grove 1967).

A more elaborate calculation of γ, specifically designed to apply to hemispherical contacts systems (point contacts) will be found in Section 3.5.1. Though that derivation is different in points of detail, the behavior of the calculated γ as a function of the applied voltage closely follows the pattern outlined above. The same is true for the variation of γ with barrier height V_D. Thus, suppose the parameters of a material are such as

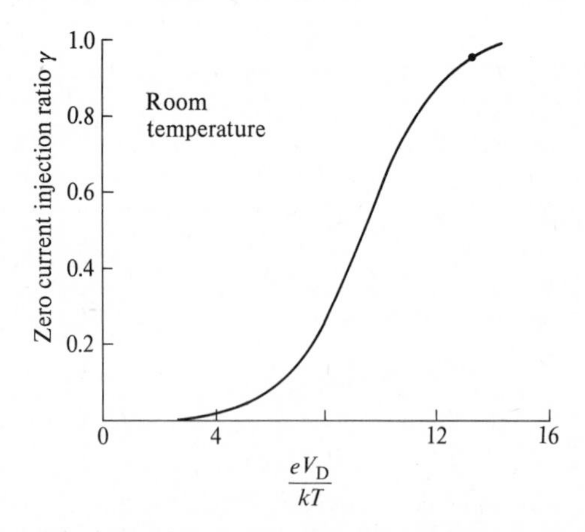

FIG. 3.8. Injection ratio as a function of barrier height; calculated 'zero current' values in accordance with eqn (3.1.29), assuming $\gamma=0.95$ for $V_D=0.4$ volts.

to make the initial (zero-current) value of $\gamma = 0.95$ for $V_D = 0.4$ V, the corresponding values of γ for lower barrier heights on the same material would be as shown in Fig. 3.8.

Once a value of the initial γ at $x = \lambda_B$ is accepted (having been calculated or assumed), it can be used as a (low current) descriptive boundary condition for calculations on injection and exclusion processes, and their effect on bulk material. This is a convenient procedure, because any assumed value of γ implies not only a definite $(dp/dx)_{\lambda_B}$ but also $(dn/dx)_{\lambda_B}$, since the field current of electrons is (also) zero at that point. Sections 3.2, 3.3, and 3.4 concern themselves with such problems. In their analysis, it will be found convenient to shift the origin to $x = \lambda_B$, thereby ignoring the barrier itself, and all its associated complications.

Further reading

On injection into heavily doped material: Slotboom 1977.

On injection under high field conditions: Dacey 1953.

On injection by barriers of a height comparable with the band gap: Huang and Crowell 1976.

On hot carriers in Si inversion layers: Hess and Sah 1974.

3.2 Injection into bulk material; lifetime semiconductors

3.2.1 *General considerations; modulation of conductivity*

The fact that injected minority carriers (e.g. produced by current flow through an interface for which $\gamma > \gamma_n$) can lead to changes of effective conductivity has already been referred to in Section 1.2.2. It was also mentioned there that the change could be an increase or a decrease, depending on circumstances. An increase is clearly expected from the fact that Δn and Δp are both positive *in a lifetime semiconductor.* When the additional carrier concentrations are reasonably uniform, they give rise to

$$\sigma - \sigma_b = \Delta\sigma = \mu_n e \Delta n + \mu_p e \Delta p \tag{3.2.1}$$

where $\Delta\sigma$ is ohmic (independent of field). When appreciable concentration gradients are present, eqn (3.2.1) ceases to be an appropriate description of the system. However, there is even then a local field F and a current density j, and it is always possible to define an *effective* conductivity

$$\sigma = j/F \tag{3.2.2}$$

where σ is now field dependent, determined by carrier diffusion as well as drift. In these circumstances σ may be smaller than σ_b, as shown by

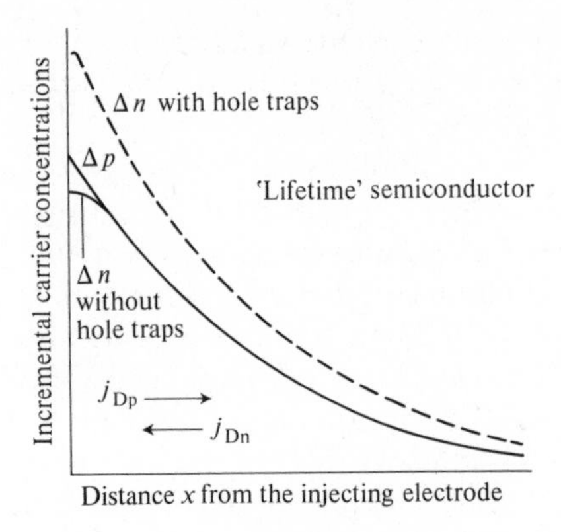

FIG. 3.9. Incremental carrier concentration contours in n-type 'lifetime' material under injection conditions. Quasi-neutrality assumed. $D_n > D_p$ makes $|j_{Dn}| > |j_{Dp}|$. Full lines: Δp and Δn without hole traps. Broken line: Δn with hole traps.

Manifacier and Henisch (1978a,b). The qualitative argument is as follows.

Minority carriers are injected and, following injection, decay as they drift and diffuse into the bulk material. Accordingly, their additional concentration $\Delta p(x)$ will follow a contour as shown in Fig. 3.9. For reasons of quasi-neutrality, the extra electron concentration $\Delta n(x)$ will be almost coincident, except close to the injecting boundary, where an appreciable difference can be shown to exist. We are not concerned here with the barrier itself, only with the consequences of its existence. The injecting boundary is at $x = \lambda_m$, renamed $x = 0$ in the quantitative considerations of the following section. As we see from Fig. 3.9, we have two virtually equal concentration gradients, and if the two diffusion constants D_p and D_n were likewise equal, their effects would cancel out. In practice, the constants are not equal, which means that one of the diffusion currents predominates. Thus, for hole injection into n-type material, electron diffusion towards the right dominates over hole diffusion in the same direction, since $D_n > D_p$. However, whereas the hole diffusion augments the drift current, the electron diffusion *opposes* it. Since the current density has to be constant everywhere, an extra field must locally exist to compensate for the electron diffusion, and an extra field in that region is equivalent to a local *conductivity decrease*. Carrier augmentation ($\Delta n(x) > 0$ and $\Delta p(x) > 0$) and carrier diffusion thus influence the system in opposing ways, with the latter dominating as long as the current density is low (and the carrier lifetime sufficiently high). Conversely, when the current density (bulk field) is high, carrier drift will dominate, and will be further augmented by the additional carrier concentrations. The local *conductivity increase* that is traditionally thought to be associated with

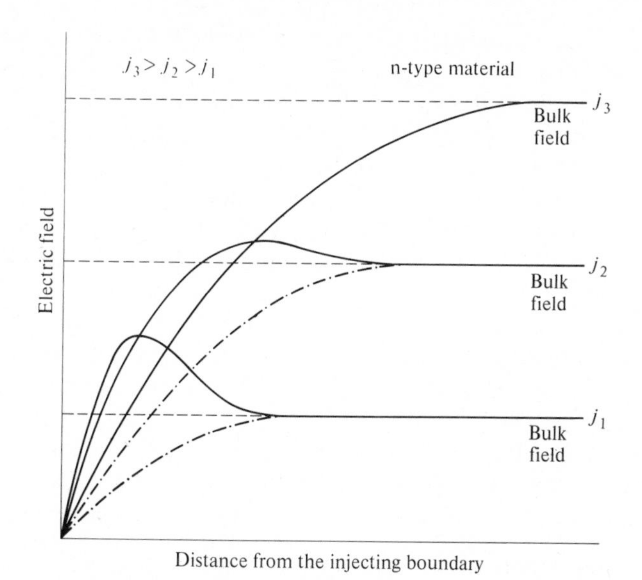

FIG. 3.10. Schematic field contours in the presence of hole injection for three current densities; n-type 'lifetime' material. Note disappearance of the field maximum (equivalent to a local resistivity increase) at higher currents. Full lines: $\mu_n > \mu_p$. Broken lines: zero injection. Chain line: $\mu_n = \mu_p$.

minority-carrier injection will then be observed. For an analysis of series resistance modulation, see Manifacier and Fillard (1976).

The situation is summarized in Fig. 3.10 for low and for high currents. If there were no injection at all, the field would be constant (broken lines). If minority and majority carriers had equal diffusion constants, the contours would be those shown by the chain lines, and their consequences would be an increase of local conductivity everywhere within the recombination region. If the majority carriers had the greater diffusion constant (e.g. as in Ge, Si, etc), the contours would be as those shown by the full lines. Immediately close to the injecting boundary, there would be an *increase* of conductivity (lower than ohmic field); beyond it, a *decrease*, as long as the current is low. A sufficiently high current density (bulk field) would, however, cause the field maximum to be eliminated as explained above.

This is one of the few cases in which the situation for n-type and p-type material is *not* symmetrical. N-type material behaves as described above, because its *majority* carriers have the higher diffusion constant. For electron injection into p-type material this would not be the case. Accordingly, no field maximum would be expected, and hence no local resistance increase.

When minority carrier traps are present, the above diffusion effects may be greatly enhanced (Popescu and Henisch, 1976b). Thus, other things being equal, the $\Delta p(x)$ contour might be essentially unchanged, but free holes would no longer be responsible for the entire positive space-charge. If a large number of holes were trapped, a large number of electrons would be attracted into the region, giving a $\Delta n(x)$ contour, as shown by the broken line in Fig. 3.9, and would give rise to a correspondingly larger electron-diffusion-current. This, in turn, would have to be compensated for by a higher field maximum, in order to keep the total current density constant. In Fig. 3.9, the $\Delta n(x)$ contours with and without traps are shown only slightly separated, but order-of-magnitude differences between them could in fact exist. Of course, the trap density cannot be arbitrarily varied in practice, without affecting other parameters of the solid, especially the carrier lifetime.

Section 3.2.2 will show in outline how these general and qualitative conclusions can be analytically derived. Since the full transport equations are not analytically soluble, one of three strategies must be adopted:

(a) The introduction of *ad hoc* simplifications, leading to analytically soluble equations, and approximate solutions for special cases.
(b) The systematic linearization of the equations, leading to general results, but only within the framework of a 'small-signal theory'.
(c) The derivation of numerical solutions for specific cases.

Strategy (a) is, of course, frequently employed in all such contexts, but whether it is really successful is often hard to ascertain. To do so would necessitate a comparison with exact solutions which, in the nature of things, are unavailable. It is also very difficult to assess intuitively which terms may safely be neglected, and attempts to do so have in the past led to a multitude of misleading results.

The importance of these aspects has only recently come to be appreciated. Thus, for instance Bardeen and Brattain (1949) and Banbury (1953b) still assumed that the majority carrier current *and* recombination can be neglected everywhere. In such circumstances, the injected minority carriers could do no other than to increase the effective local conductivity everywhere. This expectation then came to be regarded as a universal feature of all such systems. Even during later work (Swanson 1954, van Roosbroeck 1953 and Landauer and Swanson 1953), it was customary to assume local neutrality, on the grounds that Δp, the departure from hole equilibrium, and Δn, the departure from electron equilibrium, are bound to be nearly equal. They are indeed, but the neglect of the small difference between them suppresses all considerations of field curvature, and thus, indirectly, the analysis of the field maximum. For all these reasons, approach (a) is avoided here, but approaches (b) and (c) will be illustrated in this chapter.

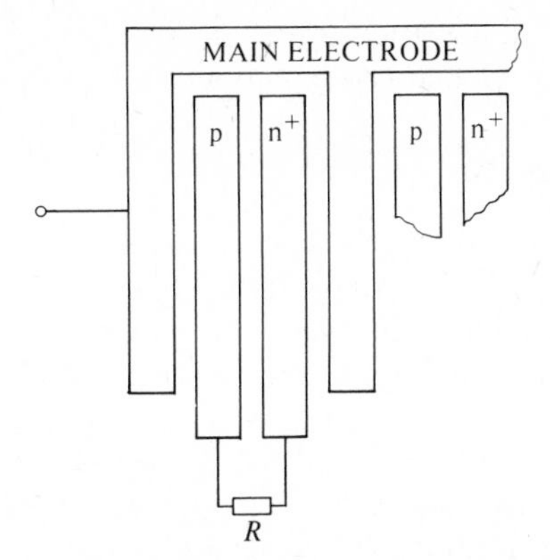

FIG. 3.11. Electrode system designed to simulate high surface recombination. After Ryvkin *et al.* (1979). Injection effects disappear as $R \rightarrow 0$.

All current-controlled non-equilibrium effects (of which minority-carrier injection is the most important example) depend, of course, on the effective carrier lifetime near the injecting boundary, determined in practice by the bulk lifetime and the surface recombination rate. When the effective lifetime is very low, all the non-equilibrium effects are suppressed. This can be achieved, when desired, by ensuring that surface and bulk contain a large number of recombination centres, e.g. as they do in a highly disordered material. It can also be done 'artificially', by providing local fields which attract holes and electrons to different segments of the electrode structure, where they can then combine. This principle was demonstrated by Ryvkin and co-workers (1979), and the corresponding electrode structure is shown schematically in Fig. 3.11. Segments of the main electrode, which carries the specimen current, are interspersed with subsidiary n^+-type and p-type strip contacts, closely spaced on the bulk n-type material. The n^+- and p-strips are connected through an external resistance. They collect non-equilibrium electrons and holes respectively, and thereby deny these carriers to the bulk material underneath. Accordingly, non-equilibrium effects tend to disappear as the external connecting resistance tends to zero. The arrangement, too complex for most practical purposes, is of interest mainly because it provides a didactic demonstration of the general principle. Ryvkin and co-workers actually demonstrated that the subsidiary electrodes could be used effectively for suppressing minority-carrier exclusion and accumulation.

A steady state is assumed to prevail in the systems discussed elsewhere in this section, but injection under transient conditions is also interesting in many contexts. As carriers are injected during a current pulse, they modify the effective conductance of the specimen as a function of time.

For a simple case, i.e. a lifetime semiconductor without traps, it is easy to show that the total conductance change will be given by

$$\Delta G = \Delta G_\infty \{1 - \exp(-t/\tau_0)\} \tag{3.2.3}$$

and this fact was used in the early days as a convenient way of measuring the lifetime τ_0 (Many 1954). A constant γ has to be assumed. It is also important to understand how an injection regime decays after cessation of the injecting pulse. Robbins and Ancker-Johnson (1971) used double-pulse techniques for investigating these matters on p-type material InSb. The first pulse injected the carriers, the second monitored the status of the electron–hole plasma after a known (but variable) delay. Though its behaviour is complex, InSb is of special interest in the matter because its carrier lifetime is strongly dependent on the departure from equilibrium. Dean (1968) has calculated approximate field profiles at different times after the onset of electron injection into highly extrinsic p-type Ge (neglecting diffusion and recombination, to make the transport equations analytically soluble).

3.2.2 *Transport relationships in a two-carrier system without traps*

For bimolecular recombination, the equations which govern transport in a one-dimensional trap-free system of lifetime τ_0 ((Manifacier and Henisch, 1978a) are, firstly, the two expressions for the electron- and hole-currents:

$$j_n = e\mu_n nF + \mu_n kT(\mathrm{d}n/\mathrm{d}x) \tag{3.2.4}$$

$$j_p = e\mu_p pF - \mu_p kT(\mathrm{d}p/\mathrm{d}x) \tag{3.2.5}$$

with, of course, $j = j_n + j_p$. Then we have two continuity relationships, here set equal to zero for the steady state:

$$\frac{np - n_e p_e}{\tau_0(n_e + p_e)} - \mu_n \frac{\mathrm{d}(nF)}{\mathrm{d}x} - \frac{\mu_p kT}{e}\frac{\mathrm{d}^2 n}{\mathrm{d}x^2} = \frac{\mathrm{d}n}{\mathrm{d}t} = 0 \tag{3.2.7}$$

$$\frac{np - n_e p_e}{\tau_0(n_e + p_e)} + \mu_p \frac{\mathrm{d}(pF)}{\mathrm{d}x} - \frac{\mu_p kT}{e}\frac{\mathrm{d}^2 p}{\mathrm{d}x^2} = \frac{\mathrm{d}p}{\mathrm{d}t} = 0. \tag{3.2.8}$$

In the last two equations, the first terms are common, as they have to be, since every electron that recombines does so with a hole. The form given here to the recombination term restricts the treatment to small departures from equilibrium, since it is not otherwise constant. Of course, other provisions can be made, e.g. in terms of Shockley–Read recombination. The second terms in eqns (3.2.7) and (3.2.8) represent the divergence of the field current, and the third terms the divergence of the diffusion current. Then, of course, we also have Poisson's equation

$$\frac{\mathrm{d}F}{\mathrm{d}x}=\frac{e}{\epsilon}(p-n+n_e-p_e) \tag{3.2.9}$$

for complete donor ionization. With appropriate boundary conditions, these equations 'contain' the solutions to all semiconductor transport problems, except those involving field-dependent mobilities; but, as mentioned above, such solutions are not analytically accessible for the general case.

The bulk transport equations can actually be formulated in various other ways. Parrott (1971, 1974), for instance, has shown that it is possible to eliminate the field, and to regard as independent variables only the space-charge density and the departure of pn from $p_e n_e$. In some contexts, this is indeed an interesting variant (though Parrott's own formulation is based on the assumption $\mu_n=\mu_p$), but the essential character of the relationships remains unaltered, nor is their solution ultimately simpler in the modified form. We shall therefore adhere to standard practice.

For computation purposes, it is convenient to normalize the equations as in Section 2.4.1, (to which we add the already implied $\boldsymbol{F}=F/kJe\lambda_D$ as well as $\boldsymbol{P}=p/n_e$) but this time expressing all distances in terms of the *two-carrier* Debye length in undisturbed material, namely

$$\lambda_{Da}=\{\epsilon kT/e^2(p_e+n_e)\}^{1/2}, \quad \text{so that} \quad \boldsymbol{j}=j\lambda_{Da}/eD_n n_e. \tag{3.2.10}$$

Normalized quantities are again identified by bold letters. The manner of normalization is, of course, arbitrary to some extent; its choice depends on the nature of the problem. With

$$b=\mu_n/\mu_e; \qquad \Delta\boldsymbol{N}=\boldsymbol{N}-\boldsymbol{N}_e=\boldsymbol{N}-1; \qquad \Delta\boldsymbol{P}=\boldsymbol{P}-\boldsymbol{P}_e. \tag{3.2.11}$$

The transport equations (3.2.4) to (3.2.9) become:

$$\boldsymbol{j}_n=\frac{b}{1+\boldsymbol{P}_e}\left(\boldsymbol{NF}+\frac{\mathrm{d}\boldsymbol{N}}{\mathrm{d}\boldsymbol{X}}\right) \tag{3.2.12}$$

$$\boldsymbol{j}_p=\frac{1}{1+\boldsymbol{P}_e}\left(\boldsymbol{PF}-\frac{\mathrm{d}\boldsymbol{P}}{\mathrm{d}\boldsymbol{X}}\right) \tag{3.2.13}$$

$$\frac{\mathrm{d}^2\boldsymbol{N}}{\mathrm{d}\boldsymbol{X}^2}+\frac{\mathrm{d}(\boldsymbol{NF})}{\mathrm{d}\boldsymbol{X}}-\frac{A_n}{1+\boldsymbol{P}_e}(\Delta\boldsymbol{N}\Delta\boldsymbol{P}+\Delta\boldsymbol{P}+\boldsymbol{P}_e\Delta\boldsymbol{N})=0 \tag{3.2.14}$$

$$\frac{\mathrm{d}^2\boldsymbol{P}}{\mathrm{d}\boldsymbol{X}^2}-\frac{\mathrm{d}(\boldsymbol{PF})}{\mathrm{d}\boldsymbol{X}}-\frac{A_p}{1+\boldsymbol{P}_e}(\Delta\boldsymbol{N}\Delta\boldsymbol{P}+\Delta\boldsymbol{P}+\boldsymbol{P}_e\Delta\boldsymbol{N})=0 \tag{3.2.15}$$

$$\frac{\mathrm{d}\boldsymbol{F}}{\mathrm{d}\boldsymbol{X}}=\frac{1}{1+\boldsymbol{P}_e}(\Delta\boldsymbol{P}-\Delta\boldsymbol{N}) \tag{3.2.16}$$

with

$$A_n = \epsilon / e\mu_n\tau_0(n_e + p_e) \quad \text{and} \quad A_p = bA_n. \tag{3.2.17}$$

The equations are now in a convenient form for linearizing (Sections 3.2.3 and 3.2.4) or for numerical treatment (Sections 3.3.2 and Section 3.4). At this stage, they are equally valid for lifetime and relaxation semiconductors. That distinction depends on the value of A_n, which will be seen to be approximately equal to the ratio of the dielectric relaxation time to the minority carrier lifetime.

Numerical solutions were first given by Popescu and Henisch (1975) but, at the time, only for $b = 1$ and $\gamma = 1$. Though the precise value of γ makes only a quantitative difference to the results, the magnitude of b has more drastic consequences. It was therefore necessary to relax these restrictive conditions. Field and concentration contours computed *without* these assumptions (and without linearization) have been reported by Manifacier and Henisch (1979).

In practice, all these bulk effects tend to be masked by resistance changes of the barrier itself (here excluded from consideration), unless special precautions are taken to isolate the barrier by potential probe techniques. This isolation is possible in materials of high lifetime and high carrier mobility. In this way, Rieder *et al.* (1980) were able to verify the theoretical predictions by direct measurement, using silver contacts on single-crystal n-germanium.

3.2.3 *Analytic treatment; linearized theory; trap-free case*

The above equations can be solved analytically within the framework of a 'small-signal theory' (Manifacier and Henisch, 1978a). As long as consideration is limited to the regime of small currents, the relationships can be linearized by assuming that $\Delta\boldsymbol{N} \ll 1$, and $\Delta\boldsymbol{P} \ll \boldsymbol{P}_e$ and by neglecting cross-products like $\boldsymbol{F}(\mathrm{d}\boldsymbol{N}/\mathrm{d}\boldsymbol{X})$ and $\boldsymbol{F}(\mathrm{d}\boldsymbol{P}/\mathrm{d}\boldsymbol{X})$ which are of second-order magnitude. Equations (3.2.12) to (3.2.16) then reduce to

$$\boldsymbol{j}_n = \frac{b}{1+\boldsymbol{P}_e}\left(\boldsymbol{F} + \frac{\mathrm{d}\Delta\boldsymbol{N}}{\mathrm{d}\boldsymbol{X}}\right) \tag{3.2.18}$$

$$\boldsymbol{j}_p = \frac{1}{1+\boldsymbol{P}_e}\left(\boldsymbol{P}_e\boldsymbol{F} - \frac{\mathrm{d}\Delta\boldsymbol{P}}{\mathrm{d}\boldsymbol{X}}\right) \tag{3.2.19}$$

$$\frac{\mathrm{d}^2\Delta\boldsymbol{N}}{\mathrm{d}\boldsymbol{X}^2} + \frac{\mathrm{d}\boldsymbol{F}}{\mathrm{d}\boldsymbol{X}} - \frac{A_n}{1+\boldsymbol{P}_e}(\Delta\boldsymbol{N}\boldsymbol{P}_e + \Delta\boldsymbol{P}) = 0 \tag{3.2.20}$$

$$\frac{\mathrm{d}^2\Delta\boldsymbol{P}}{\mathrm{d}\boldsymbol{X}^2} - \boldsymbol{P}_e\frac{\mathrm{d}\boldsymbol{F}}{\mathrm{d}\boldsymbol{X}} - \frac{A_p}{1+\boldsymbol{P}_e}(\Delta\boldsymbol{N}\boldsymbol{P}_e + \Delta\boldsymbol{P}) = 0 \tag{3.2.21}$$

$$\frac{\mathrm{d}\boldsymbol{F}}{\mathrm{d}\boldsymbol{X}} = \frac{1}{1+\boldsymbol{P}_e}(\Delta\boldsymbol{P} - \boldsymbol{P}\boldsymbol{N}). \tag{3.2.22}$$

The last three equations yield:

$$\frac{d^2\Delta \boldsymbol{N}}{d\boldsymbol{X}} - \Delta \boldsymbol{N}\left(\frac{1+\boldsymbol{P}_e A_n}{\boldsymbol{P}_e}\right) - \Delta \boldsymbol{P}\left(\frac{A_n - 1}{1+\boldsymbol{P}_e}\right) = 0 \tag{3.2.23}$$

and

$$\frac{d^2\Delta \boldsymbol{P}}{d\boldsymbol{X}} + \Delta \boldsymbol{N}\left(\frac{\boldsymbol{P} - \boldsymbol{P}_e A_p}{1+\boldsymbol{P}_e}\right) - \Delta \boldsymbol{P}\left(\frac{A_p + \boldsymbol{P}_e}{1+\boldsymbol{P}_e}\right) = 0. \tag{3.2.24}$$

We now have two simple differential equations which can be solved in the general form:

$$\Delta \boldsymbol{N} = \frac{\Gamma_1}{\boldsymbol{P}_e}\exp(\boldsymbol{X}) - \frac{\Gamma_2}{\boldsymbol{P}_e}\exp(-\boldsymbol{X}) + \Gamma_3(1-A_n)\exp(\boldsymbol{X}\sqrt{}A) + \Gamma_4(1-A_n)\exp(-\boldsymbol{X}\sqrt{}A) \tag{3.2.25}$$

$$\Delta \boldsymbol{P} = \Gamma_1\exp(\boldsymbol{X}) + \Gamma_2\exp(-\boldsymbol{X}) + \Gamma_3(1-A_p)\exp(\boldsymbol{X}\sqrt{}A) + \Gamma_4(1-A_p)\exp(-\boldsymbol{X}\sqrt{}A) \tag{3.2.26}$$

where

$$A = \frac{L_D^2}{\tau_0 D_a}, \quad \text{and} \quad D_a = \frac{kT(n_e + p_e)\mu_n\mu_e}{e(\mu_n n_e + \mu_p p_e)}. \tag{3.2.27}$$

The integration constants B, C, R, and S have to be determined by reference to the boundary conditions. Since $\Delta \boldsymbol{N}$ and $\Delta \boldsymbol{P}$ must both go asymptotically to zero as $\boldsymbol{X}$ increases, we must have $\Gamma_1 = \Gamma_3 = 0$. Since, in this case, we take the origin to be at the energy minimum where the field is zero, the current at that point is given by

$$\boldsymbol{j} = \frac{1}{1+\boldsymbol{P}_e}\left(b\frac{d\Delta \boldsymbol{N}}{d\boldsymbol{X}}(0) - \frac{d\Delta \boldsymbol{P}}{d\boldsymbol{X}}(0)\right) \tag{3.2.28}$$

and, for reasons of continuity, this is also the current everywhere else.

For the current composition ratio in the undisturbed bulk, we have: $\gamma_n = \boldsymbol{P}_e/(b + \boldsymbol{P}_e)$ and, at $\boldsymbol{X} = 0$,

$$\gamma = \frac{d\Delta \boldsymbol{P}}{d\boldsymbol{X}}(0) \bigg/ \left\{\frac{d\Delta \boldsymbol{P}}{d\boldsymbol{X}}(0) - b\frac{d\Delta \boldsymbol{N}}{d\boldsymbol{X}}(0)\right\}. \tag{3.2.29}$$

In terms of the present symbols, the electric field far from the injecting interface becomes:

$$\boldsymbol{F}(\infty) = \boldsymbol{j}(1+\boldsymbol{P}_e)/(b+\boldsymbol{P}_e) \tag{3.2.30}$$

Since

$$\boldsymbol{N}(\infty) = \boldsymbol{N}_e = 1 \quad \text{and} \quad \boldsymbol{P}(\infty) = \boldsymbol{P}_e. \tag{3.2.31}$$

Under these conditions, the integrating constants Γ_4 and Γ_2 can be shown to be

$$\Gamma_4 = \boldsymbol{j}\{\gamma(b+\boldsymbol{P}_{\mathrm{e}}) - \boldsymbol{P}_{\mathrm{e}}\}/b(1-A)\sqrt{\mathscr{A}} \tag{3.2.32}$$

and

$$\Gamma_2 = \boldsymbol{j}\boldsymbol{P}_{\mathrm{e}}\{(1-A_{\mathrm{p}}) - \gamma(1-b)\}/b(1-A). \tag{3.2.33}$$

In this form, Γ_4 and Γ_2 can be substituted into eqns (3.2.25) and (3.2.26) to yield two equations which appear to be complicated, but are in fact structurally simple, being of the form

$$\Delta\boldsymbol{N} = A_1 \exp(-\mathbf{X}\sqrt{\mathscr{A}}) - A_2 \exp(-\mathbf{X}) \tag{3.2.34}$$

and

$$\Delta\boldsymbol{P} = A_3 \exp(-\mathbf{X}\sqrt{\mathscr{A}}) + A_4 \exp(-\mathbf{X}). \tag{3.2.35}$$

Similarly, a field contour is obtained by adding eqns (3.2.18) and (3.2.19). It is in the form

$$\boldsymbol{F} = \boldsymbol{j}\left\{A_5 \exp(-\mathbf{X}\sqrt{\mathscr{A}}) + A_6 \exp(-\mathbf{X}) + \frac{1+\boldsymbol{P}_{\mathrm{e}}}{b+\boldsymbol{P}_{\mathrm{e}}}\right\} \tag{3.2.36}$$

A quick check shows that this agrees with eqn (3.2.30) for $\mathbf{X} = \infty$. The proportionality of $\boldsymbol{F}$ and $\boldsymbol{j}$ arises, of course, directly from the linearization process. Equation (3.2.36) disposes of a popular misconception, to the effect that all injection effects 'disappear' as the current goes to zero. Not so; since A_5 and A_6 depend on γ, the $\boldsymbol{F}/\boldsymbol{j}$ ratio is injection-dependent right down to the zero current limit.

Though the above equations are limited to the low current range, they have the merit of generality, in as much as they make it easy to examine how the various transport constants of the material affect the situation. Manifacier and Henisch (1978a) and Popescu and Henisch (1975,1976a) have given concentration and field contours, as well as voltage–current characteristics of the system as a whole for a variety of conditions, and have *inter alia* confirmed that a field maximum can indeed arise, as suggested by the schematic and intuitive representation in Fig. 3.10. This happens only when $b = \mu_{\mathrm{n}}/\mu_{\mathrm{p}} > 1$ and therefore does *not* apply to injection into *p-type* material. Though the local field may have a maximum value at a small distance from the injecting electrode, other regions (closer to the metal) have a lower-than-normal field. The overall resistance of the system can therefore increase or decrease with injection. In the trap-free case here under discussion, it increases, as shown in Fig. 3.12(b). The voltage–current relationship is surprisingly linear over a wide voltage range, and then becomes superlinear. Inevitably, there is a cross-over which may, however, occur far outside the range of the

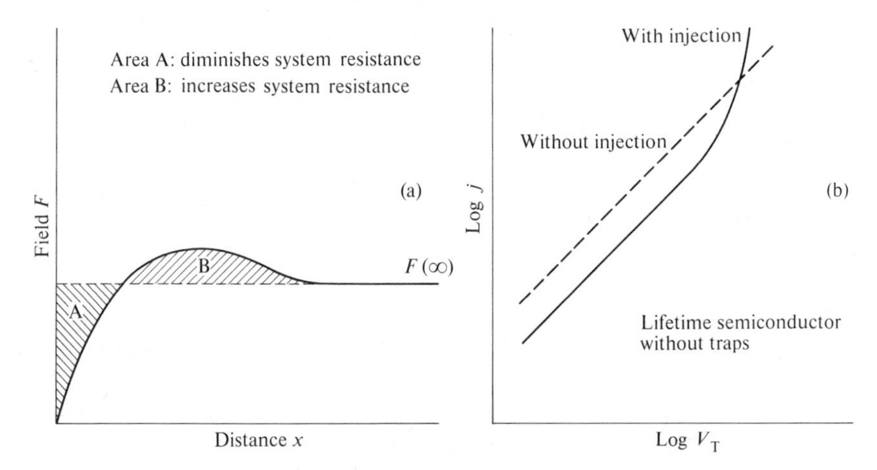

FIG. 3.12. Behaviour of trap-free lifetime systems under injection. (a) Field contour and its contributions to systems resistance. In the absence of traps, area B is more important here than area A at low current densities. (b) Voltage–current relationship of the system as a whole, which depends, of course, on the system thickness. Minority-carrier injection diminishes the total resistance at high current densities (only). Barrier resistances are ignored here.

linearized model. It can be calculated by applying numerical methods to the complete (non-approximated) equations. Manifacier and Henisch (1978a) have provided an analytic estimate.

The value of the maximum field increases with γ, of course, and γ itself is likely to be current dependent, as discussed in Section 3.1.3. The resistance increase at low current densities is not technically important, but it can influence measurements, and lead to unexpected 'anomalies'. Its existence and origin are also basic to an understanding of the injection process as such (Manifacier and Henisch 1978c).

The equations of Section 3.2.2 are applicable to the analysis of transport situations inside and outside the barrier. In contrast, the equations of the present section are inapplicable to the barrier region itself, because fields and concentration gradients are very high there, which means that many of the cross-products neglected above are not, in fact, negligible. The equations do, however, apply in other respects to all situations which involve departures from carrier equilibrium, whether they have anything to do with contact behaviour or not. For an application to Hall-effect analysis, see Manifacier and Henisch (1980b).

3.2.4 *Analytic treatment; linearized theory; case with traps*

It will be clear that the type of analysis outlined above can also be performed for the case in which recombination centers are present that act as traps. Two aspects of the analysis must be modified for this purpose. One concerns Poisson's equation which now becomes, in normalized terms:

$$\frac{\mathrm{d}\boldsymbol{F}}{\mathrm{d}\boldsymbol{X}}=\frac{1}{1+\boldsymbol{P}_{\mathrm{e}}}(\Delta\boldsymbol{P}-\Delta\boldsymbol{N}-\Delta\boldsymbol{Q}_{\mathrm{t}}). \tag{3.2.37}$$

This replaces eqn (3.2.22). $\Delta\boldsymbol{Q}_{\mathrm{t}}$ is here the normalized net charge density of carriers in traps, and in eqn (3.2.37) it can be the dominating term. It can be written as $\Delta\boldsymbol{Q}_{\mathrm{t}}=\boldsymbol{M}_{\mathrm{e}}-\boldsymbol{M}$ where $\boldsymbol{M}_{\mathrm{e}}$ is the (normalized) charge density under injection. When only a single recombination level is present, $\boldsymbol{M}_{\mathrm{e}}$ and $\boldsymbol{M}$ can be simply calculated from Shockley–Read theory, as shown by Popescu and Henisch (1976b) and Manifacier and Henisch (1978b).

The other change concerns the recombination term, assumed to be bimolecular for the purposes of Section 3.2.3. Of course, when the steady state prevails, recombination is always bimolecular in effect, since every electron recombines with a hole, but this is not actually the issue here. The issue is whether a single descriptive constant τ_0, the lifetime, is capable of describing the system. In the truly bimolecular case (e.g. when direct electron–hole recombination prevails), τ_0 is indeed a valid descriptive constant. However, when recombination takes place via centers, this is no longer true. Though it is always possible to define an *equivalent* τ_0, that quantity would no longer be a constant, but would depend on the excitation level. It is then better to describe the phenomena in terms of the separate electron and hole lifetimes τ_{n} and τ_{p}. τ_0 is then called the *diffusion length lifetime* (van Roosbroeck 1961). It is defined by:

$$\text{recombination rate}=\frac{\Delta np_{\mathrm{e}}+\Delta pn_{\mathrm{e}}}{\tau_{\mathrm{n}}(p_{\mathrm{e}}+p_1)+\tau_{\mathrm{p}}(n_{\mathrm{e}}+n_1)}=\frac{\Delta np_{\mathrm{e}}+\Delta pn_{\mathrm{e}}}{\tau_0(n_{\mathrm{e}}+p_{\mathrm{e}})} \tag{3.2.38}$$

The lifetimes τ_{n} and τ_{p}, in turn, depend on the (normalized) trap density $\boldsymbol{M}_0$, and on the energetic position of the trapping level, via n_1 and p_1, two standard parameters of the Shockley–Read (1952) model. With these modifications, the equations of Section 3.2.3 can be rewritten and again analytically solved, leading to new concentration and field contours. The details are not so important here, but the general nature of the effects instructive. When minority carriers are injected, the trap occupancy changes, and hence also the field contour. It can be shown that, with plausible recombination states, the centres act in the first instance as hole traps, and only later (at larger distances x) as electron traps, when the free-electron concentration dominates over the free-hole concentration. Within the recombination region, the positive space charge now consists of two parts, the charge density in free holes, and the charge density of trapped holes. As far as Poisson's equation is concerned, these two charge densities act as field modifiers in the same way but, of course, the trapped holes are not mobile. Electrons are attracted into the injection region for neutralization purposes in numbers which can substantially exceed the *free*-hole concentration. The situation is still qualitatively represented by the broken line in Fig. 3.9, except for the fact that the prevailing Δn values may be much higher than shown. The corresponding concentration gradients would likewise be greater, and so would the local compensating field. The overall consequence is a further augmentation of the resistance increase seen at low current densities. There are also some reactive consequences; Green and Shewchun (1973) have shown that the system behaviour can become inductive for small forward currents.

The effect of hole traps is therefore qualitatively similar to that of an increase in the mobility ratio but its quantitative scope is much greater, since the trap density can be high. On the other hand, increasing the trap density would, in general, reduce the carrier lifetime, and thereby reduce the resistance enhancement. The practical outcome is a matter of balance between these two effects.

Theoretical expectations of the way in which the presence of minority carrier traps modifies the transport relationships have been experimentally verified by Rieder *et al.* (1982), on the basis of comparative measurements on single-crystal n-Ge, before and after neutron bombardment (and annealing). This treatment introduces traps which capture injected holes. Accordingly, the local resistance increase is far more prominent than it is for the highly ordered crystal. The results represent a demonstration of principle, but in most materials the lifetime and mobility are too small to make *direct* probe measurements of the effect feasible.

Further reading

On injection with fast pulses: Slowik 1981a.

On injection into CdSe: Freeman and Slowik 1981.

On the effect of injection on series resistance: Stafeev 1958.

On the analysis of double-injection systems: Dmitriev *et al.* 1978.

3.3 Injection into bulk material; relaxation semiconductors

3.3.1 Relaxation and Lifetime Regimes

As long as electronic equilibrium prevails, there is, of course, no distinction between lifetime and relaxation semiconductors. Under non-equilibrium (e.g. in the vicinity of current-carrying contacts or under non-uniform illumination) the distinction is of fundamental importance, as was first pointed out by van Roosbroeck and Casey (1970) (see also

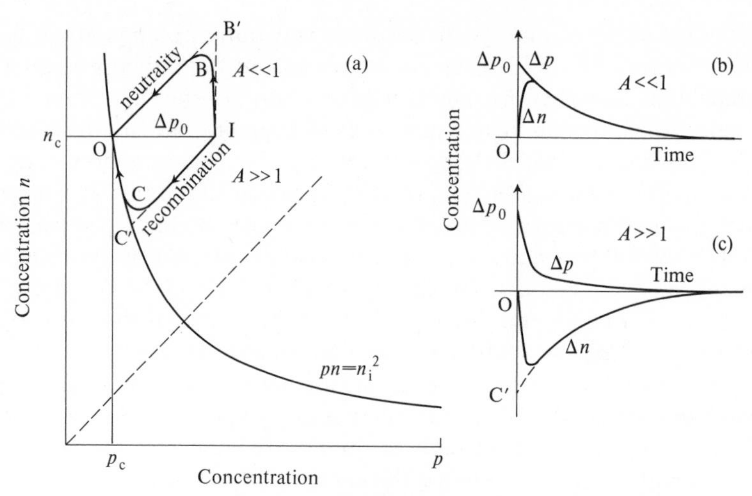

FIG. 3.13. Hole injection into n-type material; departures from equilibrium in the 'homogeneous case'. $A = A_n = \tau_D/\tau_0$. Initial hole increment: ΔP_0. (a) Curve B ($A \ll 1$): lifetime behaviour; curve C ($A \gg 1$): relaxation behaviour. (b), (c) Time-dependence of the carrier concentration increments. After Popescu and Henisch (1975).

Queisser, Casey, and van Roosbroeck (1971), and van Roosbroeck and Casey (1972)). The characteristically different reaction of the two classes of material to additional minority carriers can be simply illustrated by reference to the 'homogeneous case' (Popescu and Henisch 1975). The homogeneous case is arrived at by considering a very small sampling volume within an injection region, and ignoring concentration gradients. It is represented by Fig. 3.13, on which the parameter A is approximately τ_D/τ_0, the ratio of the dielectric relaxation time to the carrier lifetime.

Figure 3.13(a) envisages an n-type material, and displays the $pn = p_e n_e = n_i^2$ *zero-recombination contour.* Neutrality prevails at the point O. Consider now a small incremental hole concentration Δp_O, no matter how produced, taking us to the point I. At this stage there are two classes of possibilities, one denoted by $A < 1$ (lifetime semiconductor), and one by $A > 1$ (relaxation semiconductor). In the limiting lifetime case, the dielectric relaxation time might be considered zero, which means that the space charge of the extra holes will be immediately neutralized by an influx of electrons from the surrounding material. This would mean that the system goes from I to B′ in zero time, providing $\Delta n = \Delta p$. Recombination can now proceed at a rate given by the minority-carrier lifetime τ_0. In terms of Fig. 3.13(a), this implies a return to the point O along a 45-degree line, since neutrality is maintained throughout. In practice, of course, some contour OIBO would be followed, since the dielectric relaxation time cannot be actually be zero, and since the two processes go on at the same time. Throughout the whole episode, except at O and I itself, there is an *excess of majority carriers*. The fact that Δp and Δn are both positive gives rise to all the injection effects discussed in Section 3.2. This is the situation which prevails in many familiar semiconductors, including germanium and silicon, so much so that other situations remained ignored for an astonishing number of years, until van Rossbroeck and co-workers drew attention to their significance.

Let us now consider the opposite situation, namely $A>1$. In the limiting relaxation case, the carrier lifetime might be considered zero, which means that recombination would proceed in zero time, taking us to the point C′ on the zero-recombination contour. However, since this transition took place (at 45 degrees) without altering the charge balance, the material is not neutral at C′. The space charge remains to be neutralized by dielectric relaxation, which is now considered slow, and leads to the transition from C′ to back to O. In practice, a contour OICO is more likely to be followed, again, because the carrier lifetime cannot actually be zero, and because the two processes (dielectric relaxation and carrier recombination) occur at the same time. Throughout the whole episode, except at O and I itself, there is a *depletion of majority carriers*. The fact that Δn is negative while Δp is positive gives rise to a set of phenomena quite different from those described above.

The case $A=1$ represents a boundary. Under such conditions, electrons would come into the test volume for neutralization purposes at the exact rate at which they are needed for recombination, giving $\Delta n=0$ throughout. Materials can thus be classified on the basis of their value of A, but the practical situation is somewhat more complicated. Figure 3.13 makes no reference to spatial variations, and thus suggests that if (say) Δn is positive somewhere, it is positive everywhere. This need not be so; oscillatory solutions are known to apply to certain cases, of which one is discussed in Section 3.4.1. It is therefore desirable to use the terms 'lifetime *regime*' and 'relaxation *regime*' for the situations $\Delta n>0$ and $\Delta n<0$ respectively, irrespective of the materials involved. Thus, it will be shown in Section 3.3.3 that a relaxation semiconductor can be driven into the 'lifetime regime' if it contains a sufficient concentration of minority carrier traps. Figures 3.13(b) and (c) give the time-dependence of Δn and Δp for $A<1$ and $A>1$ respectively.

The transport equations which govern the relaxation case are, of course, the same as those which govern the lifetime case, except for the fact that we now have $A>1$. The dilemmas discussed in Section 3.2.1 are also the same and, accordingly, one has the choice between analytic solutions within the framework of a 'small-signal theory', and numerical solutions covering specific situations without restrictions. The analytic methods of Sections 3.2.3 and 3.2.4 are also valid here, subject only to the appropriate choice of $A_n(\approx A)$ and A_p.

3.3.2 *Solutions for materials with and without traps*

When the transport equations are solved, either analytically in linearized form (Manifacier and Henisch 1978a), or numerically in complete form, they show that majority-carrier depletion occurs within a non-equilibrium region near the contacts. Figure 3.14 gives an overview of numerical calculations by Popescu and Henisch (1975) for a situation corresponding to a trap-free relaxation semiconductor at low current level. The assumptions $\gamma = 1$ and $b = 1$ were made in this case, but are not critical for this particular demonstration: n is clearly below normal, as is $n + p$. Van Roosbroeck originally predicted a resistance increase on this basis, but this would be true only if the currents were essentially field currents. The opposite is actually the case: the currents are entirely carried by diffusion at $\mathbf{X} = 0$, and become field currents only gradually as bulk equilibrium is approached. The net result can be shown to be a *decrease* of local resistivity, as in the lifetime case, but for different reasons. Kiess and Rose (1973) have also pointed this out. (See, however, Section 3.4.1 for an exception to this rule: hybrid case.)

The majority carrier depletion produced by minority carrier injection

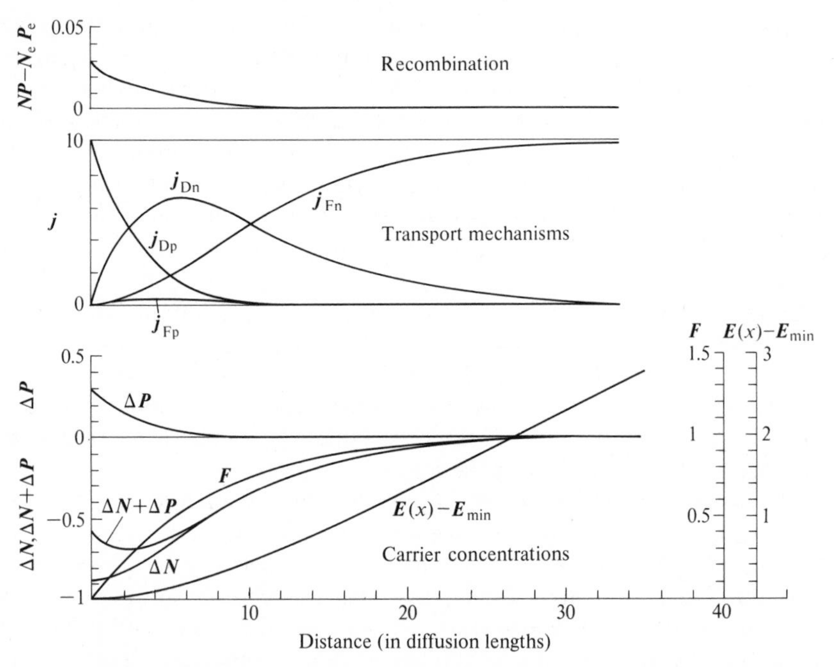

FIG. 3.14. Hole injection into a trap-free n-type relaxation semiconconductor characterized by $A = 99$ and $b = 1$. $\mathbf{j}_{Dn}$, $\mathbf{j}_{Dp}$: diffusion current densities. $\mathbf{j}_{Fn}$, $\mathbf{j}_{Fp}$; field current densities. $\gamma = 1$ assumed. $\mathbf{N}_e \equiv 1$; $\boldsymbol{E}_{min}$ = energy at $\mathbf{X} = 0$. After Popescu and Henisch (1975).

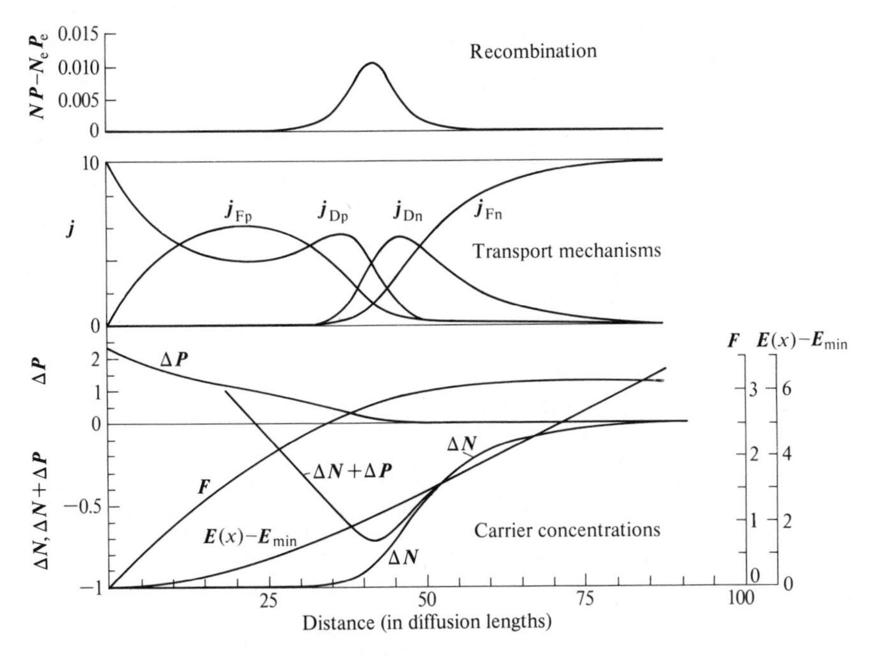

FIG. 3.15. Hole injection into a trap-free n-type relaxation semiconductor characterized by $A=990$ and $b=1$. $\boldsymbol{j}_{Dn}$, $\boldsymbol{j}_{Dp}$; diffusion current densities. $\boldsymbol{j}_{Fn}$, $\boldsymbol{j}_{Fp}$; field current densities. $\gamma=1$ assumed. $\mathbf{N}_e \equiv 1$; $\boldsymbol{E}_{min}$ = energy at $\mathbf{X}=0$. After Popescu and Henisch (1975).

has three consequences:

(a) The electron concentration gradient is now reversed, and electron diffusion is now additive with electron drift.
(b) The electron depletion increases the local space charge density, and thereby promotes the build-up of the field.
(c) The recombination rate is reduced.

All these features become more pronounced in materials with a greater value of A; this is shown in Fig. 3.15. Over a certain distance, the majority-carrier depletion is virtually complete, meaning that *no* recombiantion can take place. Far away from the contact, there is also no recombination, because equilibrium prevails. The recombination rate thus must show a maximum at an intermediate distance from the contact, and the numerical solutions reflect this. This can be demonstrated even for lower values of A, but at increased current densities. It will be seen from Fig. 3.15 that the total carrier concentration $n+p$ is below normal only over a small range of distances. The net result is still a decrease of local (effective) resisitivity, as above. Nevertheless, the total resistance of the system depends almost *linearly* on voltage, as shown in Fig. 3.16, a fact

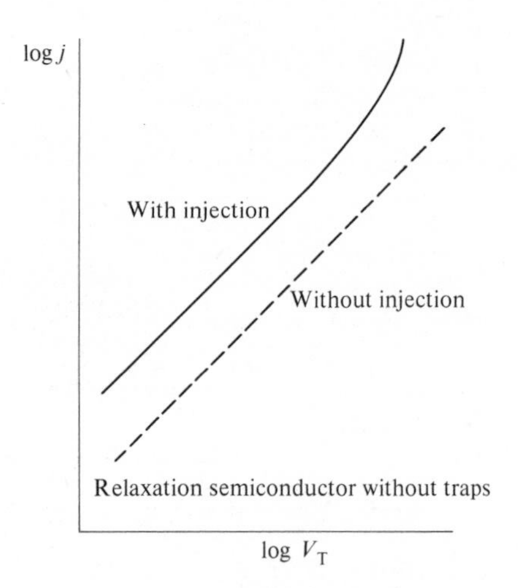

FIG. 3.16. Behaviour of trap-free relaxation system under injection. The exact shape of the contour depends, of course, on the distance from the injecting boundary at which it is measured. Barrier resistance ignored here. Compare Fig. 3.12(b).

which might so easily be misinterpreted as a symptom of 'ohmic' (equilibrium) conduction. As a diagnostic test for injection, voltage–current linearity is thus a highly unreliable criterion (see Manifacier and Henisch (1981) for examples).

Figures 3.14 and 3.15 demonstrate that the effects of injection penetrate into the material over a considerable number of diffusion lengths, but the diffusion length in a relaxation semiconductor tends to be small, because of the small lifetime. It should also be borne in mind that a relaxation semiconductor without traps, as assumed here, is a somewhat artificial construct. Most poorly conducting materials do in fact contain carrier traps (and *are* as poorly conducting as they are by virtue of that very fact); see Section 3.2.4 and the comments below.

The situation in the presence of traps is a little more complicated. A linearized 'small-signal' theory has been given by Manifacier and Henisch (1978b), as described in Section 3.2.4. Numerical solutions for more general situations have been provided by Popescu and Henisch (1976b) and Manifacier, Henisch, and Gasiot (1979). These results can be summarized as follows:

(a) As long as the trap concentration is low, injection into a relaxation semiconductor continues to lead to a relaxation regime ($\Delta p > 0$ and $\Delta n < 0$).

(b) Majority-carrier depletion under (a) is reduced compared with the trap-free case. This is because the traps store charge as explained in Section 3.2.4, and thereby augment the local electron concentration.

(c) For high trap concentrations, the space-charge effect of the traps can outweigh all other considerations in importance, and lead to a situation similar to that shown in Fig. 3.9. As a result, minority-carrier injection can lead to a 'lifetime regime', even though the material, on the basis of its bulk properties, may be classified as a relaxation semiconductor. The overall resistance at low current is then *increased* as a result of injection, just as it is for the lifetime case of Figs 3.9 and 3.12.

(d) Even in the presence of traps, the voltage–current relationship is *linear* over a wide range. Illegems and Queisser (1975) have experimentally verified this (for voltages up to 100 V!) by measurements on GaAs.

In Section 3.3.1 materials were classified into lifetime and relaxation semiconductors, according as $\tau_D < \tau_0$ or $\tau_D > \tau_0$. For trap-free material this classification works reasonably well, though even then the equations can be shown to have complex oscillatory solutions when $\tau_D \approx \tau_0$ (see Stoica and Popescu (1978)). When traps are present, this form of classification becomes less and less appropriate. It can be 'rescued' by taking the boundary to be not $\tau_D/\tau_0 = 1$ but $\tau_D/\tau_0(\boldsymbol{M}_0 + 1)$, as Popescu and Henisch (1976b) have done. However, it has since become clear that a classification based upon characteristic *lengths* would be more appropriate, e.g. as described by Manifacier and Henisch (1980a). In particular, the lengths involved are the ambipolar diffusion length L_{Da} and the screening length L_{sc}. When free carriers alone are responsible for the space charge, then $L_{sc} = \lambda_D$, the Debye length; but when space charge is also held in traps, the screening length becomes substantially smaller than the Debye length, in accordance with

$$L_{sc} = \lambda_D \left\{ 1 + M_0 \frac{n_e p_e}{(n_e + n_1)(p_e + p_1)(n_e + p_e)} \right\}^{-1/2} \tag{3.3.1}$$

where M_0 is the actual (un-normalized) trap concentration and n_1, p_1 are again the familiar Shockley–Read parameters. It can be shown that the proper classification criterion in the presence of traps should be based on the boundary given by

$$\frac{\tau_D}{\tau_0} \approx \left(\frac{L_{sc}}{L_{Da}} \right)^2 \tag{3.3.2}$$

where

$$L_{Da} = \left\{ \frac{kT}{e} \frac{(n_e + p_e)\mu_n \mu_p \tau_0}{(\mu_n n_e + \mu_p p_e)} \right\}. \tag{3.3.3}$$

Relaxation behavior (i.e. majority-carrier depletion in response to minority carrier injection) is expected when $L_{sc} > L_{DA}$.

The most drastic consequence of minority-carrier injection into a relaxation semiconductor is evidently majority-carrier depletion, but this can be of practical significance only as long as there are enough majority carriers to deplete. As we go from semi-insulating materials to 'true' insulators, majority carriers play an ever-diminishing role, and eventually none at all. The situation then becomes entirely dominated by the trapping of injected carriers, and relaxation semiconductor concepts as outlined above cease to matter. We then have to deal with space-charge controlled conduction in the conventional sense (see Lampert and Mark (1970), and Section 1.2.4).

Stoisick, Wolf, and Queisser (1971) have shown that, in some circumstances, the injection-depletion relaxation regime can give rise to low-frequency oscillations. These can be explained in terms of states which change their trapping role with increasing occupation. Oscillations are also a feature of many double injection systems, i.e. systems in which the non-equilibrium regions associated with contacts on opposite sides overlap. Such oscillations are always associated with negative (differential) resistance regimes (e.g. Migliorato *et al.* 1976), but instances have also been reported of fluctuations occurring while the differential resistance was positive ascribed to unstable filament formation. Dudek and Kassing (1979) have reported such behaviour in Au-compensated n-type silicon (see also the work of Brousseau *et al.* (1970) on p–i–n structures).

Further reading

On high injection levels in quasi-intrinsic semiconductors: Ancker-Johnson *et al.* 1970, Popescu 1981.

On neutrality and quasi-neutrality in semiconductors: Greebe and Van der Maesen 1963, Warner 1979.

On the relaxation of mobile space charge in solids: Lemke 1970.

On majority-carrier injection into insulators: Montojo and Sanchez 1974, O'Reilly and DeLucia 1975, Slowik 1981b.

3.4 Minority-carrier exclusion and extraction; bulk material

3.4.1 Carrier exclusion

At a contact carrying forward current, we have minority carrier exclusion when $\gamma < \gamma_n$. Minority carriers then enter the semiconductor at an initial rate smaller than that required for undisturbed bulk conduction. The additional carriers needed for this purpose are thermally generated within a non-equilibrium region close to the 'excluding' interface. The above equations can again be used to analyse the situation. In extrinsic material the effect of exclusion is bound to be trivial, since a small negative Δp

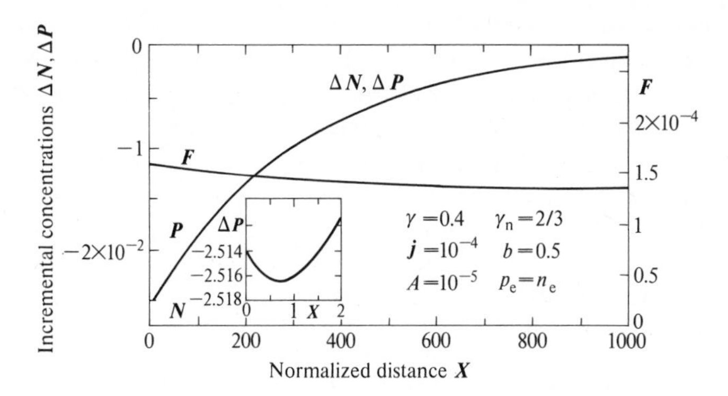

FIG. 3.17. Minority-carrier exclusion from intrinsic, high-lifetime material; concentration and field contours (Manifacier and Henisch 1979).

represents a negligible departure from the total carrier concentration. In near-intrinsic and intrinsic material the effect could be substantial, and would be greatest, of course, for $\gamma = 0$. Contours for these and other conditions have been provided by Manifacier and Henisch (1979), and some are shown in Fig. 3.17. For strong lifetime material, the field is only slightly increased at the excluding boundary, and then decreases monotonically into the bulk. One of the conditions for the appearance of the field maximum is $b < 1$, and Fig. 3.17 makes that assumption. The concentration contours are then almost identical (quasi-neutrality) for the two carriers, except that $\mathbf{\Delta P}$ shows a small dip near $\mathbf{X} = 0$. This arises here from the imposed boundary conditions, which demand that the hole current be positive, and carried only by diffusion at the boundary.

Figure 3.18 shows a hypothetical case of maximum exclusion in an intrinsic semiconductor, assuming not only $\gamma = 0$ where $F = 0$ but also

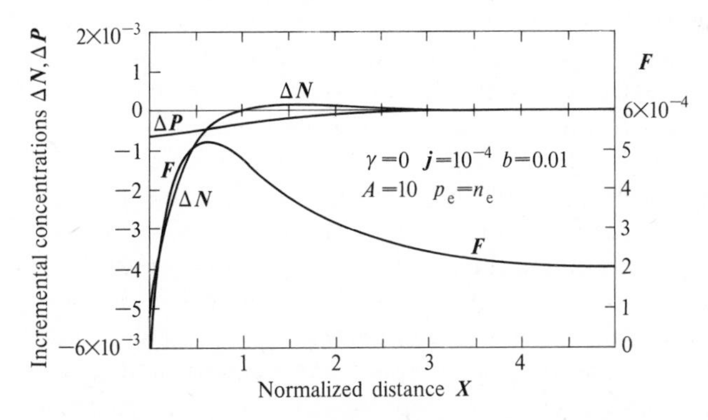

FIG. 3.18. Total minority-carrier exclusion from intrinsic relaxation material; concentration and field contours for the hypothetical case of $b \ll 1$ and $\gamma = 0$ (Manifacier and Henisch 1979).

$b \ll 1$. This is not necessarily realistic, but serves to illustrate a principle. Under such conditions, there would be a substantial field maximum (resistance increase). The overall operating regime may be classified as 'hybrid', in as much as it is characterized neither by a uniform $\Delta n > 0$ (lifetime regime), nor by a uniform $\Delta n < 0$ (relaxation regime). This demonstrates also that it is possible, at any rate in principle, to envisage a resistance *increase* in association with majority-carrier depletion, despite the arguments advanced by Kiess and Rose (1973), as long as there is minority-carrier depletion as well. On the other hand, the present assumptions make this model rather artificial.

3.4.2 Carrier extraction

The analysis of injection and exclusion processes is facilitated by the fact there is always (in the forward direction) some point close to the metal–semiconductor interface at which the field is zero. This point then serves as a convenient boundary for the injection (exclusion) region. Minority-carrier *extraction*, which occurs for reverse current flow, presents more problems precisely because there is no such point. However, a linearized analysis has been performed (Rahimi *et al.* 1981) by defining $x = 0$ as the point close to the edge of the barrier at which the field has some particular (arbitrary) small value. In this way, the relaxation effect described in Sections 1.2 and 3.3 can be quantitatively assessed. As in the case of minority-carrier injection, two processes are actually at work, one involving the unequal diffusion by carriers of different mobilities, and one involving modified carrier concentrations and correspondingly modified drift currents. However, the phenomenon of extraction is of greatest interest when its effects are of sufficient prominence to take it outside the realm of a linearized, small-signal model. For such situations, only computer solutions are feasible but, as far as is known, these have not yet been provided.

The diffusion and drift relationships discussed in Section 3.2 also control the extraction case. Thus, when a reverse current flows and $\gamma_2 > \gamma_n$, a concentration profile of minority carriers (here, holes) is set up as shown in Fig. 3.19. In a good lifetime semiconductor, quasi-neutrality will prevail over most of the material, and the electron gradient will therefore be practically identical. If the diffusion constants of the two carriers were also equal, the effects of these gradients would cancel. They do not cancel in n-type material, because the electron mobility is higher. Electron diffusion therefore dominates at low fields, and since it is opposed to the direction of the prevailing electron field current, a higher local field is necessary to keep the total current constant (compare Section 3.2.1). This, in turn, implies an increase in the effective local resistivity, and this has actually been observed by Rieder *et al.* (1980) by probe measurements on near-intrinsic germanium. Other things being

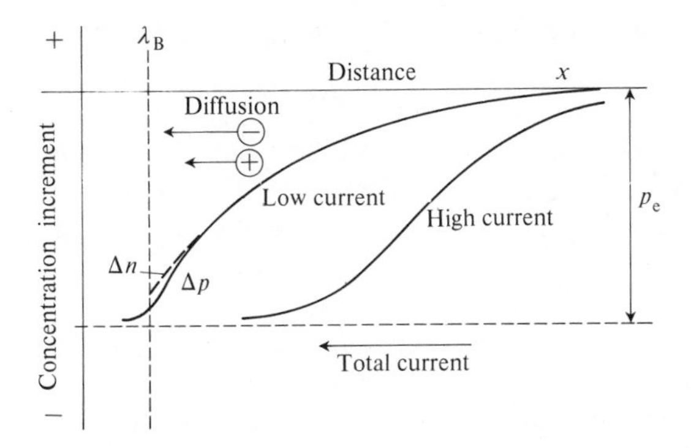

FIG. 3.19. Minority-carrier extraction; schematic concentration contours near the extracting contact for low and high reverse currents.

equal, the corresponding effect in *p-type* material should be a local resistivity *decrease.*

In spite of many parallels, there is one respect in which injection and extraction differ profoundly. There is no obvious limit to the number of minority carriers that can be injected into a material, but the number that can be extracted is obviously limited to the number originally present. When all the carriers are extracted, we have $\Delta p = -p_e$, and, for reasons of quasi-neutrality in a lifetime semiconductor, also $\Delta n = -p_e$. The local resistivity then becomes constant in the region in which these circumstances prevail, and greater than the resistivity in the undisturbed bulk. This is the effect expected and observed at higher current densities, as Fig. 3.20 shows. Assuming quasi-neutrality and total extraction, the

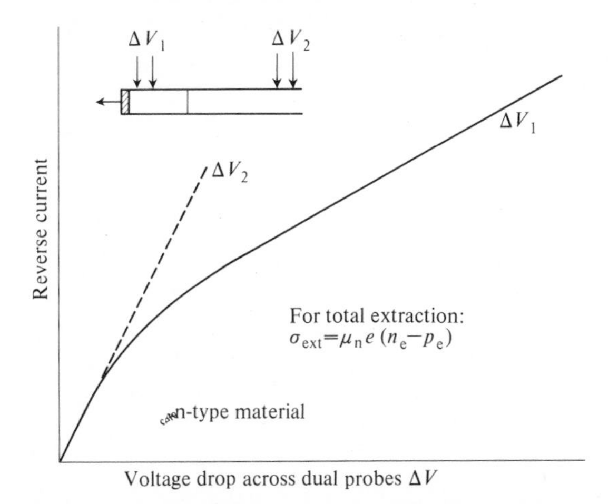

FIG. 3.20. Effect of total minority-carrier extraction on the effective local resistivity near (ΔV_1) and for (ΔV_2) from the contact. Schematic representation of observation on near-intrinsic Ge (Rieder *et al.* 1980).

conductivity in the extraction region should be

$$\sigma_{\text{ext}} = e\mu_{\text{n}}(n_{\text{e}} - p_{\text{e}}) \tag{3.4.1}$$

as compared with $\sigma_{\text{b}} = e(\mu_{\text{n}}n_{\text{e}} + \mu_{\text{p}}p_{\text{e}})$. For the near-intrinsic Ge which is appropriately used in such investigations, this implies a $\sigma_{\text{b}}/\sigma_{\text{ext}}$ value of about 3.5, very much of the same order as actually measured. The effect can be greatly intensified by the presence of hole traps, as Rieder *et al.* (1982) have shown.

The existence of an extraction region in the high-current regime was first demonstrated by Banbury (1953a) in the course of experiments on lead sulfide. Banbury did this not by measuring local resistivity, but by monitoring the characteristics of point contacts placed within the exhaustion region. The effect of extraction on such point contacts is the opposite of that produced by illumination (see Chapter 5).

We have been concerned here with the extraction of *resident* minority carriers by a reverse-biased contact. In addition, such a contact will, of course, extract any other (non-equilibrium) minority carriers which may be present, for whatever reason. This is indeed what we mean when we say that such a contact acts as a 'collector'. Illumination is one possible mechanism whereby the minority-carrier concentration can be augmented; injection by a second contact in the vicinity is another. There is also a 'passive' effect, which arises from the proximity of surfaces with high surface generation velocities; this is illustrated by Fig. 3.21. There is evidence that this effect has an appreciable influence on the detailed shape of the reverse characteristics; the proximity of high generation

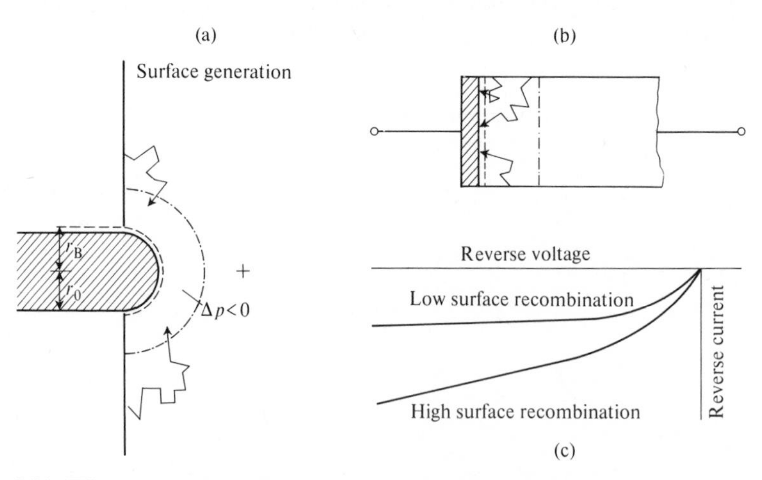

FIG. 3.21. Effect of a high surface generation rate on minority carrier extraction. Hole concentrations more nearly constant. (a) Point contact. (b) Planar contact. (c) Voltage–current relationships. Chain line marks boundary of extraction region.

surfaces 'softens' the reverse characteristic and prevents it from saturating properly. Of course, the precise influence depends on geometry.

Under dynamic contact operation, e.g. with an alternating applied voltage, minority carriers are injected during the forward half-cycle, and extracted ('collected') during the reverse half-cycle. Depending on the survival rate (carrier lifetime), we thus expected the reverse current to be higher when it follows a forward half-cycle, than it would be if a unidirectional sinusoidal voltage were applied. The rectification ratio is obviously diminished by this minority-carrier contribution.

The exhaustion of $p(\lambda_B)$ takes a certain amount of time, which means that a suddenly applied reverse voltage gives rise to a time-varying current transient. This extraction transient can be clearly observed even without augmentation (i.e. when $p(\lambda_B)=p_e$) but is, of course, augmented by an immediately preceding forward pulse. This situation is schematically summarized in Fig. 3.22. Results of this kind were first reported by Meacham and Michaels (1950). The duration of these transients depends on the carrier lifetime (here, actually, the generation time), and it follows at once that low-lifetime material is required for accurate pulse response (see also Florida *et al.* (1954) and Pell (1953)).

It will be clear from Fig. 3.22(a) that the current composition ratio γ (here, the 'extraction ratio') must necessarily vary with time throughout (say) a constant current pulse. The hole participation in the reverse current may be high at the onset of the injection process but, as more and more minority carriers are extracted, the electrical environment in which the contact functions changes drastically. Holes will then carry an ever diminishing fraction of the total current. This is qualitatively clear

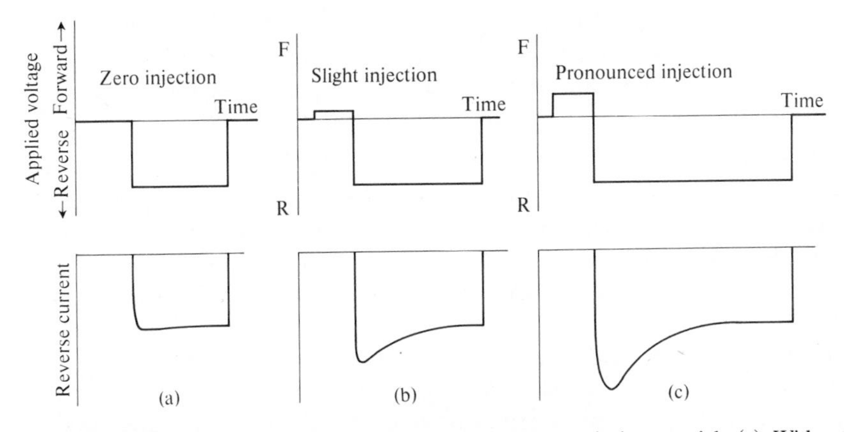

FIG. 3.22. Time dependence of the extraction current; extrinsic material. (a) Without preceding forward current. Extraction of resident minority carriers. (b)(c) With preceding forward current. Extraction of resident and injected minority carriers.

enough, but computer solutions of the time-dependent transport equations are not yet available.

3.5 Two-carrier contact relationships; lifetime semiconductors

In the sections which follow, attention is confined to lifetime semiconductors; corresponding treatments for relaxation semiconductors are not yet available.

All practical devices are finite, of course, which means that the presence of surfaces cannot be safely ignored. In an essentially one-dimensional system, it is comparatively simple to take them into account. One does this by considering the lifetime to be not τ_0 (the lifetime in infinite bulk) but τ_f, the *filament lifetime*, given (Shockley 1950) as

$$\frac{1}{\tau_f} = \frac{1}{\tau_0} + \nu_s \tag{3.5.1}$$

where ν_s is a term which depends on geometry as well as on the surface recombination velocity. The concept is based on the fact that an injected carrier can recombine *either* in the bulk material *or* at the surface, but not at both. Accordingly, the two recombination probabilities must be additive. In three-dimensional systems, the corresponding relationships are very complicated. Even for one dimension, the appropriate analysis has not yet been satisfactorily performed for contacts (with their additional boundary), but has long been available for junctions. In the literature, the use of τ_0 for τ_f is widespread and is a default measure, when better constants are unavailable.

3.5.1 *Analytic approximations; hemispherical, semi-infinite Systems*

In Chapter 2 it was shown how relatively simple situations involving only one type of carrier can be handled by analytic treatments, leading to approximate solutions. Two-carrier systems differ in two ways:

(a) The approximations that have to be made to 'secure' analytic solutions tend to be more complex, and their consequences less easy to predict. It is in any event difficult to select assumptions which are at once realistic and mathematically helpful.
(b) The simplifying separation of barrier and bulk effects is no longer feasible. This is because injected carriers survive over distances which (in the presence of a field) may be much greater than the diffusion length, and thus enormously greater than the barrier width. Analysis therefore proceeds often in terms of semi-infinite models, e.g. with the injecting barrier at $x = 0$, and infinite bulk to the right, but such a system does not permit the definition of a finite 'applied voltage' (V_B); and this is an

added complication. Fortunately, it is one that can be readily eliminated (bypassed would be a more accurate term) by confining attention to a particular situation, namely systems of hemispherical geometry.

The bulk material (*outside* the radial distances that give rise to the main spreading resistance) then contributes very little to the total voltage. Injected carriers affect only the spreading resistance itself, and can thereby have a significant influence on the total voltage–current relationship in the forward direction. For purposes of analysis, hemispherical analysis has the added advantage of permitting recombination to be neglected in many plausible cases. The essential spreading region is small, and injected carriers, drifting in the relatively high field of that region, may be deemed to pass through it without recombining. This is an enormous simplification, which is why attempts at solving the transport equations fare best with hemispherical contacts on semi-infinite bulk material. However, they are, even then, only partially successful.

Many attempts at formulating acceptable approximations have been made, notably by Bardeen and Brattain (1949), Bardeen (1950), Banbury (1953b), Simpson and Armstrong (1953), Swanson (1954), Gossick (1956, 1963), and Beneking (1958). We will outline here a later procedure employed by Braun and Henisch (1966a) which yields not only voltage–current relationships, but concentration contours and an estimate of the injection ratio.

Figure 3.23 shows the nomenclature employed. The total current will be denoted by $i(=i_n+i_p)$. The diffusion equations (3.2.4) and (3.2.5) for electrons and holes must be rewritten for spherical geometry, and adapted to handle the currents i_n and i_p, rather than current densities. With these modifications, we have

$$\frac{i_n}{2\pi r^2} = -en\mu_n\left(\frac{\mathrm{d}V}{\mathrm{d}r}\right) + eD_n\left(\frac{\mathrm{d}n}{\mathrm{d}r}\right) \tag{3.5.2}$$

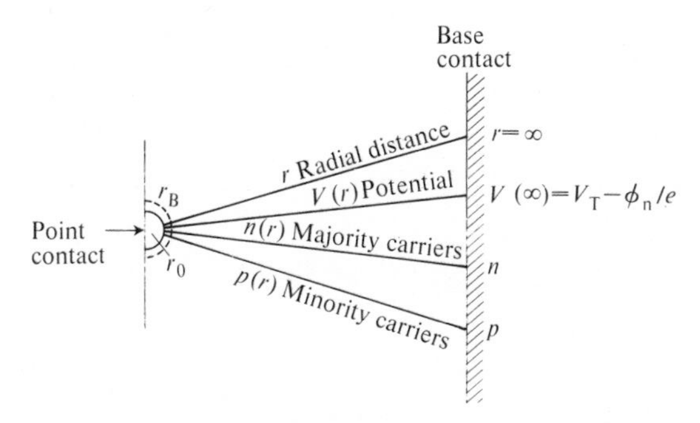

FIG. 3.23. Geometrical relationships at hemispherical contact, and definition of symbols.

and

$$\frac{i_p}{2\pi r^2} = -ep\mu_p\left(\frac{dV}{dr}\right) - eD_p\left(\frac{dp}{dr}\right) \tag{3.5.3}$$

where n, p, and V are functions of r.

As long as recombination is neglected, the two expressions can be treated independently. Each can be shown to yield an implicit expression for the carrier density, namely

$$n(r) = N_c \exp\left\{\frac{-E(r)}{kT}\right\} - \frac{i_n W_n(V_T, r)}{2\pi e D_n} \tag{3.5.4}$$

and

$$p(r) = P_c \exp\left\{\frac{E(r)}{kT}\right\} + \frac{i_p W_p(V_T, r)}{2\pi e D_p}. \tag{3.5.5}$$

Here, V_T is as usual the total voltage across the system (here clearly an asymptotic limit) and $E(r)$ is the conduction band contour. N_c and P_c are integration constants which remain to be fixed. W_n and W_p are 'field integrals', defined by

$$W_n(V_T, r) = \int_{\xi=r}^{\xi=\infty} \frac{\exp\left\{\frac{E(\xi)}{kT} - \frac{E(r)}{kT}\right\}}{\xi^2} d\xi \tag{3.5.6}$$

and

$$W_p(V_T, r) = \int_{\xi=r}^{\xi=\infty} \frac{\exp\left\{\frac{E(r)}{kT} - \frac{E(\xi)}{kT}\right\}}{\xi^2} d\xi. \tag{3.5.7}$$

This represents a development of eqn (2.2.17) for the case in which the integration distances involved are *not* small compared with the barrier width (see also Spenke (1955)). For very small applied voltages, W_n and W_p integrate approximately as $1/r$, which would make eqns (3.5.4) and (3.5.5) identical with corresponding expressions derived by Banbury and Bardeen, *provided* the injection ratio is taken as unity. No such assumption is made here.

With the boundary condition $V(\infty) = V_T - \phi_n/e$, eqns (3.5.4) and (3.5.5) become

$$n(r) = n_e \exp\left\{-\frac{1}{kT}(eV_T + E(r) - \phi_n)\right\} - \frac{i_n W_n}{2\pi e D_n} \tag{3.5.8}$$

and

$$p(r) = p_e \exp\left\{\frac{1}{kT}(eV_T + E(r) - \phi_n)\right\} + \frac{i_p W_p}{2\pi e D_p}. \tag{3.5.9}$$

The terms involving the integrals can now be seen as current-dependent corrections to the Boltzmann quasi-equilibrium relationship which is represented by the first terms. The last two expressions are not, of course, truly independent. They are linked via W_n and W_p, which both depend on $V(r)$. Since equations (3.5.8) and (3.5.9) are true for *any* value of r, we can assume a particular value, namely $r = r_0$, the contact radius, for which $n(r)$ and $p(r)$ are known. More accurately, $n(r_0) = n_s$ and $p(r_0) = p_s$ are assumed to be equilibrium values, which is equivalent to assuming that the metal–semiconductor interface has an infinite surface recombination (generation) velocity. When the sign conventions of Section 2.2.2 are adopted, $I = -i$, (i.e. with I_n corresponding to J_n, not to j_n) this procedure yields:

$$I_n = \frac{2\pi e D_n n_s}{W_n(V_T, r_0)}\left\{1 - \exp\left(\frac{-eV_T}{kT}\right)\right\} \tag{3.5.10}$$

and

$$I_p = \frac{2\pi D_p p_s}{W_p(V_T, r_0)}\left\{\exp\left(\frac{eV_T}{kT}\right) - 1\right\} \tag{3.5.11}$$

with $n_s p_s = n_i^2$ and $n_s = n_e \exp(-eV_D/kT)$, meaning Boltzmann equilibrium at the boundary. Since W_n and W_p depend on V_T, the terms involving them should not be regarded as saturation currents, all (tempting) appearances notwithstanding. Indeed, those appearances suggest that electron and hole currents rectify in opposite directions, but when the magnitudes of W_n and W_p are taken into account as a function of V_T, this is not the case.

The remaining task is to calculate W_n and W_p themselves, and this must be done by a process of successive approximation, equivalent to the use of numerical methods. At one time this need was regarded as a serious stumbling block, but it is no longer that. Indeed, we should now be inclined to resort to computational techniques at an earlier stage, without dealing with W_n and W_p explicitly. However, the first stage of this process can be pursued analytically, and yields a useful approximation by itself. The problem is, of course, that the energy profile $E(r)$ of the conduction band is not known. The procedure is therefore to assume a profile that satisfies some reasonable expectations, and to use the resulting values of W_n and W_p for calculating I_n and I_p in accordance with eqns (3.5.10) and (3.5.11). These currents, in turn, give the carrier concentrations as a function of r via eqns (3.5.8) and (3.5.9), and thus the current-dependent space-charge density. When used in Poisson's equation, the latter yields a more accurate potential contour which could, in principle, be used as the basis of the next step in the process, and so on. For a rough assessment of W_n and W_p, it is convenient to divide the space into two ranges: $r<r_B$ and $r>r_B$. In the former, the potential varies rapidly, but the field may be taken as constant as a first approximation. In the latter the potential varies slowly, which means that the equations can be linearized. The crucial 'matching point' r_B is, of course, initially unavailable, as is $E(r_B)$. For convenience a new *variable* α is introduced, such that

$$E(r_B) = -\alpha e V_T + \phi_n \tag{3.5.12}$$

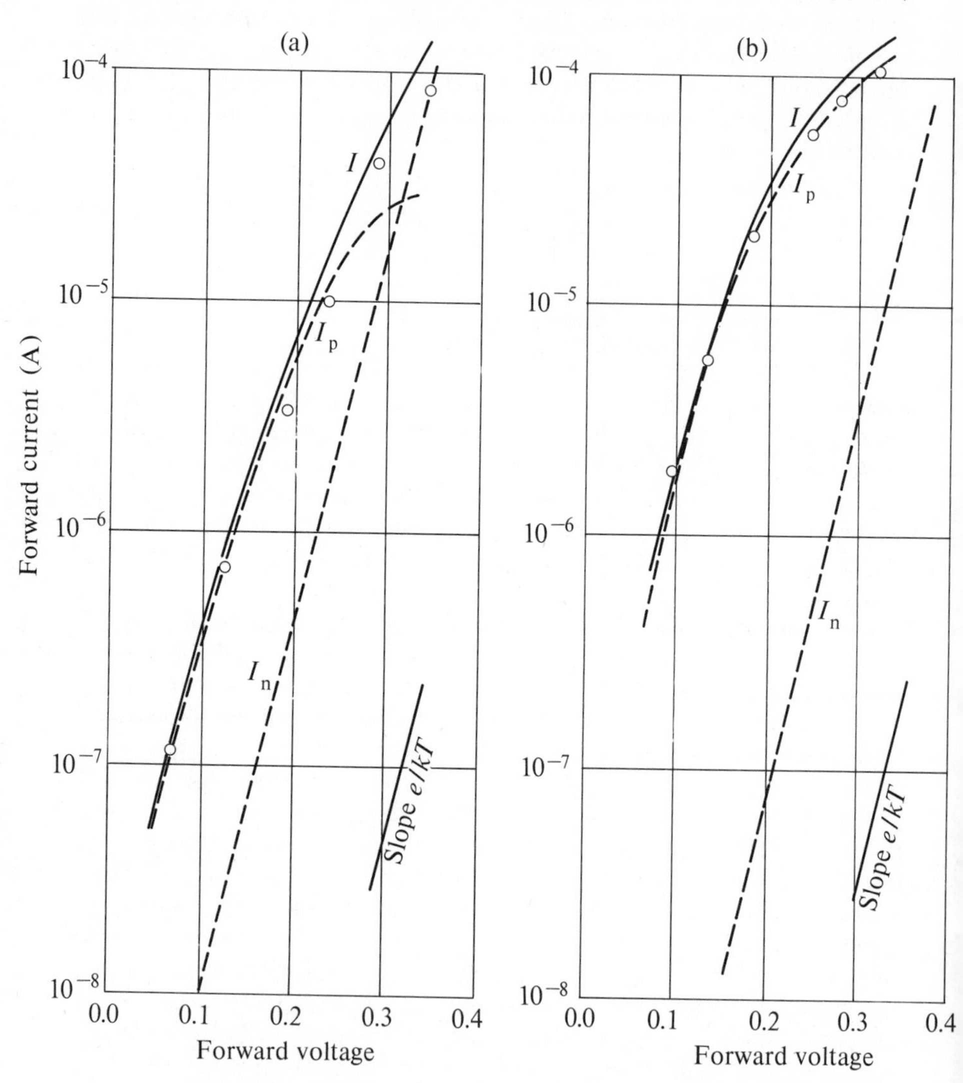

FIG. 3.24. Forward characteristics of point contacts on n-type germanium; comparison of theory (full lines) with Banbury's measurements. (a) $V_D = 0.4$ V, $n_e = 3\times10^{15}$ cm^{-3}, $p_e = 1.90\times10^{11}$ cm^{-3}. (b) $V_D = 0.4$ V, $n_e = 5\times10^{14}$ cm^{-3}, $p_e = 1.14\times10^{12}$ cm^{-3}. $r_0 = 3\times10^{-4}$ cm; room temperature. Results after Braun and Henisch (1966a).

V_T being negative in the forward direction. (It will be clear that $\alpha=1$ reduces everything to the conventional picture of a Schottky barrier, for which $V_B=V_T$.) I_n and I_p can then be calculated in terms of α, r_B, and V_T. For the general case, the results are too complex to be useful, but if barriers are high and applied voltages reasonably small, one obtains relatively simple expressions.

Though α is a variable in eqn (3.5.12) it may, for non-critical and illustrative purposes, be regarded as a constant between 0 and 1, and usually much closer to unity. It governs the ratio in which the applied voltage V_T is divided as between barrier and bulk. For increasing currents in the forward direction, the barrier resistance decreases (implying a smaller share for V_B), but the spreading resistance decreases also, because of injection (implying a large share for V_B). Since the two tendencies oppose each other, the value of α remains more nearly constant. It can be determined (along with r_B) by reference to the boundary conditions at the matching point, where energy and its slope must be continuous. In this way, Braun and Henisch (1966a) arrived at voltage–current relationships which were in satisfactory agreement with experiments (on Ge) over about four orders of (current) magnitude (Fig. 3.24). No curve-fitting was involved (see also Noble *et al.* (1967)).

One of the points of interest brought out by the use of this method was that (with $I=I_n+I_p$) the $\log|I|$ versus V_T relationship in the forward direction exhibited slopes which differed significantly from e/kT (see Section 2.2.5). Along similar but more approximate lines, Banbury (1953b) first predicted that the non-ideality factor η should be unity for low currents and 2 for high currents in a hemispherical system. Another result, derived from the above equations, shows that within limits the rectification ratio is improved by increased doping. This may be seen by reference to Figs. 3.24 and 3.25 (image-force and tunnel effects neglected).

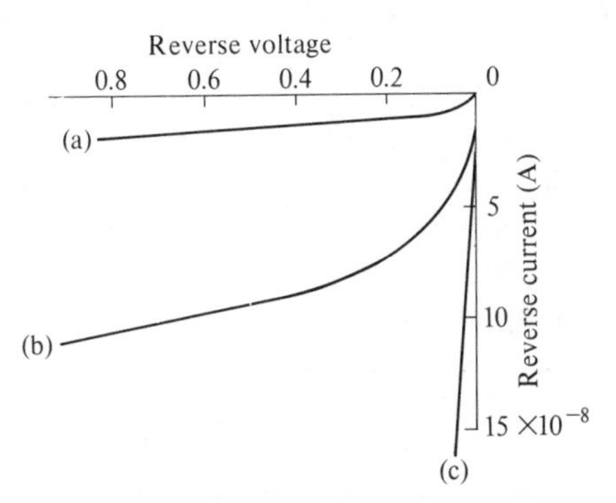

FIG. 3.25. Calculated reverse characteristics of point contacts on n-type germanium. (a) $V_D=0.4$ V, $n_e=3\times10^{15}$ cm^{-3}, $p_e=1.90\times10^{11}$ cm^{-3}. (b) $V_D=0.4$ V, $n_e=5\times10^{14}$ cm^{-3}, $p_e=1.14\times10^{12}$ cm^{-3}. (c) $V_D=0.2$ V, $n_e=3\times10^{15}$ cm^{-3}, $p_e=1.90\times10^{11}$ cm^{-3}. $r_0=3\times10^{-4}$ cm; room temperature. After Braun and Henisch (1966a).

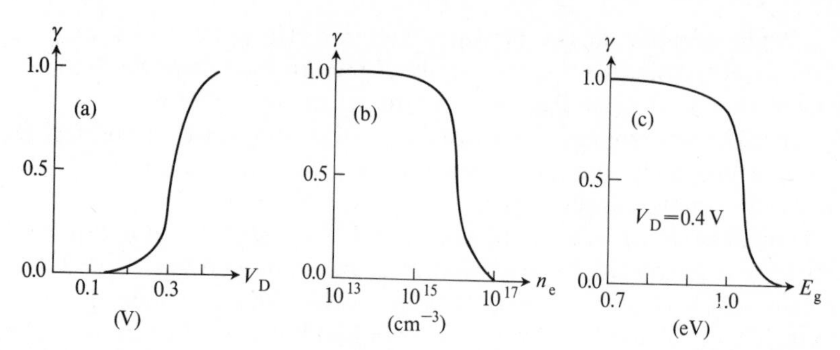

FIG. 3.26. Calculated injection ratios (at zero current) as a function of barrier height, doping level, and band gap. (a) Barrier height V_D: for $E_g=0.7$ eV, $n_e=3\times10^{15}$ cm^{-3}, $p_e=1.90\times10^{11}$ cm^{-3}. (b) Doping level n_e: for $E_g=0.7$ eV, $V_D=0.4$ V, $p_e=1.90\times10^{11}$ cm^{-3}. (c) band gap E_g: for $n_e=2.4\times10^{13}$ cm^{-3} or $\phi_n=0.35$ eV. $r_0=3\times10^{-4}$ cm; room temperature. After Braun and Henisch (1966a).

Lastly, the equations yield a current composition ratio as a function of r and V_T. For $V_T=0$ and $r=r_0$, eqns (3.5.10) and (3.5.11) ultimately (and after a good deal of manipulation) lead to

$$\gamma=1\Big/\left\{1+\frac{D_n n_s}{D_p p_n}\frac{e}{kT}V_D\left(\frac{r_0}{r_{B_0}-r_0}\right)\exp\left(-\frac{eV_D}{kT}\right)\right\} \tag{3.5.13}$$

which may be compared with the previously derived results or linear geometry, by putting $V_B=0$ in eqn (3.1.29). The calculated injection ratios were found to be in excellent agreement with experimental values reported by Banbury and Houghton (1954). Figure 3.26 shows how the calculated γ values depend on various parameters. The agreement between Figs. 3.26(a) and 3.8 is particularly satisfactory.

3.5.2 *Metal–semiconductor–metal planar systems*

The discussions of back-to-back barrier systems in Section 2.2.9 envisaged only one type of carrier, namely electrons, this situation applying if the band gap were sufficiently wide. If, on the other hand, the gap were small, then minority carriers would have to come into play at some stage as V_T increases; this is schematically shown in Fig. 3.27. The current would depend almost exponentially on V_2, but its dependence on V_T would be more complicated. Sze and co-workers (1971) have given estimates on the assumption that there is no perturbation of the relationship between V_1 and V_2 due to the advent of holes from the right, and that the hole and electron currents can therefore be treated as independent of one another. Indeed, it is only on that assumption that the transport equations can be analytically solved; but, for many practical situations, this condition is excessively restrictive. One must expect hole participation, to increase the *electron* current through the reverse-biased

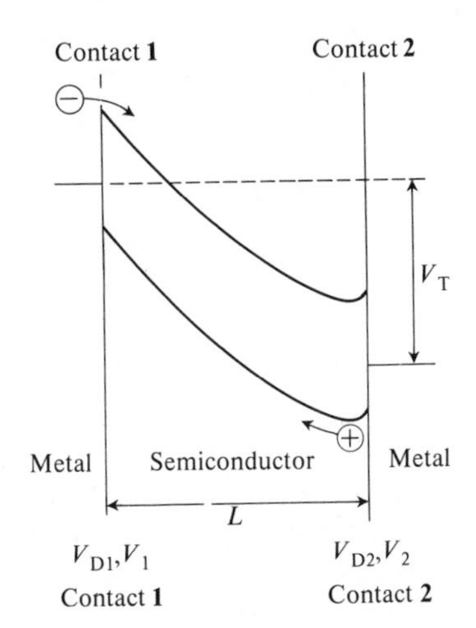

FIG. 3.27. Minority-carrier injection into a two-barrier system.

contact, via the collector amplification effect familiar in transistor physics. The onset of hole current is bound to be very rapid, once $V_T \geqslant V_{RT}$. As V_T increases further, the barrier height 'seen' by holes at contact **2** decreases, and the thickness of the field-free diffusion region soon becomes very small, which means that the entering holes are rapidly removed (towards the left in Fig. 3.27) by the field. When this stage is reached, the hole current is expected to dominate completely over the electron current, since the latter is limited by the reverse-biased contact. The consequences are therefore likely to be very much like those of avalanche breakdown (Section 6.6.2). Indeed, the two processes might easily be confused with one another. Whether the onset of hole current or breakdown will occur first must depend on the band gap. Once again a complete assessment of the system would have to be carried out by computer methods and is not yet available for contact systems. It is available for p^+–n–n^+ and p^+–p–n^+ junction structures in GaAs (Manifacier *et al.* 1980; Manifacier and Henisch 1981), and these treatments give field contours for different specimen widths L. The results show that even when the junctions are of 'low resistance', the behaviour of the system as a whole is dominated by injection effects. These often simulate ohmic behavior; compare Sections 3.3. and 3.4. The behaviour of systems with contacts (as distinct from junctions) is not likely to be very different, but remains to be computed. Meanwhile, we have some approximate analyses, e.g. by Harutunian and Buniatian (1977), who calculated the system impedance as a function of trap concentration and frequency. A different pattern of assumptions was used by Baccarani *et al.* (1974), who took thermionic emission to be the limiting factor for the hole as well as the electron currents, and had to assume complete depletion (barrier overlap) as well as zero recombination in order to make explicit solutions feasible.

3.5.3 *Contacts on intrinsic material*

There is just one case in which the two-carrier problem can be analytically solved with relative ease, namely that of a barrier on intrinsic material. Figure 3.28 shows the situation. For zero current, the accessible solution is exact (Section 3.1.3), under other plausible conditions, it represents a very good degree of approximation (Manifacier *et al.* 1983).

It is convenient to consider the problem in two parts: one dealing with the region between $x=0$ and $x=x_j$, and one dealing with the semi-infinite region beyond. x_j marks the energy minimum (potential maximum) in the forward direction, and thus also the end of the 'principal space-charge region'. λ_B, the width of the orthodox Schottky barrier, has no significant meaning in this context, since free-carrier space charges are important here. Of course, there has to be current continuity for electron and holes at $x=x_j$. If it is clear that the hole concentration at the edge of the barrier is increased by injection in the presence of a forward voltage; the question is by how much. If Boltzmann statistics were to prevail for holes over the entire barrier region, then the concentration at the edge of the barrier would be given by

$$p(x_j) = n_i \exp(-eV_a/kT) \tag{3.5.14}$$

where V_a is the applied voltage, as defined in Section 2.4.1. (Note that V_B, used elsewhere in this book, cannot serve here, because it assumes negligible voltage drop across the bulk material.)

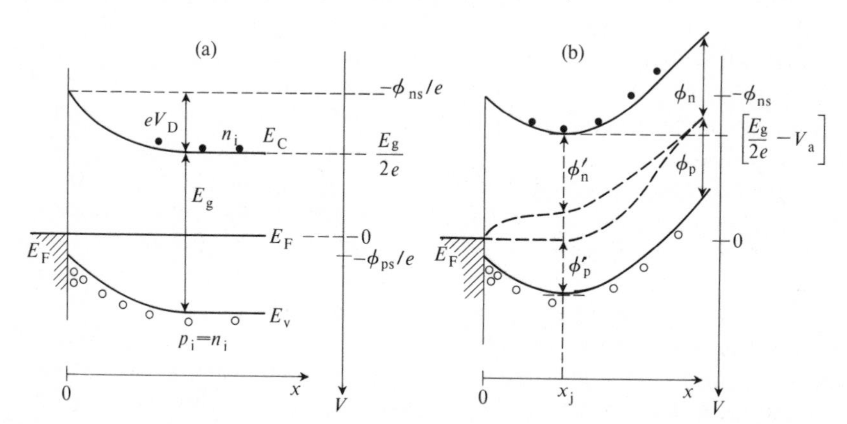

FIG. 3.28. Contact on intrinsic material; energy profiles: (a) for $j=0$, (b) for forward current. Applied voltage V_a negative in the forward direction. Surface equilibrium assumed. ϕ'_p and ϕ'_n denote quasi-Fermi levels. Compare Fig. 3.30. After Manifacier *et al.* (1983).

In practice, the equilibrium envisaged here can never be fully maintained, but in the presence of a prominent inversion layer, which constitutes an abundant hole reservoir, it is *almost* maintained for all reasonable currents. This is an intuitive expectation which computed results amply confirm.

In a lifetime semiconductor (e.g. like Ge at room temperature) the electron concentration at $x = x_j$ would adjust itself in response to the hole injection to conform to charge neutrality, as already envisaged by eqn (3.1.27). We therefore have in this case $n(x_j) = n_i + \Delta n(x_j)$ and

$$\Delta n(x_j) = \Delta p(x_j) = n_i \exp(-eV_a/kT). \tag{3.5.15}$$

In that sense, the method used here is similar to that used in Section 3.1.4, for the calculation of injection ratios. Charge neutrality is not, of course, an absolute condition, but is a very satisfactory approximation here, as computations can also show. The last equation is in fact the conventional boundary condition for a p–n junction. Unlike the hole concentration, the electron concentration is *not* 'tied' to its value at $x = 0$, because $n(0)$ is very small, whereas $p(0)$ is very large (though assumed here to be still non-degenerate). $n(0)$ shall be taken as current-independent, maintained constant by the electron dynamics in the metal, though correction terms could easily be introduced, e.g. along the lines suggested by Crowell and Sze (1966b).

Some reference must now be made to recombination. In an uncompensated intrinsic semiconductor, the lifetime can indeed be expected to be high, but it can be shown that even for bulk semiconductor lifetimes as low as 10^{-7} s, the amount of recombination which takes place within the barrier is quite negligible. On these grounds, recombination will in fact be neglected here. As a consequence, the electron and hole currents can be treated separately within the barrier, as indeed they were in Section 3.1.4.

The new electron concentration at $x = x_j$ can now be used as the boundary condition for an analytic solution, along the lines already described in Section 2.2.4. However, allowance must be made for the fact that the barrier contour differs from that of a conventional Schottky barrier, because an inversion layer is now present. This affects the evaluation of the integral in eqn (2.2.14), but is easily done, because the main contributions to the integral come from very small values of x. In this way, the electron current can be shown to be

$$j = \frac{eD_n\mathcal{N}_c}{\lambda_D} \sinh\left(\frac{eV_D}{kT}\right)\exp\left(-\frac{\phi_{ns}}{kT}\right)\left\{\exp\left(\frac{-2eV_a}{kT}\right) - 1\right\}. \tag{3.5.16}$$

The corresponding hole current is actually zero within the barrier under the conditions assumed (Boltzmann equilibrium). Within the barrier, the holes thus make their contribution to the proceedings by virtue of their space charge, not by virtue of their transport. This is equivalent to taking the quasi-Fermi level for holes as horizontal. Outside the barrier, of course, ϕ_p' changes gradually to ϕ_p, signifying hole current and, by implication, an electron current which differs from that given by eqn (3.5.16). However, at $x = x_j$, eqn (3.5.16) represents not only the electron current, but the *total* current.

The most notable feature of this result is the last term. With $J = -j$, eqn (3.5.16) could be written as

$$J = J_s\{1 - \exp(-eV_a/\eta kT)\} = J_s\{1 - \exp(-2eV_a/kT)\} \quad (3.5.17)$$

(compare eqn 2.3.31), giving a non-ideality factor $\eta = 0.5$. Though eqn (3.5.17) does not directly reflect this, the arguments on which it is based apply only in the forward direction.

Moreau *et al.* (1984) have shown that these predictions agree well with computed results obtained *without* the above approximations. It should, however, be remembered that V_a is not an operationally accessible quantity. The only voltage we ever measure is V_T, and that can differ from V_a quite substantially. Since we are here dealing with intrinsic

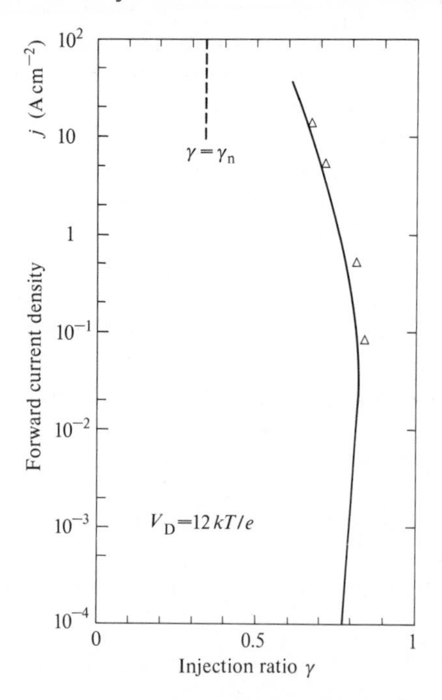

FIG. 3.29. The injection ratio for a contact on intrinsic germanium, as a function of forward current density. Continuous line: $L = 10^4\,\lambda_{Di}$. Points marked: $L = 50\,\lambda_{Di}$, where λ_{Di} is the Debye length in intrinsic material. After Manifacier *et al.* (1984).

(rather than doped) material, the bulk resistance of the system can never be neglected. This makes it virtually impossible to verify such a relationship with any degree of precision. Meanwhile, eqn (3.5.17) provides just one more example of the fact that non-ideality factors should be handled with sceptical care (if at all) (see also Henisch *et al.* (1982)).

The computed results also yield the current composition ratio γ at $x = x_j$, shown in Fig. 3.29 as a function of current density; it faithfully exhibits the initial increase mentioned in Section 3.1.4.

3.5.4 *Contacts on extrinsic material with an inversion layer*

It will be clear that the ultimate solutions must again be computer solutions. Nevertheless, some prediction can be made on the basis of the above considerations. In the first instance, the same analytic approach could be used but the necessary integration would no longer be simple Indeed, that part would in any event have to be done numerically.

For extrinsic material, the Boltzmann relationship would be

$$\Delta p(x_j) = p_e\{\exp(-eV_a/kT) - 1\} \tag{3.5.18}$$

(V_a being negative in the forward direction) and one can show (by comparison with computation) that actual Δp values differ only a little from this estimate, even when the lifetime is quite short (e.g. 0.1 μs). However, the fact is that even a small departure from Boltzmann supports a substantial injection current. In other words, as long as attention is focussed on carrier concentrations within the barrier and at the edge of it, the use of the Boltzmann approximation is perfectly in order; when attention is focussed on current, it is not. Treatments found in the literature which ignore this fact accordingly remain undiscussed here. Figure 3.30 gives a schematic representation of the quasi-Fermi level contours. They are the same as those in Fig. 3.28(b) except for the fact that now $\phi_n \neq \phi_p$, and in this case surface equilibrium is not assumed.

In as much as the hole concentration differs in fact from the Boltzmann value in the presence of hole current, it must always be *smaller* than that suggested by eqn (3.5.18), which means that Δn must be smaller also.

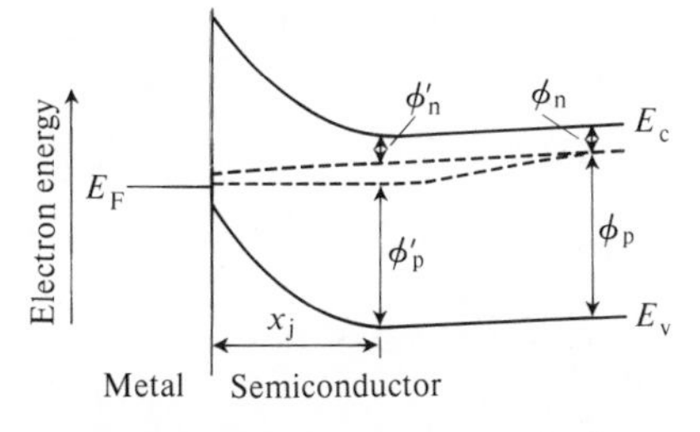

FIG. 3.30. Schematic representation of quasi-Fermi level contours under injection conditions. Electronic equilibrium at the surface not assumed. Compare Fig. 3.28.

This, in turn, will reduce the electron current, since that is based mainly on the electron concentration gradient within the barrier. Thus, an additional current due to injected holes tends to be balanced by a diminished electron current, and computations have shown that the total current for a given V_a is almost the same, with and without injection. So, also is the shape of the voltage–current characteristic in approximate terms; certainly, there is no particular feature of that relationship that would enable us to distinguish readily between contacts of the injecting and non-injecting kind.

Figure 3.31 gives a typical set of computed voltage–current characteristics, based on the assumption that equilibrium prevails at the metal–semiconductor interface. Of course, this amounts to an approximation.

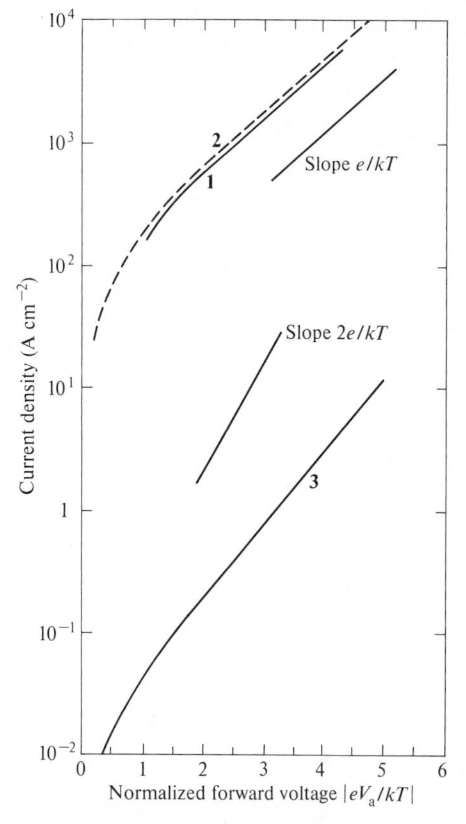

FIG. 3.31. Voltage–current relationships for a contact on extrinsic material with an inversion layer. Curve **1**: computed for highly extrinsic Ge, with $V_D = 12kT/e$, $n_e = 2.4 \times 10^{17}$ cm^{-3}, $\lambda_D = 9.6 \times 10^{-7}$ cm, $\tau_0 = 2.55$ μ. Injection terms included. Curve **2**: calculated on the basis of the classic Schottky model, neglecting injection (eqn (2.2.25), same parameters). Curve **3**: computed for slightly extrinsic Ge, with $V_d = 12kT/e$, $n_e = 1.7 \times 10^{14}$ cm^{-3}, $p_e = 3.1 \times 10^{12}$ cm^{-3}, $\lambda_D = 3.6 \times 10^{-5}$ cm, $\tau_0 = 2.55$ μ. After Manifacier *et al.* (1983).

Just because the electronic disequilibrium, which in fact prevails, cannot extend far into the metal, we are, at the interface, in a region of steep gradients, which makes the judicious guessing of boundary concentrations particularly hazardous. Gossick (1963) was the first to point this out.

For highly extrinsic material, curves **1** and **2** in Fig. 3.31 represents computed results (full line) which take injection into account, and also results calculated from the classic (on-carrier) Schottky model (eqn 2.2.25) which do not (broken line). The agreement is remarkably close, as predicted above. The logarithmic forward slope is now much closer to e/kT than to $2e/kT$ which applies to the intrinsic case. Though injection is present here, its effects (in terms of voltage–current relationships involving V_a rather than V_T) are negligible.

For a material that is only slightly extrinsic, the results are very different. Curve **3** (Fig. 3.31) shows computed characteristics with a slope considerably greater than e/kT, but not yet $2e/kT$, and this intermediate situation is again what we expect to find (see also Demoulin and Van der Wiele (1974)).

Green and Shewchun (1973) avoided the assumption of boundary equilibrium, and assumed instead that electrons and holes each have a fixed emission velocity (eqn (2.2.39a)) at the boundary. Via eqn (2.3.2) and an equivalent expression for holes, this implied definite non-equilibrium concentrations $n(0)$ and $p(0)$, given in terms of boundary velocities, and thus represented a substantial break with the conventional drift-diffusion treatment of barriers. Even then, problems relating to the surface state profile and its variable occupancy had to be ignored.

The equations were solved numerically by Newton–Raphson iteration. Amongst other things, the solutions confirm the validity of concentration contours as shown in Fig. 3.2 in the presence of forward current. They also reflect the existence of a maximum value in the current-dependence of the injection ratio, as in Fig. 3.29. For the decrease of γ, following the initial increase, there are actually two distinct reasons:

(a) the increasingly hole-rich environment into which injection takes place, and
(b) the limited hole generation rate at the metal–semiconductor interface.

Green and Shewchun dealt with finite systems of 5–100 μm thickness, with equilibrium conditions 'enforced' by the second electrode, which was assumed to be of zero resistance. The behaviour of γ under such conditions was documented, but voltage–current characteristics as such were not presented.

3.5.5 Metal–insulator–semiconductor systems

Single-carrier systems of this kind have already received brief consideration in Sections 1.3.4 and 2.4.3, Two-carrier systems present more problems because electrons and holes can cross the 'insulating' barrier by different transport mechanisms. In principle, both types of carrier can be thermionically emitted, and both types can tunnel, but one process or another generally predominates, often by a huge margin.

Though M–I–S structures are complicated, some properties are at once predictable. For instance, the ratio of electron emission to electron tunnelling (and similarly, the ratio of hole emission to hole tunnelling) must necessarily vary with the thickness λ_a of the insulating layer (typically an oxide). This is one of the factors which ultimately make the injection ratio γ dependent on λ_a. Indeed, it will be shown below that there generally *is* a barrier thickness λ_a for which the injection ratio is a maximum. Moreover, the tunnelling probability of charge carriers depends not only on the oxide thickness, but also on the energetic position of the tunnelling charge (see eqn 2.1.35). It is therefore different for electrons and holes, and is in any event a function of the applied bias voltage. (When 'the injection ratio' is under discussion without further specification, the term is always taken to refer to the current composition ratio at the edge of the barrier, determined at zero current.) An externally applied voltage V_T has a component V_a (well defined only in the forward direction) across the barrier system, and that component is in turn divided between the oxide layer and the Schottky layer proper. That division depends, of course, on the effective resistances of those two layers. Each has its own voltage–current relationship, but the series combination calls for *current continuity* at the insulator–semiconductor interface, and this restraint controls the *carrier concentrations* at that point. Following a treatment by Card and Rhoderick (1973), it is convenient to distinguish the situations arising for very thin and for moderately thick films.

Very thin oxides. The transmission probability is high, which means that holes in the inversion layer are in excellent 'communication' with the carrier population in the metal. Accordingly, they tend to maintain something like their equilibrium concentrations. High injection rates are then possible, because they are not limited by the rate of thermal hole generation at the interface.

Equation (3.1.25) gives the hole current on the assumption of thermal equilibrium at the boundary. If this equilibrium did not prevail, the quasi-Fermi level for holes would be displaced by $\Delta\phi_p$, and eqn (3.1.25) would, at the edge of the barrier, become

$$j_p = p_e e(D_p/L_p)\{\exp(-eV_a/kT - \Delta\phi_p) - 1\}\exp(-\lambda_a/L_p) \quad (3.5.19)$$

where V_a is again used in place of V_B. (Note that usually $L_p \gg \lambda_a$.) The evaluation of V_a from V_T is in general a more complex matter and

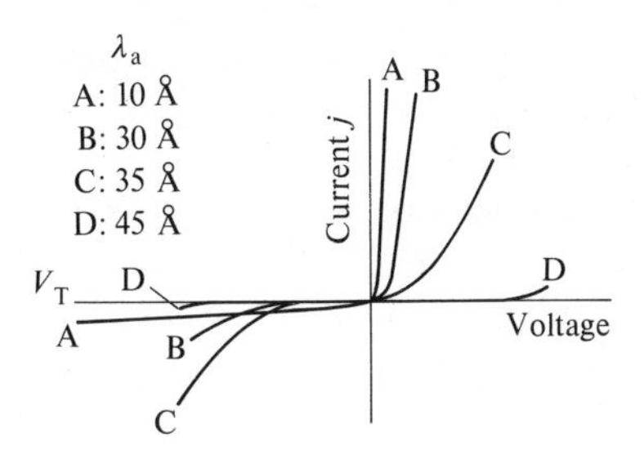

FIG. 3.32. Characteristics of metal-oxide-semiconductor diodes; schematic representation of findings by Card and Rhoderick (1971b) for gold contacts on single-crystal n-Si.

demands, of course, an explicit calculation of the tunnelling current, in terms of the voltage drop across the insulating barrier. Here, in the limit, one might consider that voltage drop to be negligible, and $V_a \approx V_T$. As the insulator thickness λ_a decreases, $\Delta\phi_p$ must decrease likewise. In the limit, it becomes zero, but $\Delta\phi_p$ tends to be negligible long before that limit is reached. Indeed, Card and Rhoderick (1971b) show that $\lambda_a <$ 20 Å yields the orthodox behaviour of a Schottky diode, because the 'insulating' barrier is not then the current-limiting feature. In contrast, drastic modifications set in for greater thicknesses, as shown in Fig. 3.32 (see also Section 5.2.3, and specifically Fig. 5.20).

Moderately thick oxides (meaning barriers which are not negligibly thin, but are still thin enough to be penetrated by tunnelling, as distinct from those described in Section 2.4.3.). When $\Delta\phi_p$ is not zero, its presence in eqn (3.5.19) obviously modifies the log j_p versus V_a relationship. $\Delta\phi_p$ is evidently governed by the tunnel current, and Card and Rhoderick (1973) used earlier results by Stratton (1962a) to establish the value of that current independently. In terms of the present conventions, with $E = 0$ at the (fixed) Fermi level of the metal, the starting point is

$$j_p = (4\pi m_x e/h^2) \int_{-\infty}^{+\infty} \{\mathcal{F}_m - \mathcal{F}_{sc}\} \int_{-\infty}^{-\phi_{ps}} T_E(E_x)\, dE_x\, dE \qquad (3.5.20)$$

where $\mathcal{F}_m$ and $\mathcal{F}_{sc}$ are the appropriate electron distribution functions for the metal and the semiconductor respectively, and E_x is the component of energy due to momentum in the x-direction (see also Harrison (1961)). Equation (3.5.20) represents a judicious simplification, in as much as it assumes spherical energy surfaces and equal effective masses on both sides. $\mathcal{F}_m$ is determined by the Fermi level in the metal, and is (here) fixed, of course. If no voltage were applied, $\mathcal{F}_{sc}$ would be the same, but in non-equilibrium it is governed (for present purposes) by the position of the quasi-Fermi level for holes at the semiconductor–insulator interface. Card and Rhoderick assumed that Boltzmann equilibrium prevails for holes, which makes their quasi-Fermi level horizontal. It was shown in Section 3.5.4 that this is actually a very good approximation even for metal–semiconductor contacts, and it is bound to be better still for contacts between semiconductors and insulators.

For the transmission coefficient $T_E(E_x)$, Card and Rhoderick used an approximation very similar to eqn (2.1.34) which may be espressed as

$$T_E(E_x) = \exp(-1.01\phi_h^{1/2}\lambda_a) \approx \exp(-\phi_h^{1/2}\lambda_a) \quad (3.5.21)$$

where ϕ_h is in electron volts and λ_a in Angstrom units (Card and Rhoderick 1971a). ϕ_h is the *average* height of the tunnelling barrier 'seen' by holes coming from the Fermi level of the metal (Fig. 3.33), and this is, of course, bias-dependent.

With these provisions it is possible to substitute for $\mathscr{F}_m$, $\mathscr{F}_{sc}$, and T_E in eqn. (3.5.20) and, though the resulting integral has to be evaluated numerically, it is possible to make a qualitative assessment of its behaviour by inspection, as Card and Rhoderick (1973) have done. As long as interface state recombination is negligible (which it may or may not be in any practical case), the values of $p(0)$, and hence $\Delta\phi_p$, can be established from a knowledge of j_p, since the currents given by eqns (3.5.19) and (3.5.20) must be equal.

For comparison with observations, we need the total current and therefore an assessment of j_n as well as j_p, either on the basis of electron tunnelling or thermionic emission or both. Card and Rhoderick assume that tunnelling alone is active (i.e. zero electron current *over* the barrier by electron emission or field-diffusion transport). For sufficiently high forward voltages ($V_a > 3kT/e$), this permits the use of

$$j_n = \left\{\frac{4\pi m_x e(kT)^2}{h^3}\right\}\exp\{-\phi_e^{1/2}\lambda_a\}\exp(-\phi_{ns}/kT)\exp(-eV_a/kT) \quad (3.5.22)$$

where ϕ_e is the *average* height of the barrier for electrons, coming from the conduction band, again expressed in electron volts (Fig. 3.33).

It will be seen from eqns (3.5.21) and (3.5.22) that the tunnelling probabilities for holes and electrons are different functions of λ_a. This is a basic asymmetry, and is additional to that arising from the different carrier supply situations discussed above (see Card 1977). Unfortunately, we have little reliable information about ϕ_e and ϕ_h and, if anything, even less about m_x for electrons and holes. The simplest (but not necessarily the most reliable) way out is to take $\phi_e = \phi_p$ and $m_x = m$, and this is often done (see, however, Duke 1969).

Card and Rhoderick (1973) obtained good agreement between injection ratios calculated on the above arguments and those observed on various silicon contacts. Card and Rhoderick have also shown that γ

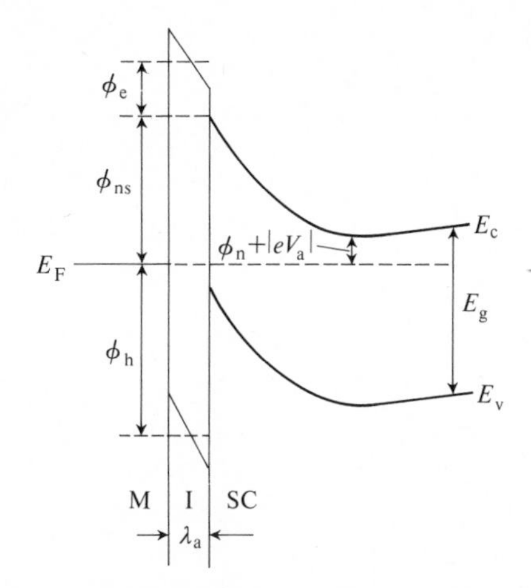

FIG. 3.33. Barrier contours in the presence of an oxide layer, with forward current flowing. Definitions of ϕ_e and ϕ_h.

increases rapidly with V_a up to the flat band condition and less rapidly after that. This is connected with a redistribution of the applied voltage. When that voltage is small, a substantial proportion of it appears across the space-charge layer in the semiconductor. However, when the applied voltage is large enough for the flat band condition to be exceeded, the semiconductor region immediately adjoining the insulator will be an accumulation region, and thus highly conductive. Nearly all the voltage drop will be across the insulator after that. Most significantly, γ was found to increase with λ_a at first, and then to reach a maximum. Indeed, the fact that oxide layers can improve injection has suggested their use in luminescent diodes which depend on radiative recombination of the injected carriers (see Buxo *et al.* 1976).

A physical explanation for the initial injection rate increase is readily available. When λ_a is very small, j_n is limited primarily by the existence of the barrier, not by the supply of free electrons from the relatively well-conducting bulk. On the other hand, the minority-carrier current j_p *is* limited by the supply rate, i.e. by diffusion from the interior of the semiconductor, while tunnelling itself may be highly probable. The presence of the oxide film thus diminishes j_n more than j_p. As λ_a increases, j_p will likewise experience tunnelling as the principal limitation, and the injection ratio γ will then increase more slowly. (This form of behaviour is easily simulated by a parallel network of resistors, with a constant voltage applied.) The subsequent decrease of γ is linked with the fact that ϕ_{ns} decreases in practice as λ_a increases; see also Gundlach and Kadlec (1972), and DiMaria (1974). Indeed, one of the principal functions of the oxide film is to permit the band structure of the semiconductor to move with respect to that of the metal. This is also why the insulating and space-charge layers cannot be considered as a simple series combination of two circuit components; they interact with each other, as Card and Rhoderick (1971b) have pointed out. The interaction takes two distinct forms, which occur simultaneously:

(a) via electron–hole recombination—essentially a process which tends to restore equilibrium, and
(b) via space charge effects—essentially a process which tends to augment disequilibrium.

When interface recombination is negligible, j_n governs $n(0)$ at the interface, and j_p governs $p(0)$. When recombination is important, then $n(0)$ and $p(0)$ are both functions of j_n and j_p. Moreover, because of the space charges involved, a small incremental hole concentration at the interface can lead not only to a proportionate modulation of the hole current, but to an increase of the local electron concentration, and thus to an increase of electron current. Minority-carrier multiplication (also called current amplification) comes about in this way; the net charge at the interface has a governing influence on the entire barrier structure profile. Green and Shewchun (1974) have reported multiplication factors of 10^2–10^3 in silicon diodes with 25 Å thick silicon dioxide layers. They have also provided a partly analytic and partly numerical treatment of such systems. The matter is, of course, most important in connection with the use of such diodes as sensitive light detectors. (see Section 5.2).

When a system is otherwise well-defined, the all-important barrier height ϕ_{ns} can be determined from observations of internal photoemission, e.g. as described by DiMaria and Arnett (1975) for contacts on silicon, with silicon nitride intermediate layers. Unfortunately, many systems are in practice not 'well-defined', and suffer from a variety of other complications. One such is the unwanted diffusion of metal atoms into the insulating film (see Kar and Dahlke 1971).

Card and Rhoderick (1971b) have also examined the effect of oxide films on the *reverse* characteristics, and found that the presence of such a film can, in some circumstances, *increase* the reverse current at high voltages, in comparison with that observed on oxide-free contacts (Fig. 3.32); see also Wu (1980).

A very substantial level of understanding has thus been achieved, but it remains true that the overall voltage–current relationship is a highly implicit entity from which the detailed aspects of the various processes involved cannot be evaluated with any degree of confidence. In this light, attempts to measure parameters as specific as, for instance, the effective masses of tunnelling charge carriers (Temple and Shewchun 1974) or *the* bulk carrier lifetime (Kar 1974) by means of voltage–current observations would seem to be based on overoptimistic expectations (see also Ashok *et al.* 1978).

When the interface states are energetically deep, non-equilibrium charge situations can be 'frozen in', and can give the system as a whole bistable or even polystable characteristics. Adán and Dobos (1980) found this for metal–SnO–Si layers, and such structures can be optically as well as electrically switched. Trapping processes within the 'bulk' of the insulator layer may also play a role in this type of behaviour, as Verwey (1972) had earlier suggested; Arnett and DiMaria (1975, 1976) made observations on silicon nitride layers which can readily be interpreted in this way. The transient (as distinct from steady-state) behaviour of MIS structures depends inter alia on the fact that interface states can act as temporary *reservoirs* of minority carriers. Some of these come to be injected when forward voltages are applied, and the limited nature of this supply can have a strong influence on the frequency characteristics (see also Laflére *et al.* (1982). Structures on Si which include not only an oxide but a nitride film are now of technological importance as long-term (e.g. one year or longer) digital storage devices (see Neugebauer and Burgess (1976), Kahng *et al.* (1977), and Pryor (1981).

Further reading

On the manner in which 'ideality' is influenced by traps: Lee and Nussbaum 1980.

On the effect of geometry on injection diodes: Gossick 1960.

On inversion layers at GaAs contacts: Bose *et al.* 1979.

On carrier mobilities within inversion layers (Si): Coen and Muller 1980.

On radiation damage in Au–GaAs Schottky barrier diodes: Borrego and Gutmann 1976, Ludman and Nowak 1976, Taylor and Morgan 1976.

On high injection levels in quasi-intrinsic semiconductors: Popescu 1981.

On the frequency characteristics of metal–oxide–semiconductor structures: Mar *et al.* 1977.

On the effect of non-parabolic energy bands on carrier tunnelling: Sarnot and Dubey 1972.

On the electronic characteristics of GaAs MIS diodes: Ashok *et al.* 1979.

On transient behaviour due to carrier generation at Si–SiO_2 interfaces: Arnold and Poleshuk 1975.

On the electrical properties of metal–polymer–Si devices: Sanchez *et al.* 1974.

On the static characteristics of M–I–S–I–M structures: Djurić and Smiljanić 1975.

3.6 Contacts with highly disturbed and amorphous semiconductors

Highly disturbed and amorphous semiconductors are not necessarily two-carrier systems, but many of the configurations that are of great practical interest do indeed involve both electrons and holes, which makes it convenient to discuss these matters in this chapter.

3.6.1 Bulk relationships; screening lengths

An amorphous semiconductor and, indeed, a highly disturbed crystalline one, is characterized by unsharp band edges, e.g. as shown in Fig. 3.34. Above a certain energy E_{cm}, the states are continuous, as they are in ordered materials; just below it they are localized in varying degree. In a similar way, valence band states are deemed to be continuous below E_{vm} and localized above it. In as much as the distinction between 'continuous' and 'localized' may be sharp, E_{cm} and E_{vm} are usually referred to as *mobility edges.* It is convenient to adopt this concept here, even though there is as yet no proof that such sharp transitions are a universal feature of amorphous systems. The band tails may be long or short, depending on the type and degree of disorder. When the tails are dense and extended, the Fermi level is itself located in a quasi-continuum of states, rather than in a state-free 'forbidden band'. The localized states below it are, of course, mostly full, and are neutral in that condition; they can act as hole traps. Localized states above the Fermi level are mostly empty, and neutral in that condition; they can act as electron traps.

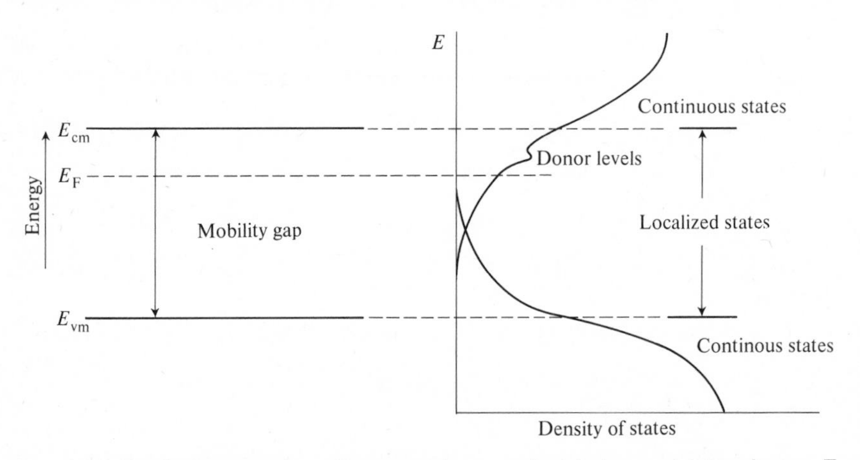

FIG. 3.34. Band representation of an amorphous semiconductor containing donors. E_F arbitrarily placed (here). E_{cm} and E_{vm} are mobility edges.

The unsharpness of the bands arises from the breakdown of long-range periodicity and is, besides, controlled by the presence of unsaturated bonds which depend, in turn, on the degree of network connectivity and on additives. Annealing tends to diminish the concentration of localized states, and the incorporation of hydrogen or fluorine in controlled amounts can cause a similar reduction by orders of magnitude. However, problems relating to the breakdown of long-range lattice periodicity, to the exact nature of the localized states, and to the relationships which govern their distribution are outside the present scope. For detailed treatments see, for instance, Tauc (1974) and Mott and Davies (1979). We shall here take the presence of such states for granted, and concern ourselves with their effects upon the behaviour of contacts. In general terms, the effects of these states depend on their ability to trap space charges. At low temperatures, the band tails could also make their presence felt by supporting hopping conduction but, as far as is known, no effect of hopping on contact properties has ever been unambiguously substantiated. The localized states also act as scattering centres, and thereby give rise to carrier mobilities which are generally much lower than those of corresponding crystalline materials.

Because the band gap is now not empty, dopants must, in a sense, 'compete' with resident localized states. The more pronounced the distribution tails, the less effective are the dopants. Because it is highly desirable to control the properties of amorphous semiconductors in this way, the suppression of band tails (by additives which have no other function) is an important technological objective. It is linked to the notion that even an amorphous material may be characterized by its degree of perfection; 'perfect' meaning (in this context) a situation in which, despite disorder, no bonds are left unsaturated.

As discussed in Section 2.1.1, a Schottky barrier width (λ_0) is a screening length, distinguished from the Debye length $L_D = (kT\epsilon/n_e e^2)^{1/2}$ by the fact that only charges in donors of fixed position are involved, whereas the Debye length refers to free carriers whose concentration can change in response to a field. When traps are present, the relationships are complicated by the fact that charges in (the immobile) traps also play a role in screening. In a semi-insulator, the Debye length may be large, but the trap concentration may be high; and since traps can hold charge, the screening distance L_{sc} in an amorphous semiconductor with substantial band tails may be quite small. Manifacier and Henisch (1980a) have analysed this problem in terms of a single trapping level, within the framework of a small-signal theory, conditions under which the linearized transport equations are explicitly solvable. The results show that systems under non-uniform excitation are governed by two characteristic lengths, the screening distance L_{sc} and the ambipolar diffusion length L_{Da}. The most significant control is exercised by the larger of the two lengths. As discussed in Section 3.3.2, the presence of traps can make L_{sc} much smaller than the Debye length. The matter is actually more serious now because, instead of a single trapping level we now have a quasi-continuous distribution of them. Moreover, whereas a linearized 'small-signal theory' is successful for bulk material, it cannot serve for the analysis of contact relationships, because the high fields prevailing within barriers make the linearization invalid. Meanwhile, one qualitative conclusion is obvious: for a given value of the free carrier concentration n_e, barrier widths are reduced by the presence of additional localized states, just as bulk screening lengths are. This was first pointed out by Rose (1956).

The position of the Fermi level in amorphous bulk material is, as always, determined by the neutrality condition. In Fig. 3.34 smooth tails are assumed for simplicity, though the density profiles can in fact be a good deal more complicated. No matter, there will always be some energy E_F such that the corresponding electron and hole distributions satisfy the neutrality condition. Exactly where E_F comes to be under equilibrium conditions depends also on doping. In the absence of dopants, E_F will be somewhere near the middle of $E_{cm}-E_{vm}$, the *mobility gap*; in the presence of doping it would be closer to one of the mobility edges. Once the position of E_F relative to E_{cm} and E_{vm} is known, it is always possible to explore what the charge density would be if E_F were displaced to some

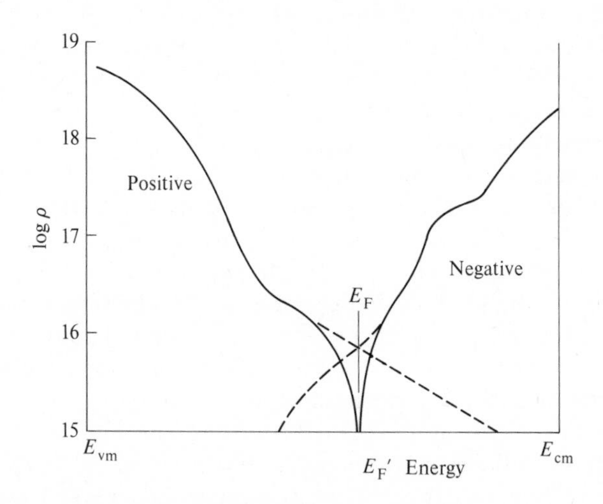

FIG. 3.35. Charge density in a typical amorphous semiconductor as a function of E'_F, the hypothetical position of the Fermi level. The actual Fermi level E_F corresponds to $\rho = 0$. Broken lines reflect the band tails, and the cusp-like contour is a consequence of logarithmic plotting. Schematically after Spear *et al.* (1978).

other position E'_F. Indeed, this is how E_F is determined in the first place,. i.e. by calculating the charge density which would result from a hypothetical Fermi level in the position E'_F, and then changing E'_F until that density is zero. That point ($E'_F = E_F$) is reached by numerical analysis. Figure 3.35 shows a charge density contour as a function of E'_F, schematically after Spear *et al.* (1978), whose classical paper describes an important line of approach to barrier analysis (albeit only in the absence of current flow). In such an analysis, contours of the kind shown on Fig. 3.35 play a key role.

3.6.2 Contact barriers in the presence of deep traps

Figure 3.36(a) shows a barrier, (there) represented by a purely schematic profile. Relative to the mobility edges E_{cm} and E_{vm}, the Fermi level comes to occupy various positions within the barrier and, for each such position, a corresponding charge density can be established from Fig. 3.36(b). Spear *et al.* did this for α-Si, and proceeded to use the values so obtained in a step-by-step numerical integration of Poisson's equation (essentially by adapting eqn (2.1.10) to finite differences). Their method was only approximate, but it yielded all the essential features of the system. Those can be qualitatively inferred from an inspection of Fig. 3.36(a). In a classical Schottky barrier, the charge density ρ is uniform; here it is not. Close to the interface, many localized states which are occupied in the bulk material (where neutrality prevails) are empty, and

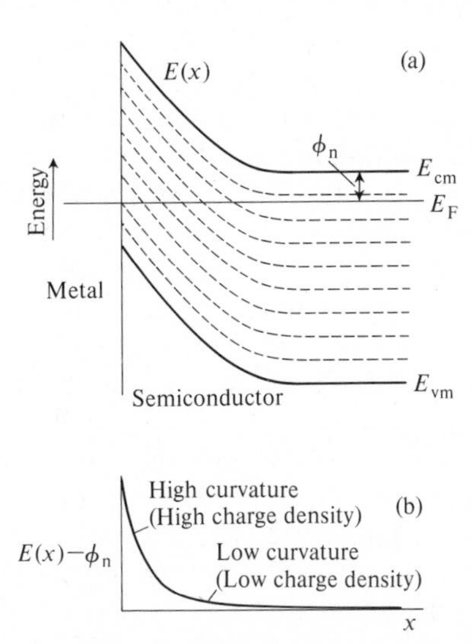

FIG. 3.36. Schematic representation of barrier profiles at zero bias. (a) The thin broken contours between E_{cm} and E_{vm} symbolize *possible* trapping states without, in this instance, taking account of space charges held in them. (b) Actual barrier curvature.

this creates a substantial positive space charge. Accordingly, the boundary field at $x = 0$ is relatively high. Towards the edge of the barrier, the charge density diminishes, which means that the curvature of the barrier profile also diminishes. The actual shape is therefore more like that shown in Fig. 3.36(b), which differs from that of a classical Schottky barrier, but is not generically different from that of a barrier with an inversion layer (Section 3.1.2). The detailed contour is established by computation. Of course, corresponding calculations can be made for inverted barriers on p-type material.

The manner in which a barrier of this kind reacts to applied voltages is complicated by the fact that the local space charge density $\rho(x)$ is now current-dependent. Indeed, small variations of $n(x)$ can be associated with large changes of $\rho(x)$. No general solution has become available to date, not even for a single-carrier system. This gap remains to be filled, notably in the context of solar cell technology. However, an approximate (non-numerical) treatment of amorphous p–n *junctions* has been given by Harris *et al.* (1980). Junctions and contacts have one important feature in common, in as much as they both offer tunnelling possibilities via gap states. Even so, a deeper analysis would be substantially more complicated, partly because the equations have no explicit solutions, and partly because amorphous semiconductors behave non-linearly *even in bulk.* Bonch-Bruevich (1979) has suggested that this non-linearity arises from internal random fields due to non-homogeneity, but on a scale small enough to leave the specimen *macroscopically* homogeneous. The bulk non-linearity shows itself as an increase in effective mobility with increasing field, which means also with increasing carrier energy above the mobility edge. This suggests that the edge is not *sharply* defined as a boundary between discontinuous and continuous states. Silver, Cohen, and Adler (1982) have observed some astonishingly high carrier mobilities in high-field regimes, at temperatures high enough to make band conduction (as distinct from hopping) possible.

As is well known, trapping states in the bulk of a material can be investigated by thermally stimulated current (TSC) and photoconductive methods, and the same is true for trapping traps within a contact barrier. Indeed, for such investigations barrier systems offer a special advantage, in as much as the application of forward and reverse voltages can fill and empty traps more or less at will. Illumination can be used as an additional tool. When filled traps within the barrier are irradiated at a suitable wavelength, they are emptied, and when the applied bias voltage is such that they are not replaced, irradiation gives rise to a photoconductive transient. Grimmeis (1974) has used such a method for investigating trapping states in GaP, and the same procedures can, of course, be applied to the investigation of amorphous materials (see also Hamilton (1947)).

Because the position of the Fermi level in relation to the bands can be changed within a barrier, the method makes levels accessible to experimentation that are not ordinarily accessible to TSC measurements in bulk material.

Smith (1972, 1974) has demonstrated the use of straightforward TSC methods for barrier analysis. In his procedure, traps were filled at a low temperature by applying a forward voltage, and then thermally released while heating in the presence of a reverse voltage. In sulphur-doped n-type GaP, as many as six peaks were detected which identified discrete trapping levels. The evaluated trap depths and trap concentrations were checked against isothermal measurements, in the course of which a reverse bias was suddenly applied, and the corresponding current transient monitored. The total amount of charge released from M_0 traps (per unit volume) at a single trapping level E_T, is obviously $Q = eM_0(\lambda_B - \lambda_T)$, where λ_T is defined as shown in Fig. 3.37(a). By using eqns (2.2.1) and (2.2.2), we find

$$\lambda_B - \lambda_T = \left(\frac{2\epsilon}{eN_d}\right)^{1/2} \left\{ (V_D + V_B)^{1/2} - \left(\frac{E_F - E_T}{e}\right)^{1/2} \right\} \tag{3.6.1}$$

which means that Q should vary linearly with $(V_D + V_B)^{1/2}$. Figure 3.37(b) gives a schematic representation of the results which Smith obtained with gold electrodes on n-type GaP. By selecting different fixed temperatures, the method can be used for each (distinctly resolved) trapping level E_T, as identified by a previous temperature run. The method and, in particular, eqn (3.5.1) does, however assume that $M_0 \ll N_d$, and that there are no secondary interactions between the charge released from traps and the barrier conductance. This limits its applicability in the above simple form.

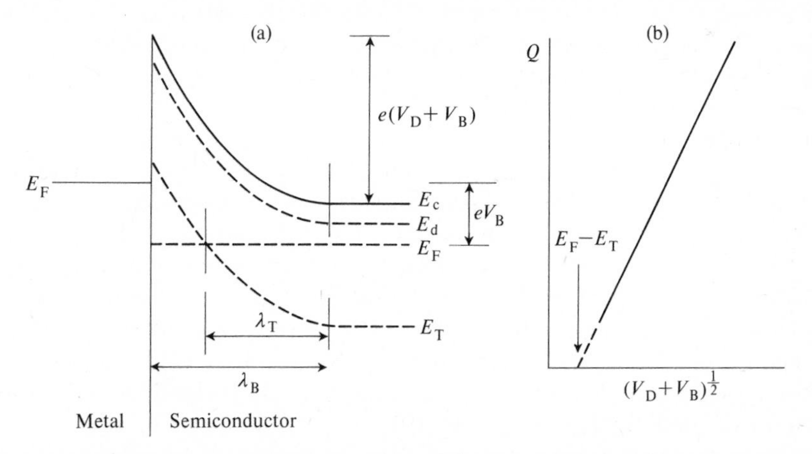

FIG. 3.37. Deep trap 'spectroscopy' of a contact barrier by the thermally stimulated current (TSC) method. Q = measured charge after sudden bias reverse. Abrupt end of barrier at $x = \lambda_B$ assumed. Schematically after Smith (1972).

It is often said that barriers formed by direct metal contact with amorphous semiconductors are characterized by low rectification ratios and low resistance and, indeed, this tends to be true. Certainly, it is what one would expect, considering

(a) the near-intrinsic character of many amorphous materials, and
(b) the small barrier thickness associated with trap screening.

However, neither expectation applies to the form of amorphous silicon that can be deposited as a thin film in a glow discharge, even without deliberate doping; e.g. see Carlson and Wronski (1976). Such films are n-type and their trap density is actually very small (optical band gap ≈ 1.55 eV; electron mobility in the extended states, 1–10 $cm^2 V^{-1} s^{-1}$; resistivity $\approx 10^7$ ohm cm). Wronski *et al.* (1976) have described the behaviour of Pd and Pt contacts with such material, and have shown that diffusion (rather than diode-) theory models are well obeyed. Barrier heights, as 'seen' from the metal side, were about 0.94 eV. Together with $E_{cm} - E_F = 0.57$ eV, this gave a diffusion potential V_D close to 0.4 volt. Surface states seem to play a little role in determining such barrier heights. This is suggested, for instance, by the observations of Wey and Fritzsche (1972), who found that the nature of the metal has a profound effect on barrier properties. Thus, on a multicomponent switching alloy, $Ge_{16}As_{35}Te_{28}S_{21}$, sandwiched between semi-transparent electrodes, contacts of Au, Al, and nichrome led to negative space charge regions, whereas contacts of Sb and Te produced almost no space charge (as assessed by photovoltaic tests). The ineffectiveness of surface states arises, of course, from the fact that trapping states which co-exist in the same band gap dominate over surface states at the boundary. There are, indeed, calculations which show that fewer than about 2×10^{12} cm^{-2} of surface states should have no detectable effect on measurements at all (Barbe 1971).

Similar conclusions apply to amorphous silicon (e.g. see Paul *et al.* (1976)). Nichrome, with a thermionic work-function close to that of intrinsic Si, yields low-resistance contacts and can therefore be used as a base electrode. Molybdenum, on the other hand, yields highly rectifying contacts on n-Si, with a forward direction corresponding to metal positive, and again $V_D \approx 0.4$ volt. Stainless steel has also been reported to yield a low-resistance contact (Wronski and coworkers 1976, 1977), and highly doped ($N_d \approx 10^{18}$ cm^{-3}) α-Si specimens with gold contacts have been found to support substantial forward current densities, e.g. of the order of 30 A cm^{-2}. (Snell *et al.* 1979).

Fritzsche (1974) has suggested that the localized states may become 'delocalized' in a high barrier field, and Bonch-Bruevich (1979) has shown that this idea would be compatible with the approximately exponential field dependence of the bulk conductivity. However, no direct evidence is available. In practice, oxide films often exist between amorphous semiconductors and the contacting metal, and these complicate the task of arriving at simple conclusions. Amorphous films are usually measured in the transverse mode, which means that they hardly ever correspond to *single*-contact systems. Their full characteristics therefore include a variety of switching phenomena in which injection and impact ionization are believed to play important roles (see Ovshinsky (1968), and Adler, Henisch, and Mott (1978), and references given there). A description of the behaviour of electrolyte contacts with amorphous chalcogenide alloys has been given by Montrimas (1976). The practical interest in contacts with highly disturbed or amorphous material derives in part from the fact that such materials tend to be less sensitive to radiation damage than well-ordered single crystals.

3.6.3 *Contacts with organic materials*

Asymmetrical conduction and sensitivity to light have been observed in many phthalocyanine (Ph) sandwich systems (e.g. see Gutman and Lyons 1967) but the amount of information obtained on really well-characterized specimens is still very small. The Cu, Mg, and Ag compounds appear to be the members of this family most frequently investigated. Using Cu as the main electrode and Au as a low-resistance base contact, Sussman (quoted by Gutman and Lyons) found rectification ratios as high as $10^5:1$ on CuPh; Delacote *et al.* (1964) have made similar observations. Fedorov and Benderskii (1971a,b) obtained pronounced rectification on Al/MgPh/Ag structures into which Al is thought to have diffused to replace Mg, thereby forming donor centres. A somewhat more detailed study of this material has been provided by Ghosh and co-workers (1974). Films were, as usual, deposited by vacuum evaporation, and measured in nitrogen atmospheres, because oxygen is known to act as a dopant; the organic material itself is ordinarily p-type. Heat treatment in vacuum at 100 °C led to a substantial improvement of rectification; the barriers were associated with an estimated diffusion potential of $V_D \approx 0.6$ volt. The photocapacitance was strongly time-dependent, which suggested the participation of deep trapping centres. The presence of such centres was also inferred from the conclusion that the contact barrier was very thin, even though the concentration of active acceptors in the bulk was low. By varying the direction of light incidence and observing the resulting effects, the Al electrode was identified as the seat of rectification (see also Barkhalov and Vidadi 1977 and Ghosh *et al.* 1974).

Another material which has been the subject of experimentation is anthracene. It has a wide band gap and contains only a few carriers in bulk, arising mostly from the dissociation of excitons (Silver 1962). Since the carrier mobility is relatively high (e.g. see Kepler 1960 and LeBlanc 1960) conduction is limited essentially by electrode processes. Kallmann and Pope (1960a,b) have demonstrated that electrolyte contacts (NaCl and NaI–I_2) tend to be associated with much lower resistances than metallic contacts. From the polarity of the observed photo-effects, it appeared that holes are injected into the organic material by a positively biased electrolyte contact. Silver (1962) ascribes the low-resistance behaviour of such contacts to ionic screening of image force effects, but the detailed mechanism is still not clear. With metal electrodes (using Inconel), Silver and Moore (1960) observed electron injection, and Boroffka (1961) found widely differing voltage–current relationships for contacts of Al, Cu, Fe, Ag, Cu, and CuI. At this stage, all interpretations must be regarded as very tentative, since no analysis appears to have taken account of the fact than anthacene is ordinarily a relaxation semiconductor.

Further reading

On rectification and photovoltaic properties of Al–α-GeSe contacts: Kottwitz, *et al.* 1977.

On tunnelling into amorphous germanium: Nwachuku and Kuhn 1968.

On computer simulations of hopping processes: Silver, Schoenherr, and Baessler 1982.

On the study of electronic structure by reverse bias transients in pin junction structures: Silver, Giles, *et al.* 1982.

On the structure of amorphous solids: Zallen 1983.

4

REACTIVE AND FREQUENCY-DEPENDENT CONTACT PROPERTIES

4.1 Capacitance of trap-free contact systems

IN Sections 4.1.1–4.1.5 we shall be concerned with single-carrier models; the effect of minority carriers on capacitance relationships is discussed in Section 4.1.6. Unless otherwise stated, 'insulating' (e.g. oxide) barriers are assumed to be absent.

4.1.1 Equivalent circuits

This section deals with the capacitive aspects of contact behaviour and, in particular, with the barrier capacitance C_B (defined per unit area). However, C_B is not equal to the measured capacitance C_M, since the equivalent circuit of a rectifying contact involves not only C_B, but also associated resistances (Schottky and Deutschmann 1929). For reasons discussed in Section 4.1.2, all measurements must be carried out with small alternating ripples which may, however, be superimposed on large, stable-bias voltages. The equivalent circuit must certainly take account of the series resistance, no matter what its origin may be in any particular case. In order to ascertain the structure of the equivalent circuit in practice, it is necessary to measure the real and complex parts of the impedance over the widest possible frequency range, and then to perform a Cole–Cole plot. If that plot yields a simple shape, e.g. as in Fig. 4.1(a), then an equivalent circuit made up of frequency-independent components can be deduced from it. More accurately, there is then a small choice of equivalent circuits which are electrically indistinguishable from one another. Thus, if the Cole–Cole plot yields a semicircle with its centre on the resistive axis, this choice is limited to the two networks shown in Fig. 4.1(b) and (c). At any frequency, the component values of one circuit can be transformed into the other, and the choice between them depends, therefore, not on electrical but on structural arguments. Circuit (b) is usually preferred, because R_B and R_b can then be simply identified as barrier and bulk resistance respectively. At zero frequency, the total resistance (per unit area) is $R_B + R_b$; at high frequency (and in the absence of further complications) the total resistance would be R_b, since R_B would be capacitively short-circuited by C_B. Various high-frequency complications actually tend to make the impedance diminish continuously, instead of tending to a constant value as $\omega \rightarrow \infty$. The complex impedance at any other frequency is given by the vector from the origin

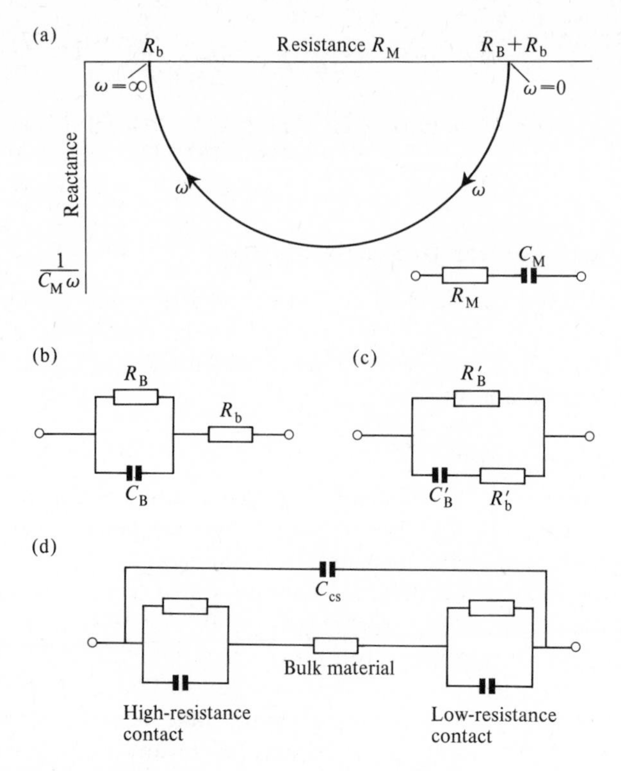

FIG. 4.1. Definition and measurement of barrier capacitance. $\omega = 2\pi$ (measurement frequency). (a) Cole–Cole plot of system impedance in a simple case. (b)(c) Functionally identical equivalent circuits of a single barrier without oxide layer. (d) Equivalent circuit of a total (two-contact) system without oxide layers. Equivalent circuits with frequency-independent components.

on Fig. 4.1(a). This can be shown to be:

$$Z = \left\{R_b + \frac{R_B}{1+R_B^2C_B^2\omega^2}\right\} - i\left\{\frac{R_B^2C_B^2\omega}{1+R_B^2C_B^2\omega^2}\right\}. \tag{4.1.1}$$

Measurements are made over the widest possible frequency range ω, for a whole series of direct bias voltages, to reveal the voltage-dependence of C_B. That is the usual procedure, but other methods have been described (e.g. Lehovec 1949, Hoffmann 1950) which involve the application of large-amplitude AC, and the observation of phase shifts.

From a structural point of view, the complete equivalent circuit of a barrier without an oxide layer would, of course, have to be that given by Fig. 4.1(d), but the additional features will, in practice, show themselves as modified values of C_B, R_B, and R_b. In any event, the errors arising from the neglect of this complication are generally very small (Rose

1951). When a Cole–Cole plot fails to yield a simple semicircle with its centre on the resistive axis, a further choice has to be made; one can adopt a more complicated equivalent circuit (still made up of frequency-independent components), or one can retain the circuit which seems justified on structural grounds, and then envisage some of its components as frequency-dependent (Burgess 1950). There is no *a priori* rule; either approach can be useful.

The situation is more complicated when an oxide layer is present. For such cases, Kar and Dahlke (1972) proposed the general equivalent circuit shown in Fig. 4.2(a), in which some of the (frequency-independent) components are identified with functions, rather than structural elements. Here C_{ox} represents the polarization capacitance of the oxide layer, and C_B, as usual, the space-charge barrier capacitance in the semiconductor, whose bulk material gives rise to R_b. The barrier resistance R_B is taken here to be associated with thermionic emission. C_{IS} is an additional

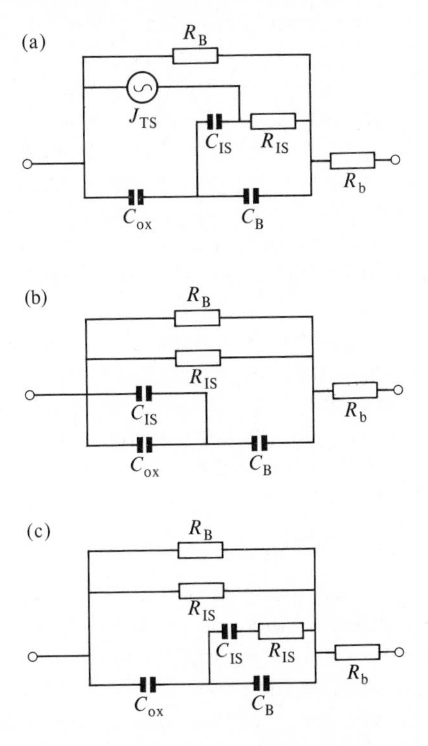

FIG. 4.2. Equivalent circuits of single barriers with oxide layers (after Kar and Dahlke 1972). (a) General. (b) Interface recombination controlled ($\lambda_a \leqslant 10$ Å). (c) Tunnelling controlled (10 Å $> \lambda_a < 40$ Å). Components: C_{ox}, polarization capacitance of oxide layer (thickness λ_a). C_B, space charge barrier capacitance. R_b, bulk series resistance. C_{TS}, additional capacitance associated with interface states. R_{IS}, interface state recombination resistance. R_{TS}, tunneling resistance.

capacitance associated with interface states (assumed to be at a single level), and R_{IS} is a recombination resistance, denoting the flow of carriers to those states from the semiconductor. J_{TS} is a current generator which represents the tunnelling process between metal and interface states (see Section 3.5.5 for comments on current continuity). Of course, the evaluation of the various components cannot now be achieved by AC ripple measurements and Cole–Cole plots alone. Kar and Dahlke (1972) and Jonscher (1983) have described some of the procedures involved.

For various limiting cases, the circuit of Fig. 4.2(a) can be simplified. Figure 4.2(b) gives a version, specially applicable to very thin oxide layers, in which tunnelling is highly probable, so much so as to permit the current-generator J_{TS} to be replaced by a short circuit. The system is then controlled (in the sense of limited) by interface recombination. Conversely, the circuit tends to the configuration shown in Fig. 4.2(c) when the system is controlled primarily by tunnelling. The current generator can then be replaced by a frequency-independent conductance $1/R_{TS}$. For thick layers (e.g. $\lambda_a > 40$ Å), R_{TS} becomes effectively infinite.

Of course, even the more complex equivalent circuits of Fig. 4.2 assume *lateral* homogeneity of the contact structure, but practical systems do not always conform to this expectation (e.g. see Day *et al.* 1970), and edge capacitances can play a significant role. Anomalies may occur at very low frequencies due to creep and forming effects, and even positive reactances are sometimes encountered under these conditions. The larger intercept on the resistive axis does not then correspond to $\omega = 0$, but to some definite small value. The physical structure of the contact precludes the existence of any appreciable inductance in the normal sense. Whatever inductive behaviour may occasionally be observed must, therefore, be simulated by some other mechanism which involves a time delay. Electrolytic migration is just such a mechanism, and so is heating. A quantitative analysis of thermal effects has been made by Oldekop (1952) (see also Section 6.2).

4.1.2 *Barrier capacitance with complete donor ionization*

'Capacitance' (here always per unit area) is ordinarily defined by two equal charges, one positive and one negative, and by an unambiguous separation distance between them. In that sense, the Schottky barrier has no capacitance. The term 'static capacitance' is sometimes applied to the ratio of the total charge ($\rho\lambda_B$) within such a barrier to the total voltage ($V_D + V_B$) across it, but such a quantity would have no operational meaning. (A 'static capacitance' C_{BS} is used in Section 2.3.5 and elsewhere as a convenient constant of proportionality in model making. Popović (1979) defined an 'integral capacitance' used for the interpretation of very low frequency measurements on systems with traps.)

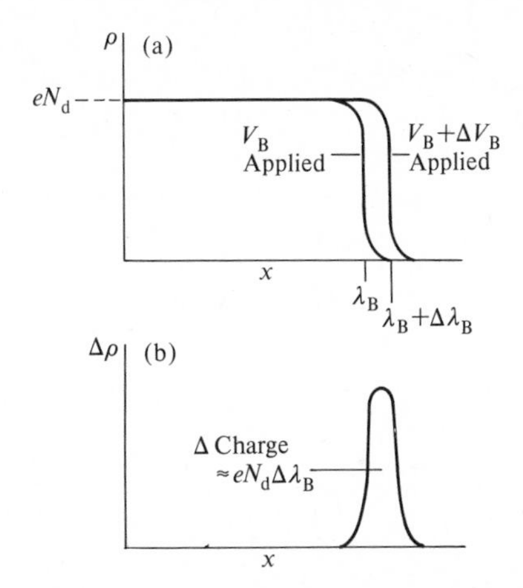

FIG. 4.3. The dynamic (differential) capacitance arising from the change of charge density at the edge of the barrier. The charge increment within the semiconductor at $x \approx \lambda_B$ is balanced by an equal and opposite increment on the metal surface.

The same difficulty is also encountered (and not always avoided) when the capacitance is defined as a differential entity (Fig. 4.3). Much depends on how the distribution of charge density within the system adjusts itself to the external voltage imposed. If the charge density changes more or less everywhere, then even the differential capacitance has no simple meaning, because the fields (and field increments) at the two electrodes are not the same; Callarotti (1981) has pointed this out. While it is true that an *effective* differential capacitance can always be *measured*, there is not necessarily any way in which that measured value can be interpreted in terms of system thickness and dielectric constant. That interpretation is simple only when the space charge density in *most* of the barrier region remains unchanged. For a classical Schottky barrier, this happens to be the case, with the space charge density fixed at $\rho = eN_d$ except near the barrier edge at $x = \lambda_B$. Only there does a change occur in response to the externally imposed ripple voltage, and the charge separation is then well defined (Fig. 4.3). Accordingly, we have for the dynamic capacitance:

$$C_B = dQ/dV_B = eN_d \, d\lambda_B/dV_B. \tag{4.1.2}$$

Substituting for dV_B from eqn (2.2.1b) then gives

$$C_B = \epsilon/\lambda_B. \tag{4.1.3}$$

The barrier width λ_B can be eliminated between equations (4.1.3) and (2.2.1b), leading to the well-known expressions

$$\frac{1}{C_B^2} = \left(\frac{2}{\epsilon e N_d}\right)(V_D + V_B) \tag{4.1.4}$$

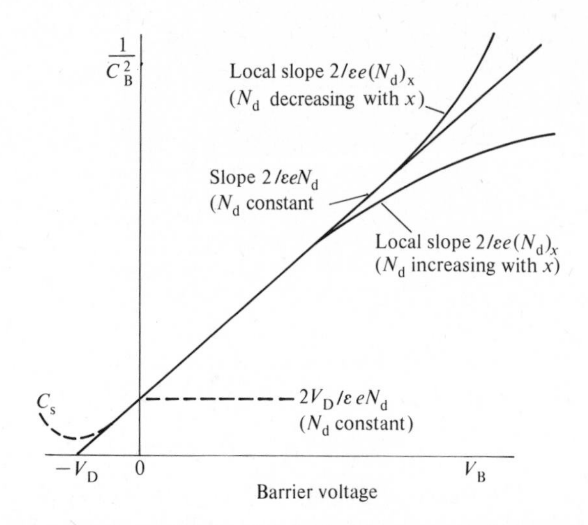

FIG. 4.4. Schottky capacitance relationship (eqns 4.1.4 and 4.1.6). Schematic representation, after Macdonald (1962). Broken line: departures from linearity which may arise from incomplete donor ionization; system capacitance, C_s (see Section 4.1.3).

which is represented in Fig. 4.4. The discovery of this relationship by Schottky in 1942, and its subsequent experimental confirmation (at any rate, in essentials) were principal factors in the general acceptance of the Schottky space-charge model. Other models might have accounted for rectification as such, but only Schottky's model offered a convincing explanation of the voltage-dependent contact capacitance. This striking success caused other features of the Schottky model (e.g. its dependence on thermionic work-functions) to be accepted also, perhaps without some of the critical examinations to which they might otherwise have been subjected.

Equation (4.1.4), by now the 'classical capacitance relationship' is in fact well-obeyed by contacts on a variety of materials; e.g. see McColl and Millea (1976) for results on InSb, Millea *et al.* (1969) on GaAs, and Szydlo and Poirier (1980) on TiO_2. Of course, such observations are routinely made on Ge and Si (e.g. see Senechal and Basinsky (1968) and Goodman (1963)). Even so, the arguments leading to eqn (4.1.4) are an oversimplification, valid only for high barriers on materials in which the donors are completely ionized at the temperature concerned. Only in such cases can the charge density be simply represented as eN_d for $z < \lambda_B$ and as zero for $x > \lambda_B$. The model thus suits material like germanium and silicon particularly well, though it was actually developed in the much less promising context of research on microcrystalline selenium and cuprous oxide rectifiers.

The practical utility of eqn (4.1.4) derives from the possibility of evaluating N_d (from the slope) and V_D (from the intercept) of the $1/C_B^2$ versus V_B relationship. It has been widely used for this purpose, and, indeed, not always within the limits of its validity. The treatment neglects not only minority carriers, but also the contribution of free majority carriers, (here) electrons. It assumes that the barrier ends abruptly at $x = \lambda_B$, but there is in fact no such discontinuity, as explained in Section 2.1.2, where it is also noted that the *effective* diffusion potential (in a non-degenerate system) is not V_D but $(V_D - kT/e)$. This also applies here. The intercept of eqn (4.1.4) will yield an 'effective' V_D in this sense.

It is interesting to note that a treatment of the same form (albeit only a coarse one) is still possible, even when N_d is non-constant. The argument is conventionally presented in the following way. When N_d is a function of x, eqn (4.1.4) is replaced by

$$C_B = dQ/dV_B = eN_d(\lambda_B)\, d\lambda_B/dV_B \tag{4.1.5}$$

with $dV_B = (e/\epsilon)N_d(\lambda_B)\lambda_B\, d\lambda_B$ from eqn (2.2.1b). Since $1/C_B^2 = \lambda_B^2/\epsilon^2$, we have

$$\frac{d(1/C_B^2)}{dV_B} = \frac{d(1/C_B^2)}{d\lambda_B} \cdot \frac{d\lambda_B}{dV_B} = \frac{2}{\epsilon e N_d(\lambda_B)} \tag{4.1.6}$$

where $N_d(\lambda_B)$ is, of course, the donor concentration at a distance $x = \lambda_B$ from the metal boundary, and $d(1/C_B^2)/dV_B$ is the local slope of the contour in Fig. 4.3. In this way, concentration profiles can be experimentally determined, since every value of V_B corresponds to a known value of λ_B.

The method, though widely used, is of course subject to all the cautions outlined above, and these cautions depend in no way on whether the attempted interpretations are analytical or numerical, e.g. as reported by Lubberts and Burkey (1975). Nor is this the only problem. A determination of N_d in the manner suggested would be crucially dependent on the exact circumstances which prevail at the edge of the barrier, where the electron concentration decays in accordance with its Debye length. What the capacitance measures is, of course, the differential electron concentration, rather than N_d itself, and that concentration is sampled in a region in which it has a distinct gradient. Kroemer and Chien (1981) have pointed out that the distances over which these charge exchanges take place can greatly exceed the $d\lambda_B$ by which the nominal Schottky barrier width changes under the voltage ripple. An averaging process is therefore involved, governed by the non-zero amplitude of the ripple used for measurement purposes. This is always true, but in the presence of donor gradients the situation is more complicated still. When $N_d(x)$ is non-constant, that variation leads to a space-charge contribution of its own, because charge is exchanged between neighboring regions of unequal donor content. Thus, even if the determination of $n(x)$ as a function of x were itself totally simple, its interpretation in terms of the $N_d(x)$ profile would not be a simple matter at all. In practice, this means that the capacitance method is applicable only to small variations of N_d, with the additional rider that we do not know reliably in any particular case just how small that variation has to be. For a more detailed analysis, see Kroemer and Chien (1981), Kennedy *et al.* (1968), Kennedy and O'Brien (1969), Johnson and Panousis (1971), and Nishida (1979).

It should be noted that the diffusion potential V_D evaluated by the method discussed above is the barrier height as 'seen' from the semiconductor side, *without* image-force lowering. The image force has no relevance to the present considerations, which is why the capacitance method generally yields greater barrier heights than does the analysis of voltage–current characteristics and their temperature dependence. For instance, Rideout and Crowell (1970) report

W–GaAs (n)	Au–GaAs (n)	
0.71 eV	0.74 eV	from voltage–current characteristic
0.77 eV	0.83 eV	from voltage–capacitance characteristic.

The systems analysed here are assumed to be non-degenerate as far as carrier concentrations are concerned. For a discussion of the degenerate case, see Dewald (1960), Goodman and Perkins (1964), and the brief discussion in Section 2.1.2.

Some of the ways in which the evaluation of N_D and V_D from capacitance measurements can fail to yield significant results are described in Sections 4.1.3, 4.1.4, 4.1.6, and 4.2.1. Goodman (1963) has given a survey of such failure modes, including those arising from the presence of interfacial layers (see also Yaron and Frohman-Bentchkowsky 1980). Coleman (1975) has discussed the effect of lateral inhomogeneities. The unwanted diffusion of metal ions has already been mentioned in Section 3.5.5 as a complicating factor in the assessment of voltage–current characteristics; it also shows itself here, e.g. as demonstrated by Neville and Hoeneisen (1975) in measurements on contacts with n-type single-crystal strontium titanate. Another factor, though rarely considered, arises from the possibility that the dielectric constant of the medium may be field dependent. There are some signs that this applies to GaAs (Pellegrini and Salardi 1978), though further evidence would be desirable to fortify the diagnosis. The effect which may be interpreted along these lines show itself as an additional and voltage-independent parallel capacitance; this has also been seen by Vasudev *et al.* (1976).

4.1.3 *Barrier capacitance with incomplete donor ionization*

Depending on the donor activation energy, the temperature, and the position of the Fermi level in the bulk, the donor may *not* be immediately ionized, or may be completely ionized only in the region immediately adjoining the metal. Such a totally ionized region is sometimes called an 'exhaustion layer'. In the limit, incomplete ionization may prevail throughout the barrier, and this is sometimes referred to as a 'reserve layer' (see Spenke 1941, 1950).

Complete ionization in the bulk means, of course, that $n_e = N_d$. Incomplete ionization means that the equilibrium between donors and free electrons is not completely biased to one limit. The law of mass action is then operative, giving

$$n_d(N_d - n_d) = K = n_e^2 \tag{4.1.7}$$

where K is a (temperature-dependent) constant, and n_d the concentration of electrons remaining on unionized donor sites. Within the barrier, n and n_d would both be x-dependent. The space charge ρ is then given by

$$\rho = e(N_d - n_d - n) = e\left\{\frac{n_e^2}{n(x)} - n(x)\right\} \tag{4.1.8}$$

as compared with $\rho = e\{n_e - n(x)\}$ which would have prevailed for complete ionization. Even if the validity of the Boltzmann relationship between $n(x)$ and $V(x)$ were assumed (which is not actually permissible when current is flowing), the integration of Poisson's equation, and thus the establishment of an equivalent barrier thickness, would not be analytically possible without an explicit knowledge of $V(x)$ versus x. Nevertheless, an approximate solution for the components of the complex barrier impedance has been obtained, the outcome of an analysis of impenetrable complexity (Spenke 1941, 1950).† The results themselves are surprisingly simple, and can be summarized as follows:

(a) The correcting terms arising from incomplete donor ionization (extreme case of $n_d \approx N_d$) are proportional to the current density and inversely proportional to the dielectric relaxation time *of the boundary region*, which is defined as $\tau_{DI} = \epsilon/\sigma_I$ where σ_I is the conductivity at the metal–semiconductor interface.

(b) Other things being equal, the real part of the impedance is doubled at low frequencies, and left unaffected at high frequencies.

(c) The complex part of the impedance is increased by an amount that is equivalent to a constant capacitance decrease. This arises, of course, from

† Spenke writes disarmingly at one point: "Eine auch nur andeutungsweise Wiedergabe der diesbezüglichen recht mühseligen Untersuchungen ist aber keinesfalls möglich. Wir müssen uns vielmehr mit der Wiedergabe der Resultate begnügen."

the increased barrier width that is bound to accompany a diminished space-charge density.

More precise requirements have to be met by numerical solutions. This applies particularly when it is desired to take account of free carriers in the calculation of the space charge and, as expected, this correction is most important in the forward direction. Macdonald (1962) has shown that the total differential capacitance conventionally interpreted, is slightly increased by the free-carrier concentration. In the same paper, calculated results are given for the *system capacitance* (differentiated from the barrier capacitance C_B) as a function of the applied voltage. Whereas the idealized C_B goes to infinity for $V_B = -V_D$, the system capacitance, C_s is expected to diminish after reaching a maximum, as schematically shown by the broken line in Fig. 4.4. However, measurements presented by Millea *et al.* (1969) for Schottky barriers on GaAs are veritable textbook examples of $1/C^2$ linearity.

4.1.4 Barrier capacitance and interface state occupancy

We have seen in Section 2.3 that varied steps can be taken to make the basic theory of voltage–current characteristics more sophisticated, and the same is obviously true for the theory of voltage–capacitance relationships. Indeed, the capacitance case is simpler, since it is not concerned with charge-transfer processes that are influenced by image forces. There remains the possibility of ϕ_{ns} changing with the occupancy of surface states, as discussed in Section 2.3.5, and this mechanism can in principle also affect the capacitance. It is sometimes invoked to interpret observed anomalies, but its potency in that context is actually small. This is obvious for contacts characterized by tightly bunched interface states, since only minor variations of barrier height (and hence, thickness) can then be envisaged. Even those are important as far as the *current* is concerned, in view of the exponential relationship between these quantities, but barrier height and barrier thickness are related via a far less sensitive square-root law. Only a minimal effect of varying interface state occupancy on capacitance can then be foreseen. However, for the more general situation one might envisage a model in which the surface states are not tightly bunched, but are thinly spread over a wide range of band-gap energies. In that case a simple formulation is possible. Of course, we are here concerned only with coarse trends; to pursue highly detailed models would be 'empty sophistication', because the opportunities for comparing theory with experiments carried out on sufficiently well-defined specimens are very few.

To make an assessment of the effect, we recall that, as discussed in Section 2.3.5, the variable surface state occupancy changes ϕ_{ns0} into ϕ_{ns}.

Using the proportionality constant α_s, as before, eqn (2.1.5) then becomes

$$\frac{\phi_{ns0}-\phi_n}{e}=\frac{N_d e\lambda_B^2}{2\epsilon}+\alpha_s eN_d(\lambda_B-\lambda_0)=V_D+V_B \tag{4.1.9}$$

where $\alpha_s eN_d(\lambda_B-\lambda_0)$ is an incremental, but nevertheless macroscopic, charge. The diffusion potential V_D is now defined by reference to ϕ_{ns0} rather than the actual barrier height ϕ_{ns}. Accordingly, we have a *differential* capacitance given by

$$C_B=\frac{eN_d\,d\lambda_B}{dV_B}=eN_d\left(\frac{eN_d\lambda_B}{\epsilon}+\alpha_s N_d e\right)^{-1}$$
$$=\frac{\epsilon}{\lambda_B+\alpha_s\epsilon},\ \text{as compared with the normal}\ \frac{\epsilon}{\lambda_B}. \tag{4.1.10}$$

The capacitance is thus *diminished*, and this result can be easily rationalized. A voltage increase dV_B ordinarily implies the inclusion of increased charge in the barrier. However, as a result of $\alpha_s\neq 0$, the barrier height would *decrease* at the same time, and that means less charge. The smaller capacitance arises from the opposition of the two tendencies.

By eliminating λ_B between eqns (4.1.9) and (4.1.10) one arrives at a modified capacitance relationship of the form

$$\frac{1}{C_B^2}\approx\left(\frac{2}{\epsilon eN_d}\right)(V_D+V_B)(1+\text{correcting terms}) \tag{4.1.11}$$

This is close to the conventional result (compare eqn 4.1.4) but leads, for $\alpha_s\neq 0$, to a slightly excessive V_D when eqn (4.1.11) is used for evaluation purposes. The effect is ordinarily very small. In special situations (e.g. low density of widely spread surface states) it would become more prominent, but nothing drastic can be envisaged. Experimental data for close comparisons would be desirable, but do not appear to be available. In any event, the use of α_s is based on the assumption that the electron concentration is in equilibrium at $x=0$, and we have already seen (Section 2.3.2) that this is not necessarily correct. Note that the error due to free electrons is in the opposite direction.

All these difficulties notwithstanding, there is reason to believe that capacitance measurements, with appropriate choice of an equivalent circuit, can still serve to extract useful information from barrier systems. Thus, Deneuville (1974) succeeded in obtaining truly stable values of V_D for Ag–SiO_2–Si, and Baccarani *et al.* (1973) have evaluated convincing interface state density profiles. Even so, the methods are straightforward only if the doping of the semiconductor is uniform.

4.1.5 Capacitance of back-to-back barrier systems

Corresponding to the resistive properties discussed in Section 2.2.9 there are, of course, capacitive relationships which arise from the fact that a back-to-back barrier system involves two voltage-dependent capacitances in series connection. The resultant total capacitance can always be calculated as long as the parameters of the system are known, bearing in mind that the reverse voltage applied to one contact generates the same current as the forward voltage applied to the other. If the bulk material can be neglected (frequency not too high), we have for the system capacitance $C_s = C_1C_2/(C_1 + C_2)$ and, in the terminology of Section 2.2.9,

$$C_1 \propto 1/(V_{DI} + V_1)^{1/2}, \qquad C_2 \propto 1/(V_{D2} - V_2)^{1/2}. \tag{4.1.12}$$

V_1 and V_2 depend, of course, on the voltage–current relationship of the contact, and can be numerically evaluated with any desired degree of correction for tunnelling, Schottky effect, and self-heating. Figure 4.5 displays calculated relationships of this kind, based on *un*corrected diode theory. Resistive bulk effects have been neglected. If the systems are structurally symmetrical, the capacitance–voltage relationship is symmetrical also (curve **A**). At zero bias, the total capacitance must evidently be half of the zero-bias-capacitance corresponding to a single barrier. For asymmetrical structures, the maximum is displaced. When one of the

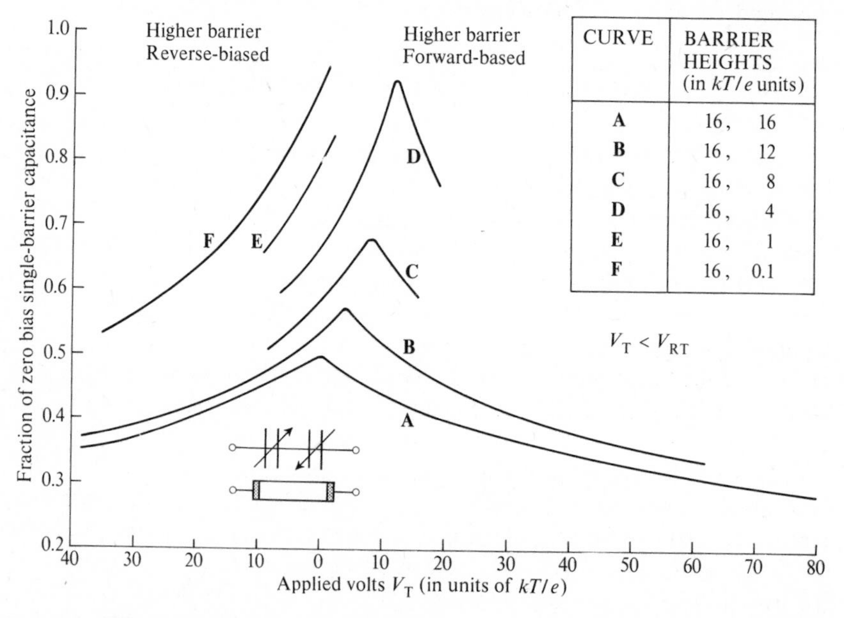

FIG. 4.5. Differential system capacitance of two back-to-back Schottky barriers for symmetrical and unsymmetrical structures of equal contact areas; bulk material neglected.

barriers is of zero height (a situation almost reached by curve **F**), we get the simple square-root dependence described by eqn (4.1.4). These curves apply as long as $V_T < V_{RT}$. When $V_T \geqslant V_{RT}$, the entire semiconductor slab is depleted and acts, thereafter, as a dielectric of constant width. From that point onward, the capacitance must be a voltage-independent constant (not shown in Fig. 4.5).

Sze and co-workers (1971) have obtained results of the kind shown on Fig. 4.5, and have also pointed out that unequal contact areas can distort the results, making a system with symmetrical diffusion potentials appear like an unsymmetrical one.

4.1.6 *Effect of inversion and interface layers on contact capacitance*

The nature of inversion layers has already been discussed in Section 3.1. It enters here, because the charges represented by the minority carriers in such a layer affect the barrier capacitance. One way of assessing the effect would be to integrate the charge density numerically for various applied voltages V_B, but even this is simple only in the absence of current flow. In any other case, the complete transport equations discussed in Section 3.2.2 would have to be solved to obtain the potential and charge-density profiles of the barrier for subsequent integration. It would certainly not be sufficient to assume that equilibrium carrier concentrations prevail throughout the barrier. Accordingly, the situation is complex and a complete solution, i.e. one valid in the presence of current flow, is not yet available, though a beginning has been made (Schwarz and Walsh 1953).

A qualitative assessment has been given by Rhoderick (1978) along the following lines. Figure 4.6 shows that the inversion layer produces a sharp cusp, followed by a space-charge region governed entirely by donor ionization, and decaying into neutrality at the edge of the barrier. When a ripple voltage is superimposed upon a bias V_B, the principal change occurs, as always, in the neighbourhood of $x = \lambda_B$. Immediately to the left

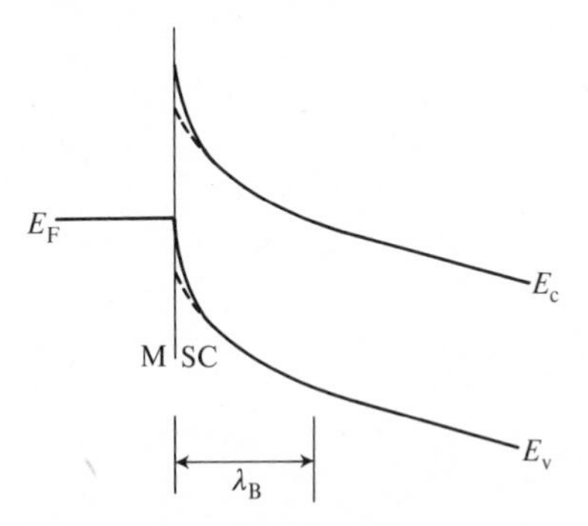

FIG. 4.6. Barrier with inversion layer under reverse bias. Image forces neglected. After Rhoderick (1978).

of $x = \lambda_B$, the space charge density is unaffected, in as much as it is controlled by N_d, but close to the metal, where minority carriers may predominate over N_d, there may be some further changes. That region is, however, very thin and also very conductive; in arguments about the capacitance it can therefore be deemed to 'belong to the metal'. In any event, as V_B increases, and λ_B grows, the inversion layer will constitute an ever-diminishing fraction of λ_B. Eventually, therefore, C_B will be proportional to $(V_D + V_B)^{-1/2}$, as it is in the case without minority carriers. However, for small values of V_B, V_D will *appear* to be non-constant. Even at high value of V_D, for which an equation of the form (4.1.4) should be obeyed, measurements will yield a misleading intercept to the $(1/C_B)^2$ versus V_B relationship, amounting to an underestimate of V_D. This is suggested by the broken E_c and E_v contours in Fig. 4.6, which represent extrapolations of the parabolic barrier contours that correspond to $\rho(x) = eN_d = \text{constant}$. Changes of applied voltage ΔV_B will cause the barrier to distort *almost* as if it had a height indicated by the extrapolation. Green (1976) has dealt with the interpretation of the 'misleading intercept' mentioned above, on the basis of earlier work by McNutt and Sah (1974). Below barrier-heights (on Si) of about 0.85 eV, the errors are minor, but above that value they become substantial. It has also been shown (Green and Shewchun 1973) that, in some circumstances, the presence of minority carriers (injected by forward bias) can actually make the system inductive.

An extreme case has been discussed by Walpole and Nill (1971), involving barrier heights greater than the band gap of their degenerate semiconductors, PbTe and InAs. Astonishing as it may be, $(1/C_B)^2$ versus V_B linearity was observed even in such systems and high (and, indeed, not so high) values of V_B.

In many ways the complications which arise from the presence of insulating interface layers are greater than those which arise from inversion layers, in part because more descriptors are involved. A detailed analysis has been provided by Crowell and Roberts (1969), and a general review by Rhoderick (1978). There are two main questions: what can we learn from the intercept of the capacitance relationship, and is it still possible to interpret $d(1/C_B)^2/dV_B$ in terms of the donor concentration. To answer these questions, it is necessary to make some assumption about the occupancy of the interface states between the semiconductor and the insulator. In that respect, one can distinguish between two limiting cases: the case in which interface state occupancy is governed entirely by the (constant) Fermi level of the metal (very thin interface layer), and the case in which it is governed by the (variable) Fermi level of the semiconductor. The first is obviously the simpler, and for that Cowley has shown that the $(1/C_B)^2$ versus V_B relationship is still that given by eqn (4.1.4), the slope being directly given in terms of N_d. However, because the applied voltage is now shared between the interface layer and the Schottky barrier, the interface is no longer interpretable as V_D. Even so, the true V_D can be calculated from the observed value, on the basis of Cowley's findings for each of the two limiting conditions. Measurements by Card and Rhoderick (1971a) on silicon diodes, with oxide films between 8 and 26 Å thick, have demonstrated once again how resistant the linear $(1/C_B)^2$ versus V_B relationship is to a variety of complicating features.

Further reading

On early bridge measurements of contact capacitance: Pfotzer 1949.

On the capacitance of contacts with oxide layers: Korol *et al.* 1975.

On procedures used in AC admittance studies: Engemann 1981.

On capacitance methods of determining impurity profiles: Anderson, Baron, and Crowell 1976, Feltl 1978, Heime 1970a and 1972, Lang 1974, Ziegler *et al.* 1975.

On measurements of photocapacitance; Au–Si contacts: Lee and Henisch 1968.

On the capacitance characteristics of Schottky barriers on diamond: Glover 1973.

On the capacitance of M–I–S–I–M systems: Djurić *et al.* 1975.

On changes of contact capacitance due to non-equilibrium inversion layers: Zakharov and Neizvestnyi 1975.

4.2 High-frequency and microwave effects

4.2.1 Relaxation of donors

When donors are completely ionized in bulk, the application of voltages does not involve ionization and recombination processes which are subject to relaxation effects. If, on the other hand, the impurity centres are only partially ionized, such relaxation effects must be taken into account, because they affect the AC behaviour of the barrier. A voltage excursion into the reverse direction then involves additional ionization, and one in the forward direction involves recombination. Neither process is instantaneous, and important time lags may occur if the frequency is sufficiently high. The effect was recognized long ago, and several early attempts at its assessment are on record, e.g. Torrey and Whitmer (1948) and Schottky (1952). They were based on simplifying assumptions which are no longer quantitatively acceptable but which, nevertheless, provided useful order-of-magnitude estimates. The following considerations illustrate this in terms of n-type material, in the absence of an inversion layer. It will be assumed that the dielectric relaxation time is negligibly small compared with other time constants involved, i.e. that the *free*-electron concentration everywhere responds instantaneously to a change of barrier voltage. Of course, this would not be so in a semi-insulating specimen.

The concentration of electrons at donor sites in the bulk material has been denoted by n_d. We shall here use n_{dB} for the corresponding values within the barrier. The rate at which n_{dB} changes is obviously

$$\frac{dn_{dB}}{dt} = -a_1 n_{dB} + a_2 n(N_d - n_{dB}) \tag{4.2.1}$$

where a_1 and a_2 are constants. The first term represents ionization, the second recombination. n_{dB} and n are, of course, functions of x. (Deeply within the barrier, we assume $n_{dB} \approx 0$ and $n \approx 0$.) In the presence of a small alternating ripple voltage (angular frequency ω) applied to the barrier, n and n_{dB} become periodic functions of time, with a phase lag between them. We can schematically put

$$n = n_0 + \tilde{n}\exp(i\omega t) \tag{4.2.2}$$

$$n_{dB} = n_{dB_0} + \tilde{n}_{dB}\exp(i\omega t - \theta). \tag{4.2.3}$$

Substituting into eqn (4.2.1) leads to

$$\frac{\tilde{n}_{dB}}{\tilde{n}} = \frac{a_2(N_d - n_{dB_0})\exp(i\theta)}{a_2 n_0 + a_1 + i\omega} \tag{4.2.4}$$

and since this must be *real*, we have

$$\frac{\tilde{n}_{dB}}{\tilde{n}_0} = \frac{a_2(N_d - N_{dB_0})\{(a_2 n_0 + a_1)\cos\theta + \omega\sin\theta\}}{(a_2 n_0 + a_1)^2 + \omega^2} \tag{4.2.5}$$

where

$$\tan\theta = \omega/(a_2 n_0 + a_1).$$

This can be further simplified to

$$\frac{\tilde{n}_{dB}}{\tilde{n}} = \frac{a_2(N_d - n_{dB_0})}{\{(a_2 n_0 + a_1)^2 + \omega^2\}^{1/2}}. \tag{4.2.6}$$

The ratio evidently goes to zero as ω increases, and this is true for all locations, but the most important effects are those which take place at the edge of the barrier, within the 'reserve layer' close to $x = \lambda_B$. There, the last equation becomes

$$\frac{\tilde{n}_{dB}}{\tilde{n}} \rightarrow \frac{a_2 n_0}{\omega} = \frac{a_2 n_e}{\omega} \tag{4.2.7}$$

at very high frequencies. Theoretical estimates of a_2 for cuprous oxide and germanium suggest that this rate constant is of the order of $10^{-9}\,cm^3\,s^{-1}$ (Schottky 1952, Burton *et al.* 1953). Goodman *et al.* (1947) have given a similar value for Al acceptors in silicon. On the basis of their model, the probability of ionization by a single phonon decreases exponentially with the depth of the impurity level below the conduction band (see also Section 4.2).

The problem of incomplete and time-dependent donor ionization is particularly important in connection with applications of GaAs. Such donors must evidently make the barrier capacitance C_B frequency dependent, which greatly complicates the proper use of equivalent circuits. The matter is all too often ignored, accepting the system capacitance for C_B; but this is a dangerous procedure. It can be shown that, to a fair approximation, the barrier capacitance itself (if only it could be readily measured) is expected to be composed of two parts in the form

$$\frac{1}{C_B} = \frac{1}{C_{BV}} + \frac{1}{C_{B\omega}} \tag{4.2.8}$$

where C_{BV} depends only on voltage and $C_{B\omega}$ only on frequency (see also Schibli and Milnes 1968, Perel' and Efros 1968, and Schultz 1971). The last term is zero at zero frequency (Hesse and Strack 1972), which means that C_B is always decreased by the frequency dependence (see also Section 4.2.2). Surprisingly, again, the $1/C_B^2$–voltage relationship is still approximately linear, although slope and intercept need to be reinterpreted.

In an attempt to cope with the time-dependent deep-donor occupancy (Au in n-Si), Senechal and Basinsky (1968) applied sudden reverse pulses to a barrier system, and monitored the subsequent capacitance change with time (see also Kaplan 1980). Other experimenters have used sawtooth voltage ramps of varying steepness for similar purposes (e.g. Tsao and Leenov 1976).

4.2.2 *Asymmetry of contact characteristics*

At very high frequencies, the shunt reactance of a rectifying contact may be comparable to the barrier resistance. When the reactance is calculated on the assumption that the barrier capacitance is constant, the resulting values are very low, and it would be concluded on this basis that rectification becomes impossible at wavelengths of (say) a few metres.

The fact that rectification can still be observed at microwave frequencies is believed to be due to relaxation effects of the type described above, since their most important consequence is a reduction of barrier capacitance. At low frequencies, the donors within the reserve layer are still able to participate in a periodic charging and discharging process. The barrier thickness which governs the capacitive behaviour is then relatively small. At very high frequencies, the charges at donor sites are almost fixed, and unable to follow the applied signal. By way of compensation, larger numbers of them are needed. The effective barrier thickness then becomes large, and the capacitance correspondingly small. In this way, the contact asymmetry is maintained. At microwave frequencies, the non-zero dielectric relaxation time of the bulk material will also begin to play a role in reinforcing the above effects.

Schottky (1952) has drawn attention to the fact that the increase of effective barrier thickness at higher frequencies should diminish the importance of accidental non-uniformities in the structure of the barrier. In this way, contacts which have an unfavourable voltage–current relationship under low frequency conditions can sometimes exhibit good rectification at high frequencies. Beringer (1944) has reported experimental evidence for this conclusion, though only on the basis of early (and thus insufficiently defined) semiconductor specimens.

In the case of injecting contacts the analysis of high-frequency behaviour is evidently much more complicated. The minority-carrier concentration in the neighbourhood of the injecting interface (and sometimes deeply into the semiconductor) may be greatly augmented during the voltage excursion in the forward direction. If there were a subsequent voltage excursion in the reverse direction before the minority carriers have had time to decay, these carriers will be 'collected' by the contact. It is as if the emitter and collector of a transistor were periodically exchanging their roles (see Fig. 3.22). At low frequencies, the consequence of collection is a temporarily increased reverse current, but at high frequencies, i.e. those comparable with $1/\tau_n$, the increase would prevail throughout the reverse half-cycle, and rectification would sharply deteriorate. A short carrier lifetime is thus essential for good high-frequency rectification. An injected concentration of minority carriers also affects the contact capacitance, but a detailed assessment of this process is not yet available.

For the sake of achieving good noise characteristics, a good deal of work has also been carried out on the high-frequency properties of contacts between semiconductors and superconducting metals. McColl *et al.* (1973, 1976) have described such experiments, using Pb as the superconductor. They report band widths far in excess of those obtainable with parametric and maser amplifiers, with comparable noise temperatures.

Further reading

On the design of microwave diodes: Kellner *et al.* 1980, MacPherson and Day 1972.

On the mechanism of RF burnout in microwave diodes: Gerzon *et al.* 1975.

On the microwave performance of injection diodes: El-Gabaly 1980.

On the high-frequency response of inversion layers: Hofstein and Warfield 1965.

On submillimetre wave detection with submicron Schottky diodes: Hodges and McColl 1977, McColl *et al.* 1977.

On submillimetre mixing in point contact diodes: Zuidberg and Dymanus 1976.

On microwave mixing in GaAs Schottky diodes: Wortman and Kohn 1975.

4.3 Capacitance of systems with traps

The introduction of localized trapping states has two sets of consequences, as far as capacitive properties are concerned. It alters the relationship between the barrier capacitance C_B and the barrier voltage V_B (Section 4.3.1), and it also makes that capacitance frequency-dependent (Section 4.3.2). Both phenomena can be used, in conjunction with photo-techniques, for deep-level spectroscopy, i.e. attempts to evaluate the distribution of trapping centres over the energy spectrum from barrier measurements.

4.3.1 Capacitance–voltage relationships

We have seen in Section 4.1.2 that the relationship between $1/C_B^2$ and the barrier voltage V_B can be interpreted in a simple way when the donor content is constant, but only with tentative caution when it is variable. A similar situation arises when traps are present, even though the donor concentration may be constant. Because traps in the barrier come to be at different distances from the Fermi level, the local charge density is again a function of x. The question is then whether differential capacitance measurements as a function of applied voltage still serve any useful purpose. The answer is that they do, albeit to a far more limited extent, and after some degree of adaptation of the model to the more complicated situation now envisaged. Even that adaptation (which follows) is possible only as long as we ignore the major (and totally general) problems associated with capacitance concepts in systems of non-uniform space charge, as outlined in Section 4.1.2 (Callarotti 1981).

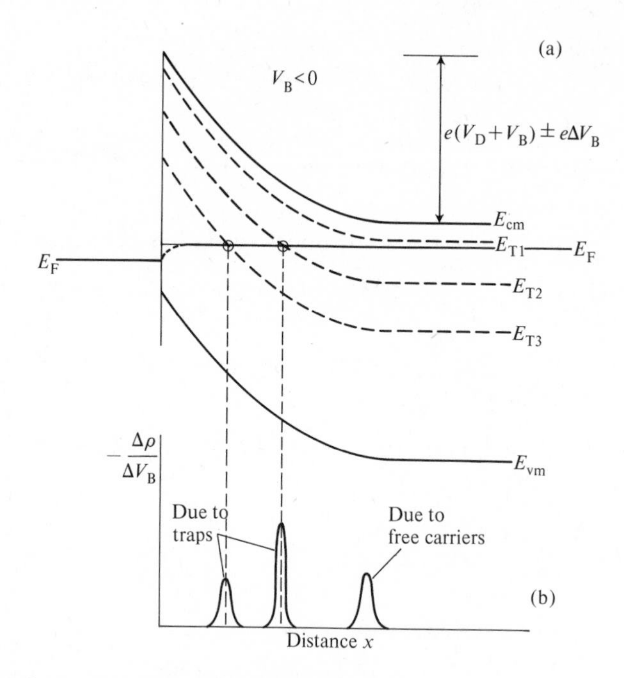

FIG. 4.7. Barrier on material with three trapping levels, E_{T1}, E_{T2}, E_{T3}, and the corresponding contributions to the capacitance. Forward voltage applied ($V_B<0$). After Roberts and Crowell (1970).

In this section, we shall be concerned *only* with low-frequency relationships. For these conditions, Roberts and Crowell (1970) have given a derivation of the capacitance for a system containing a series of localized states, uniform in x, but distributed in terms of energy. The qualitative consequences can be ascertained by reference to Fig. 4.7(a) (which is arbitrarily presented for a system of just three discrete levels). As the barrier is deformed by the application of V_B, e.g. in the reverse direction, new levels of the E_T series will come to be below the Fermi level. Conversely, deformation in the forward direction will bring these states above E_F, and since the entire barrier profile is affected, all the cross-over points (marked with circles) are involved. In the diagram, ΔV_B represents an infinitesimal barrier voltage involved in the differential capacitance measurement, and because the cross-over points are discretely defined, the space charge density will show abrupt changes at the corresponding values of x. In practice, many factors lead to a certain amount of smoothing, and Fig. 4.7(b) shows this. The relative magnitudes of the (here, only two) peaks is, of course, related to the relative trap concentrations at E_{T1} and E_{T2}. The analysis can be extended to a continuous distribution of states, e.g. like that shown in Fig. 3.34.

The differential capacitance is, of course, governed by the total change of charge within the barrier, resulting from the application of ΔV_B. $\Delta\rho/\Delta V_B$ or, in the limit, $d\rho/dV_B$, is already an incremental capacitance, and needs only to be integrated from $x=0$ to the edge of the barrier to yield the total effect. For a single carrier system, and following Roberts and Crowell (1970), this may be done as shown below, beginning with Poisson's equation. Accordingly, we have

$$d^2V/dx^2 = -\rho(x)/\epsilon. \tag{4.3.1}$$

For its integration, it is convenient to observe that

$$\frac{d^2V}{dx^2} = \frac{\left(\frac{d}{dx}\right)\left(\frac{dV}{dx}\right)^2}{2(dV/dx)} \tag{4.3.2}$$

and since the local potential $V(x)$ is single-valued, $\rho(x)$ can be alternatively defined as $\rho(V)$. Substitution of (4.3.2) into (4.3.1) and rearrangement then yields

$$d\{(dV/dx)^2\} = -(2/\epsilon)\rho(V)\,dV$$

or

$$dV/dx = \left\{-\frac{2}{\epsilon}\int \rho(V)\,dV\right\}^{1/2} \tag{4.3.3}$$

which, except for sign, represents the local field. The entire objective of the manipulation is to avoid having to integrate with respect to x. One can do this by noting that the integrated charge over the whole barrier region must equal the compensating surface charge Q_s (both per unit area). Thus

$$Q_s = \epsilon F(0) = -\epsilon(dV/dx) = \left\{2\epsilon\int_{V(\lambda_B)}^{-\phi_{ns}/e} \rho(V)\,dV\right\}^{1/2}. \tag{4.3.4}$$

The differential capacitance C_B is then given by

$$C_B = dQ_s/dV_B = \epsilon\rho(V_s)/Q_s \tag{4.3.5}$$

where Q_s is already known from (4.3.4), and $V_s = -\phi_{ns}/e$. For an experimental verification of the predicted deep-trap effects, see Roberts and Crowell (1973).

There are benefits from expressing the results in this form, and Spear *et al.* (1978) have taken full advantage of those. Thus, the integral in eqn (4.3.4) can be directly evaluated from information of the kind contained in Fig. 3.35, as long as some further simplifying assumptions are made (see below). With the cautions already discussed, it is then possible to calculate the differential capacitance of the barrier as a function of the applied voltage for any given distribution of (spatially constant) localized states.

A simplifying assumption is necessary, because the use of Fig. 3.35 assumes that we know exactly where the Fermi level is within the barrier, relative to the mobility edges, and thus relative to the levels E_{T1}, E_{T2}, etc. Without extensive numerical computation, we do not actually know this and, faced with this impasse, most authors have assumed that it is permissible to take the Fermi level as flat, not only at the semiconductor end of the barrier, but *throughout* its thickness. The problem has already been encountered elsewhere in this book: a flat Fermi level implies *balance* between field and diffusion, as one can easily prove by solving eqn (2.2.8) for zero current. To say that the slope of E_F (while perhaps not zero) is 'small' over most of the barrier does not solve this problem because, even if that were true, there would have to be a place close to the interface where the slope is large if a current is to flow at all. The assumption of flatness is not, therefore, realistic. Nevertheless, a useful approximation may be based on it, as long as attention is limited to the *capacitive* properties of the barrier; along those lines the procedure has proved itself capable of reflecting measurements on amorphous silicon with gold contacts rather well. Spear *et al.* have demonstrated this, though only for reverse voltages below 0.1 volt. The departures observed at higher voltages were not fundamental, but had to do with the frequency effects discussed in the following Section. Along those lines, corrections can be introduced which greatly widen the range of agreement.

The most important result of the analysis concerned the behaviour of C_B, and thus of $1/C_B^2$, as a function of V_B, which, as Spear *et al.* (1978) have pointed out, is 'diametrically opposite' to that expected for crystalline material (Fig. 4.8(a)). This finding is also intimately linked with the limitation of the above arguments to low frequencies, i.e. frequencies at which all localized states are capable of reaching equilibrium occupancy, following the imposed changes of potential. How low is 'low'? In order to achieve their agreement (with the above model as it stands), Spear *et al.* had to make their measurements by a method which implied a frequency of 10^{-3} Hz. The measurements were actually performed by keeping the system at a uniform temperature for a prolonged period, to ensure equilibrium occupancy of all states, and then applying $V_B = 20$ mV, and integrating the resulting current changes over 10^3 s. With corrections for the time-dependent relaxation effects discussed below, satisfactory agreement between theory and measurements was extended up to 10 Hz. At such frequencies, the behaviour of $1/C_B^2$, though not actually linear with V_B, became much more like that for crystalline material (Fig. 4.8(b)). (See also Wronski 1977.) The corrections ignored the deepest traps, on the grounds that these could not be charged and recharged within each cycle.

Of course, the proper evaluation of barrier capacitance (and, indeed, of

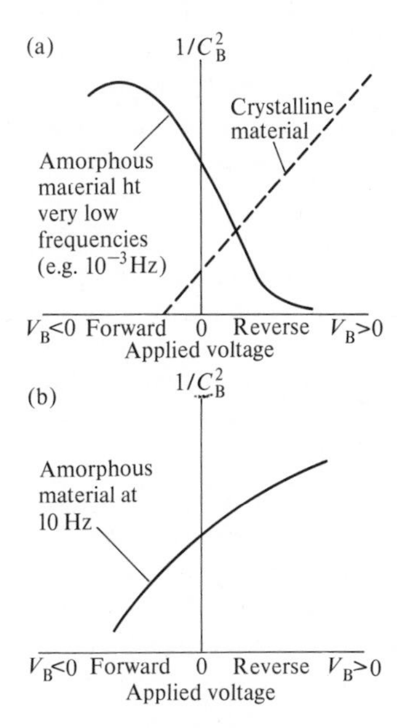

FIG. 4.8. Barrier capacitance as a function of applied voltage in the presence of traps: (a) At very low frequencies, and (b) at low audio-frequencies. Schematically after Spear *et al.* (1978).

many other barrier properties) from measurements calls for the use of an equivalent circuit, as explained in Section 4.1.1. The problems involved in its choice are not trivial by any means but, even if that choice were carefully made, the use of such a circuit would call for measurements over a wide frequency range; and this requirement is bound, sooner or later, to lead to a very real impasse when we are dealing with traps which are themselves associated with time constants (see Snell *et al.* 1979). Despite a great deal of work in this field, these problems remain to be satisfactorily resolved; meanwhile, it would be wise to limit oneself to modest expectations. Complications also arise from other causes. Thus, estimates of barrier *width* from capacitance measurements tend to be distinctly unreliable in the presence of traps, in view of the barrier shape shown in Fig. 3.36. Indeed, the 'width' lacks a clearcut operational definition. The situation is even more complicated when an oxide layer is present. The interface states associated with such a system constitute carrier traps, and since these do not acquire and release their charges instantaneously, they lead to a marked hysteresis in the C–V relationship (Kaplan 1981). The effect is most important in the forward direction, as one would expect, and causes substantial departures from the ideal straight-line behaviour of $1/C_B^2$ versus V_B (Deneuville 1974).

The change of capacitance under illumination can also be used for clarification. Each system tends to have its own complex series of responses, but the principle is simple enough: when a trap is emptied within a barrier, it tends to remain empty because the high barrier field suppresses recombination. In any event, the barrier region is almost free of electrons which might otherwise take part in recombination. As a result, illumination is a highly effective means of altering charge density, and hence of modulating the barrier capacitance. Marfaing *et al.* (1974) have demonstrated this for gold contacts on n-type CdTe. Of course, trap-emptying by irradiation takes time, which means that photocapacitive changes are never instantaneous; they may typically continue for several hundred seconds. Radiative trap-filling by electrons makes the space charge more negative, and thereby decreases the capacitance; radiative trap-emptying has the opposite effect. In simple cases, the sign of the photocapacitance can thus give information about the energetic position of the traps. Grimmeis (1974) has shown that the nature of the $1/C_B^2$ versus V_B relationship, when measured below the freezout temperature, depends on the wavelength and intensity of previous illumination, and this is indeed what one must expect.

4.3.2 *Frequency dependence of the barrier capacitance*

The basic problem has already been encountered in Section 4.2. It takes time for traps (and, indeed, for donors and acceptors) to capture or release charge, and if the frequency of the applied ripple voltage is too high, the traps cannot charge and recharge in response. As a result, the capacitance must decrease. Thus, for instance, Beichler and his co-workers (1980) have observed such a decrease for Pt contacts on amorphous Si, amounting to a factor of 10, as the frequency changed from 10^{-3} to 10^2 Hz. In the present context, we can write eqn (4.1.6) in the form

$$\frac{d(1/C_B^2)}{dV_B} = \frac{2}{\epsilon\bar{\rho}(\lambda_B)} \tag{4.3.6}$$

where $\bar{\rho}(\lambda_B)$ is the *average* charge density in a barrier width λ_B. The higher the frequency, the lower $\bar{\rho}$ and hence, other things being equal, the greater the slope of the $1/C_B^2$ versus V_B relationship. For an amorphous semiconductor with a mobility edge E_{cm}, the average thermal capture time of an electron by a trap of capture cross-section S_T at an energy E_T is given by

$$\begin{aligned}\tau_T &= (\mathcal{N}_c\bar{v}S_T)\exp(E_{cm}-E_T)/kT\\ &= 1/\nu_T\end{aligned} \tag{4.3.7}$$

and ν_T could be called a trap escape frequency. For amorphous silicon, Spear *et al.* (1978) have used

$$\nu_T = 2\times10^3\cdot\exp\{-(E_{cm}-E_T)/kT\} \tag{4.3.8}$$

Thus, at frequencies higher than ν_T, traps at E_T and below will play a diminishing role. A trap at a depth of 1 eV would (at room temperature) fail to respond to frequencies greater than $2 \times 10^{13} \cdot \exp(-40) \approx 10^{-4}$ Hz! Of course, the E_{cm} used here would be simply E_c in a crystalline material. If the trapping states were in fact discretely defined, as assumed in Fig. 4.7, then the frequency spectrum would likewise show pronounced capacitance steps at the corresponding frequencies. However, even for a continuous trap distribution the frequency can always be calculated from eqn (4.3.8), as long as the trap concentration profile is known. By simple analogy with the classical Debye argument, the effective density of traps at an (angular) measurement frequency ω would be

$$M_{0\,\text{eff}} = \frac{M_0}{1 + \omega^2 \tau_T^2} \tag{4.3.9}$$

which is plotted in Fig. 4.9 for amorphous silicon. The curves show how rapidly successive layers of traps are rendered ineffective as the frequency increases. By simply subtracting these traps from the known trap-distribution profile. Spear *et al.* (1978) were able to reach their agreement with experiment. It is also obvious that the capacitive behaviour must tend to that of trap-free crystalline systems as all traps become inactive, i.e. at high frequencies. Correspondingly, the capacitive behaviour of amorphous systems tends to become more like that of crystalline systems, as the trapping states in the gap are eliminated.

Roberts and Crowell (1970) have likened a multitrap system to a parallel combination of R–C series circuits, each with its own time

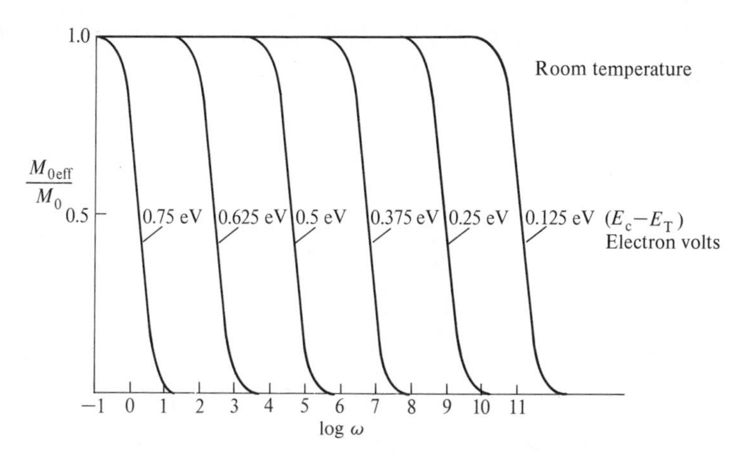

FIG. 4.9. Trapping efficiency as a function of applied frequency for traps of identical cross-section at various trapping levels E_T. As the frequency (ω) increase, fewer traps can 'follow' the signal. Deep traps have the longest time constants, and are therefore eliminated first from participation in the charge exchange.

constant, to reflect a particular trapping time. The capacitive component ΔC_T may be calculated as above for each trapping level, and the resistive series component is simply $\Delta C_T/\tau_T$. In this way, the problem can be schematically converted into one of circuit analysis. The curves presented by Roberts and Crowell reflect all the aspects of the observed behaviour, including the 'corrugation' of the $1/C_B^2$ versus V_B curves in systems with discrete E_T states. (See also Crowell and Nakano (1972) for a theoretical development, involving a single deep donor level, partially ionized, and also a set of completely ionized shallow impurities.)

Further reading

On the use of capacitance–voltage measurements for the exploration of band profiles: Diligenti *et al.* 1980, Salardi and Pellegrini 1979.

On the use of capacitance–voltage measurements in the presence of insulator layers, for the investigation of traps in the semiconductor: Cook *et al.* 1980.

On the characterization of multiple deep-level trapping systems by admittance measurements: Beguwala and Crowell 1974.

On the capacitance of GaP systems: Vincent *et al.* 1975.

On capacitance spectroscopy of deep impurity levels at contacts, with ZnTe: Losee 1972; with Ge: Pearton 1982.

On measurements of photocapacitance, with ZnS: Lee and Henisch 1969; with GaP: Vincent *et al.* 1975.

On the space-charge capacitance of M–I–S systems: Grove *et al.* 1964.

5

METAL–SEMICONDUCTOR CONTACTS UNDER ILLUMINATION

5.1 Opto-electronic principles; application to simple barriers

THOUGH some contacts in darkness can evidently be represented as single-carrier systems (see Chapter 2), contacts under illumination always involve electrons and holes, and are therefore governed by the considerations of Chapter 3. It was found there that numerical solutions are needed even in the absence of light and, of course, they are required here. In principle, one could take the same equations and adapt them by adding terms for the optical electron–hole pair generation and for recombination at the contact interface. New solutions could then be obtained. Indeed, only this procedure can be relied on to give sophisticated answers, always assuming that algorithm and boundary conditions have been properly chosen. Even so, the computation is not simple, and it is often desirable to seek analytic approximations into which the various descriptive parameters of the system enter with greater clarity. In as much as there exists a body of knowledge under this heading, it is a collection of such attempts at simplification, and some of these (but, of course, not all) will be described below. Just how satisfactory these approximations are is hardly ever a straightforward question, since we cannot be sure that any actual situation under review corresponds in all respects to the assumptions made by the model. This is true even in darkness, but under illumination the uncertainties are actually greater. New parameters enter into the picture, in particular the optical reflectivity and absorption constant as a function of wavelength. It is therefore wise to approach models with realistically modest expectations when direct comparisons with experiment are intended. Even so, the field has already provided much scientific insight, as well as technologically important guidance for the design of electro-optical devices in general, and solar cells in particular.

The modern interest in all energy-related matters has motivated a great deal of research into solar cells of various types and, accordingly, the subject now has a vast literature of its own, primarily addressed to the design specialist (see literature 'for further reading' at the end of this section). Substantial advances in solar cell engineering have been made in recent years, both, in connection with crystalline and amorphous structures. Important as they are, most of these matters will not be our concern here. The overview which follows will confine itself to essential ideas that relate to contacts (as distinct from junctions) and to the relationship between analytic and numerical approaches.

5.1.1 *Qualitative survey of photo-effects; terminology*

Though complex in terms of detail, the action of light on a contact is simple in principle, and consists of three parts:

(a) The bulk concentrations of charge carriers are augmented, e.g. from n_e and p_e to (n_e+n_L) and (p_e+p_L), where n_L and p_L depend on the light intensity. This alone is sufficient to give the system a new, modified voltage–current characteristic, as shown in Fig. 5.1.
(b) Additional electron–hole pairs are generated within the barrier itself and are separated there from one another by the high field, as shown in Fig. 5.2. If these separated charges manage to reach the boundaries of the barrier before recombining, they contribute to the current and, indeed, the barrier may be considered as a current-generator.
(c) Since light is absorbed as it penetrates (in a manner governed by the bulk band-structure) shallow depths of the system will experience more electron–hole pairs than deeper regions. Accordingly, a (distributed) *Dember effect* voltage V_{Dem} will be generated within the bulk components of the structure. In comparison with *externally* applied voltages, V_{Dem} is usually small; but in the absence of such voltages it can become prominent (Fig. 5.3).

Mechanisms (a) and (b) are clearly distinguished from one another only at zero current. When current is flowing, it leads to non-equilibrium

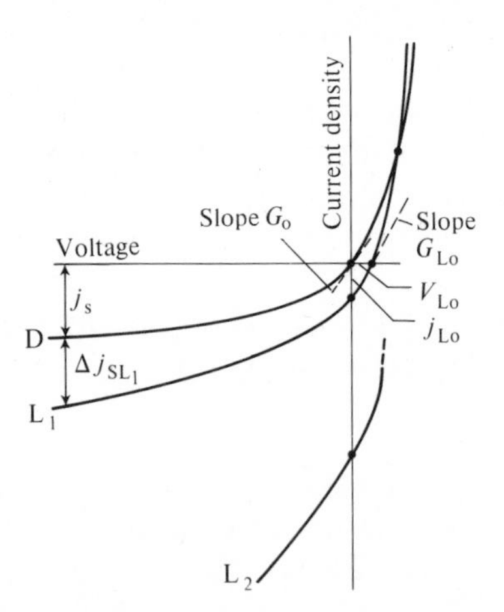

FIG. 5.1. Effect of light on a metal–semiconductor barrier structure; schematic characteristics. D: in darkness; L_1, L_2: under increasing light intensities, L_1: $p_L/p_e \approx 10$, L_2: $p_L/p_e \approx 100$; j_{LO} = short circuit current density; V_{LO} = photovoltaic e.m.f.; reverse currents positive, photovoltaic e.m.f. negative.

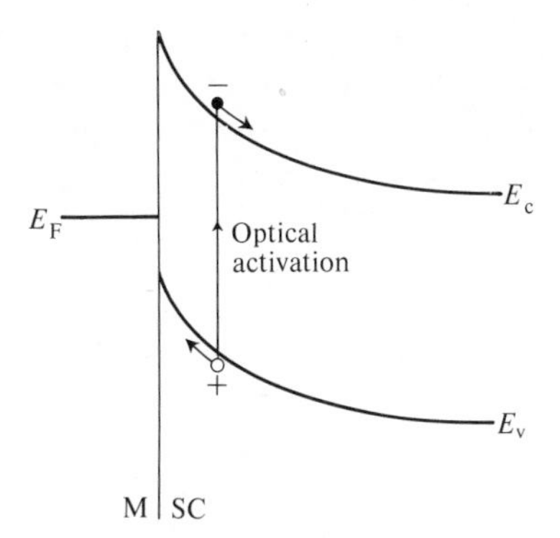

FIG. 5.2. Separation of light-generated electron–hole pairs in the barrier field. (Recombination not shown.)

carrier concentrations on its own at the edge of the barrier, and these can no longer be separated from p_L and n_L. Even at zero current, the relative importance of these effects may vary.

A numerical solution of the transport equations in the presence of light excitation (varying with depth in accordance with the absorption law) would take care of all these effects at the same time, but such solutions are not easily obtained. It is therefore instructive to look at partial, analytic solutions (under the heading of 'illustrative models'), although a good deal of the published literature fails to identify the underlying assumptions with sufficient clarity. Processes (a) and (c) are obviously most important when the light penetration is deep compared with the barrier thickness, so that only a few carrier pairs are generated within the high field region. The opposite applies to (b), which will evidently be most prominent for light of a wavelength that is quickly absorbed. Excitation

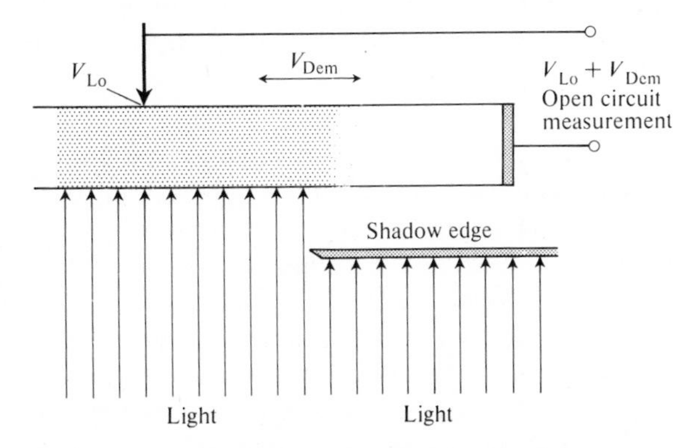

FIG. 5.3. Generation of a Dember effect voltage in a non-uniformly illuminated semiconductor.

by particle fluxes also comes under this heading; e.g. see Cornet *et al.* (1970) for an account of Au–CdTe contacts under 5 MeV α-particle flux at low temperatures. Of course, contact systems are not *stable* under intense bombardment, as Darley and Christopher (1974) have shown by experiments on In–Ge contacts with 3 MeV proton irradiation.

Within a contact barrier in darkness, and in the absence of current flow, diffusion and field effects are exactly balanced. Illumination upsets this balance and an open-circuit voltage (*photovoltaic e.m.f.*) V_{Lo} then appears, because such an e.m.f. is necessary to guarantee zero current. It will be seen from Fig. 5.1 that it is essentially a 'forward' e.m.f., and this is easily explained. In a typical situation (transparent contact on an extrinsic n-type semiconductor) the incoming light might augment n_{e} by a few per cent, and thereby cause a very small electron diffusion current to flow in the forward direction. However, this effect is greatly exceeded by the current due to hole flow. That current (holes into the barrier) is ordinarily controlled by p_{e} and (under extraction conditions) by the rate at which holes at the edge of the barrier can be replaced. The augmentation of p_{e} by p_{L} thus leads to an almost proportionate increase in hole current, and that is in the reverse direction. The net current is, therefore, a reverse current, and if that current is to be zero, a forward voltage must necessarily be applied. The photovoltaic e.m.f. is that forward voltage. It serves to ensure that there is a forward injection current which exactly balances the drift of light-generated holes towards the metal (which the diffusion of light-generated electrons in the same direction cannot balance on its own).

Alternatively we may consider the short-circuit conditions, under which the very same mechanisms cause a *short-circuit current* (density) j_{Lo} to flow, as shown in Fig. 5.1. Open-circuit and short-circuit conditions represent two fixed points on the voltage–current characteristic under light but, of course, all other contact properties are sensitive to illumination, e.g. the barrier capacitance (see below), and the current composition ratio. All can in principle serve as measures of the incoming light intensity.

When a substantial forward voltage is applied to a barrier in darkness, a stage is reached when the total resistance is bulk-controlled, rather than barrier-controlled. Under illumination, the system as a whole is then governed by the photoconductive properties of the bulk material. Accordingly, the current must be higher than in darkness. This fact is responsible for the cross-over of the D and L curves in Fig. 5.1. However, over a certain voltage range, just before cross-over, the current density is actually *reduced* by illumination, and this is sometimes confused with negative photoconductive effects which can (in highly specialized situations) arise in bulk material. It is here a barrier phenomenon.

The differential conductance at zero current under illumination, and the corresponding conductance *change* ΔG_{Lo} are easily measurable parameters, and can be shown to be sensibly proportional to the light intensity. ΔG_{Lo} can therefore be used to provide at least an approximate measure of p_L, whether arising from light (as here) or from other mechanisms (e.g. minority-carrier injection by some contact nearby).

Of course, a barrier system does not react to light as such, but to the extra carriers generated, and reacts in the same way to extra carriers generated by some other mechanism, e.g. by injection. The equivalent of the photovoltaic e.m.f. in response to *injected* carriers is called a *floating potential* (V_f). Such a floating potential may be large or small but its presence, as such, is inevitable. This is a matter which affects the confidence we can place in (currentless) potential probe measurements of all kinds. Probes placed on a solid acquire the local potential *only* if that solid is in electronic equilibrium. If it is not, then the probe potential will differ from the lattice potential by an amount which eepends on the nature of the contact barrier and on the degree of non-equilibrium (see Section 5.1.6).

On n-type material (assuming a conventional barrier), the floating potential makes the metal contact positive with respect to the semiconductor. This is once again due to the fact that additional holes will be drawn into the barrier by the barrier field, to decay on the metal–semiconductor interface. The metal thus acquires a positive charge, and the resulting potential difference between semiconductor and metal represents a forward bias on the barrier. Indeed, this is *how* the forward bias is generated that is needed to make the total current zero.

Appearances to the contrary, neither the calculation nor the measurement of floating potentials is a completely simple problem. Difficulties arise in part from the fact that the very presence of the surface on which a potential probe rests gives rise to a local concentration gradient. The magnitude of this gradient is, of course, determined by the degree of non-equilibrium in bulk, and by surface recombination. For this reason surfaces are often sand-blasted before probing, a method which creates (in effect) an infinite surface recombination velocity. The hope is that this will not only 'standardize' the surface, but will make V_f small. In contrast, careful etching of the surface would diminish the concentration gradient by diminishing surface recombination, but would increase the floating potential by giving rise to a higher barrier.

A second practical problem arises from the Dember voltage described above. The Dember e.m.f. exists (inevitably) between illuminated and unilluminated parts of the solid. It can be calculated on the basis of simplified models (e.g. see van Roosbroeck 1950), or else numerically computed (e.g. Moreau, Manifacier, and Henisch 1981). If the contact barrier in series with it is sufficiently prominent, we have $V_{\mathrm{Lo}} \gg V_{\mathrm{Dem}}$. The externally measured open-circuit e.m.f. then equals the e.m.f. across the barrier to a good approximation. However, when V_{Lo} is also small, the external measurement ($V_{\mathrm{Lo}} + V_{\mathrm{Dem}}$) has no simple interpretation. The problem cannot be eliminated by the removal of the shadow edge in Fig. 5.3; that would remove the Dember e.m.f. from the bulk material, only at the cost of replacing it by a new floating potential at the base contact. At zero current, we would then have two floating potentials in opposition; differences in contact structure tend to make them unequal. Floating potentials have a disconcerting habit of entering into many types of measurements, in which their presence is unsuspected. They deserve more serious attention than they generally tend to get (see Section 5.1.6).

It will be clear from the present considerations that it matters not only *whether* an incoming photon is absorbed (resulting in the generation of an electron–hole pair), but *where*. On these grounds, the spectral sensitivity of the barrier response should be similar to that of the bulk (photoconductivity), but not identical. The dependence of response on the *locus* of absorption can, at times, have spectacular consequences, of which two should be briefly mentioned. Thus, Seib (1971) has described structures (made by diffusing Cu into heavily n-type GaAs) which exhibit a reversal in the polarity of the photovoltage with increasing photon energy, the reversal being strongly bias dependent. A different situation has been the subject of investigations by Ng and Card (1981), namely one arising from the potential profile created by the image force. Reference to Fig. 2.8 shows the field reversal close to the metal interface very clearly, and any hole–electron pair generated *within* that boundary region will, of course, be field-separated in the wrong way. In that region, the metal collects *majority* carriers, and a loss of efficiency at short wavelengths is bound to arise from this effect, just as it does ordinarily from the surface recombination promoted by minority-carrier influx. The width x_{m} of the field-reversal region is, of course, easily calculated, as in Section 2.1.5, and on this basis Ng and Card found the bias dependence of the photocurrent in good agreement with the image-force model for Au–n-Si contact systems.

A set of observations by Carlson (1977) is of fundamental as well as practical interest. Carlson has demonstrated that, in Schottky barrier cells based on amorphous silicon, the open-circuit voltage increases systematically with the work-function of the contacting metal. This shows that 'Fermi-level pinning' is not a significant process in such systems, as one might intuitively expect.

As mentioned above, illumination of a barrier affects not only conductive, but also its capacitive properties (*photocapacitence*). Since the barrier-width changes under light, the barrier capacitance must also change and should generally increase. Other factors, besides thickness, are involved (e.g. changes of dielectric constant due to the changed trap occupancy) but they are generally orders-of-magnitude smaller. Thus, many observations which purport to detect the photocapacitance in bulk samples are believed to be in fact due to contact barriers, either at the electrodes or between grains or both. Lee and Henisch (1969), for instance, reported (differential) capacitance increases by factors of the order of 8 which are not plausibly interpretable in any other way.

Light actually changes the barrier in three different manners:

(a) by changing the boundary carrier concentrations on the bulk side,
(b) by the photoexcitation of traps within the barrier region, and subsequent removal of the liberated carriers by the high field, and
(c) by the adjustment of trap occupancies within the space charge region to the new majority- and minority-carrier distributions.

These effects have not yet been clearly isolated from one another in the analysis of experimental results, and attempts to do so are indeed full of difficulty. Effect (b) should, of course, be sensitively field-dependent; and a field-dependent capacitance creep (over a period of many seconds) has been reported (Lee 1969) on Au–collodion–Ge contact systems. Important electrode edge effects complicate all such measurements (see also Lee and Henisch (1968), and other references given there). (Sah and co-workers (1969) were able to use effect (b) for impurity level spectroscopy in p–n *junctions* by measurements of photocapacitance.)

5.1.2 *Illustrative models; analytic and quasi-analytic solutions*

The search for analytically solvable models is the search for judicious approximations but, as a common experience, approximations which may seem judicious in one context may seem misguided in another. Earliest treatments take the concentration increments n_{L} and p_{L} as equal and uniform throughout the bulk material. Recombination is neglected which makes it possible to deal with the hole- and electron-currents independently. The adoption of a hemispherical contact structure helps to make this plausible, because it ensures that the rapid passage of carriers through the region which is the principal seat of resistance. In the same way, hemispherical structures avoid some of the complicating effects due to series resistance. These simplifications are adopted below. To make the equations tractable, Bardeen (1950) and Banbury (1953a) assumed, in addition, an injection ratio of unity ($\gamma = 1$), as well as $p_{\mathrm{L}} \ll n_{\mathrm{e}}$. The resulting model yielded two widely used conclusions, namely expressions for the open-circuit voltage V_{Lo} and the total short-circuit current I_{Lo}.

$$V_{\mathrm{Lo}} = -\frac{kT}{e}\log\left(1+\frac{p_{\mathrm{L}}}{p_{\mathrm{e}}}\right) \tag{5.1.1}$$

$$I_{\mathrm{Lo}} = 2\pi r_{\mathrm{B}}\mu_{\mathrm{p}} kT p_{\mathrm{L}}. \tag{5.1.2}$$

(In terms of the present sign conventions, I_{Lo} corresponds to J in planar systems, being positive in the reverse direction. In Fig. 5.1, j_{Lo} represents the true current *density*.)

Some of the difficulties involved were soon recognized: even if electron currents could be neglected in darkness, they would have to play some role under light. Interaction terms between electron- and hole-currents were schematically introduced under the heading of 'current gain' (as observed in transistors), but could not be calculated from the model itself. For this reason, a somewhat different approach is adopted below. Meanwhile, let us note that for $\gamma = 1$ (only!), as envisaged above, the diode saturation current would be roughly proportional to p_e in darkness, and of the same structure as eqn (5.1.2). This means that eqn (5.1.1) could be rewritten in terms of observables as

$$V_{Lo} = -\frac{kT}{e}\log\left(1+\frac{I_{Lo}}{I_s}\right) \tag{5.1.3}$$

though I_s is not always as well defined as one might wish. An equation of that type can actually be arrived at on thermodynamic grounds alone (e.g. see Rose 1960 and Müser 1957.

There are several possible strategies for going beyond the simplest notions outlined above, of which three shall be mentioned. The first is to adapt the treatment (for darkness) outlined in Section 3.5.1, e.g. as described by Braun and Henisch (1966b). Geometry and terminology are again as shown in Fig. 3.23, escept for the replacement of n_e by $(n_e + p_L)$, and of p_e by $(p_e + p_L)$. This model assumed neither the neutrality condition, for $\gamma = 1$, but retains the idea that recombination can be neglected, and for a hemispherical system this is particularly plausible. For this degree of additional flexibility there is, however, a heavy price to pay: most of the relationships obtained are implicit, and therefore involve numerical solutions in part. This is not a penalty of the hemispherical geometry; it is a general feature of such models. In the model outlined below, the additional assumption $p_L \ll n_e$ limits the status of the treatment to that of a 'small-signal theory' (low levels of illumination). The carrier concentrations at the metal–semiconductor surface are taken to be current-independent, as before, and (now) also as illumination independent. With these provisions, eqns (3.5.10) and (3.5.11) become

$$I_{nL} = \frac{2\pi e D_n n_s}{W_n'(V_T, r_0)}\left\{1 - \exp\left(\frac{-eV_T}{kT}\right) - \frac{n_L}{n_e}\,\mathrm{esp}\left(\frac{-eV_T}{kT}\right)\right\} \tag{5.1.4}$$

and

$$I_{pL} = \frac{2\pi e D_p p_s}{W_p'(V_T, r_0)}\left\{\exp\left(\frac{eV_T}{kT}\right) + \frac{p_L}{p_e}\exp\left(\frac{eV_T}{kT}\right) - 1\right\} \tag{5.1.5}$$

bearing in mind that $n_L = p_L$. Implicit in this treatment is the assumption that mechanism (a) of Section 5.1.1 is active on its own, equivalent to assuming that the contact is exposed to deeply penetrating light. V_T is, as before, the total voltage applied. For the hemispherical system, it has the status of an asymptotic limit. The terms W_n' and W_p' are V_T-dependent functions of the same kind as those used in Section 3.5.1, except for the fact that the barrier profile is slightly changed by illumination. That change makes it impermissible to regard the dark-current and the photo-current as additive. In practice, the two currents interact via the space charges that they generate, and does so even while recombination is neglected. W_n' and W_p' can be numerically evaluated, and this involves either some judicious guessing of the barrier shape (in which case the evaluation takes only one stage) or else an iterative process of successive approximation. The process is tedious, but does permit useful comparisons between theory and experiment *without* curve fitting.

When I_{nL} is negligibly small (and as long as we neglect the Dember effect), we can calculate V_{Lo} by putting $V_{Lo} = V_T$ and $J_{pL} = 0$. This yields eqn (5.1.1) again, independent of barrier height and shape, whereas the complete voltage–current characteristic does depend on these parameters. When I_{nL} is not negligibly small, V_{Lo} must be found numerically from $I_{nL} + I_{pL} = 0$. It is then always smaller than the value given by eqn (5.1.1), and is no longer independent of barrier parameters. Accordingly, the photovoltaic e.m.f. V_{Lo} is the highest and most reliable measure of p_L when the injection ratio of the contact is unity. However, since p_L appears under the logarithm, V_{Lo} is never a *sensitive* measure of optical excitation.

Figure 5.4 shows how the V_{Lo} computed in this way varies with barrier height. The corresponding short-circuit current ($I_{nLo} + I_{pLo}$) can be shown

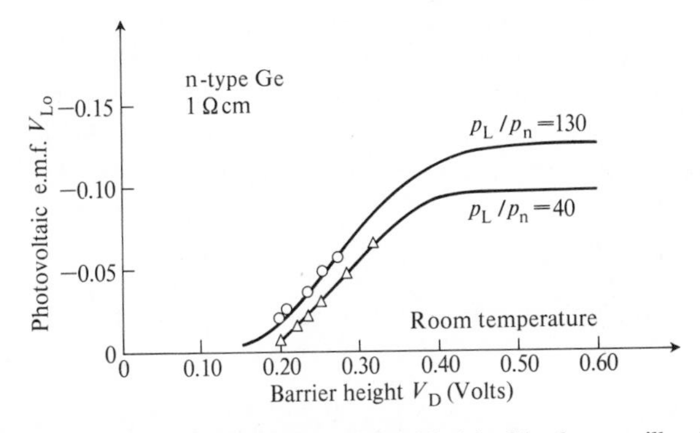

FIG. 5.4. Photovoltaic e.m.f. as a function of initial height V_D, for two illumination levels. Contacts on germanium. Calculations (full lines) and experiments (○, △) after Noble, Braun, and Henisch (1967).

to be almost proportional to p_L, at any rate for high barriers. For low barriers, it is less than proportional, but even then mostly composed of holes. On this basis, Braun and Henisch (1966b) have also calculated the voltage–current relationships for electrons and holes, as well as their sum, which is the total voltage–current relationship (I_{nL} and I_{pL} being connected through their field integrals W_n' and W_p', which are governed by the common barrier profile). This has shown that incremental currents due to light are no longer proportional to p_L in the presence of applied voltages. The very fact that the electron- and hole-currents are coupled makes the system non-linear.

Noble, Braun, and Henisch (1967) have used this model for the evaluation of voltage–current characteristics measured on freshly cleaved Ge surfaces (under oxygen-free liquids) by means of mercury-jet micro-contacts. This technique was employed to obtain contacts which were mobile over the surface, highly localized, but incapable of deforming or damaging the crystal surface. Barrier heights were evaluated from the dark-characteristics, and p_L from the light-characteristic. The barrier heights determined in this way behaved in accordance with expectations as a function of doping, which helped to generate confidence in their validity. Other things being equal, the heights increased with increasing crystalline disorder, presumably due to the appearance of additional interface states.

A completely different strategy and approach (due to Gutkowicz-Krusin (1981)) to the problem of photocurrents is outlined below, but some of the more obvious complicating features must first be discussed. Note, for instance, that all the quantitative aspects considered so far are linked to the action of deeply penetrating light. Under more general conditions (and reverting to planar geometry), one might consider that there is a normal dark current (density), augmented by a current density J_L due to light. As mentioned above, the two quantities are not inherently additive, but are often so regarded, a practice which is occasionally elevated to the ‘principle of linear superposition’. On the other hand, there do indeed appear to be some instances in which this idea has led to useful agreements with observations. Proceeding along those lines, one could write

$$J = J_s\{1 - \exp(-eV_B/kT)\} + J_L(V_B). \tag{5.1.6}$$

The open-circuit voltage ($V_B = V_{Lo}$ when $J = 0$) can then be obtained in the form

$$V_{Lo} = -\frac{kT}{e}\log\left\{1 + \frac{J_L(V_{Lo})}{J_s}\right\}. \tag{5.1.7}$$

Compare eqn (5.1.3). The inclusion of a non-ideality factor η would simply replace kT/e by $\eta kT/e$ and would thus, in the ordinary way, make V_{Lo} larger (see Section 5.1.3). J_L itself would remain to be determined and, since it is itself a function of V_{Lo}, it cannot be truly proportional to the incident light intensity (see also Tarr and Pulfrey (1980) for a general discussion of linear superposition). If it were not for the fact that interface recombination creates concentration gradients, J_L would be zero under flat-band conditions, since those correspond to field-free homogeneity. In the presence of reverse voltages, the situation is more complex. The barrier width increases, thereby increasing the region within which light-generated electron–hole pairs are separated by the high field. At the same time, their chances, of reaching the barrier boundaries before recombination, may decrease. The net effect is generally a small increase of J_L, and the situation is then as shown in Fig. 5.5, where $J_L(V_{Lo})$ is smaller than J_{Lo}, but not necessarily much smaller. The difference has been emphasized here for purposes of illustration (see also Section 5.2.3). At any rate, eqn (5.1.7) could be coarsely tested on the basis of $J_{Lo} \approx J_L(V_{Lo})$ assuming that J_s is reasonably well-defined. A reliable value of $J_L(V_{Lo})$ can only be obtained by numerical methods (always assuming that the parameters fed into the computation are themselves reliable).

The non-linear interaction between dark- and photo-currents can give rise to *photoconductive gains* greater than unity, by which is meant a (bias-dependent) rate of carrier flow across the boundary which is greater

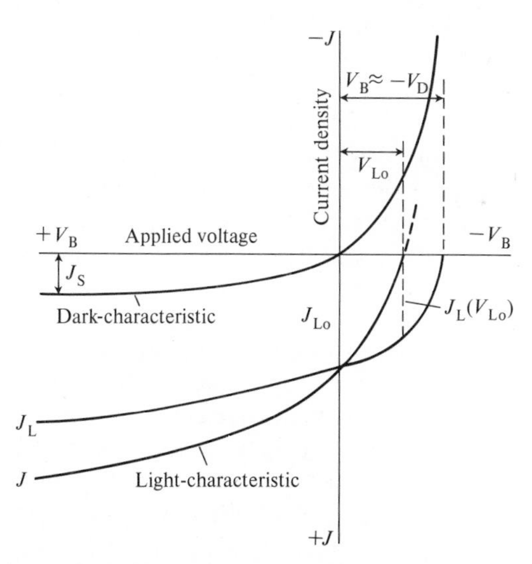

FIG. 5.5. Schematic representation of photovoltaic and photoconductive effects at a metal–semiconductor barrier under illumination. Construction of the photoresponse by linear superposition of dark current and light-generated current increments. Reverse currents, positive; photovoltages, negative.

than the optical generation rate. The barrier then acts in two ways, as a current-generator, and as a current-limiter. Gains as high as 10^6 have been reported by Bücher *et al.* (1973) for Au–CdS contacts with narrow Cu-compensated layers, and values only up to 300 or so by Mehta and Sharma (1973) for contact systems on CdSe. The reasons for this enormous difference remain unclear. Similar observations have been made by Gill and Bube (1970) on Cu_2S–CdS heterojunctions, in which the Cu_2S played the role of the metal. The gains diminished with increasing photon flux, as they are bound to do, with electron and hole-populations gradually equalizing.

When the semiconductor involved has a high dark-resistivity, the fact that the dark-current component may be (and often is) injecting has an important bearing on the overall behaviour of the system, and this is confirmed by the observations of Mandelkorn and Lamneck (1973) on high-resistivity silicon. In all simple models, this injection is necessarily neglected, because there *is* no simple way of treating it. However, Berry (1974) has shown that the prevailing minority-carrier concentrations can in fact be orders-of-magnitude greater than those projected by the assumption of equilibrium. This is entirely reasonable, *inter alia* because the density of photo-generated carriers can greatly exceed that of thermally generated carriers, and often does so. Any injecting component of the dark current is, of course, opposed to the minority-carrier 'collecting' function of the contact, and this diminishes efficiency. Correspondingly, the free majority-carrier concentration within the barrier may also be substantially higher under illumination than it is in equilibrium (Heasell 1981), and this also diminishes the efficiency by enhancing the majority-carrier forward current. As one would expect, this efficiency loss is most serious for illumination by light of short wavelength, since this results in a high rate of majority-carrier generation close to the interface, where their chance of escape is a maximum. For calculations of this effect (based on the assumption of a Schottky barrier of unperturbed shape) see Lavagna *et al.* (1977).

There are circumstances in which any one of the complicating features outlined above may have an essential role, making numerical treatment mandatory; but there are also others in which useful solutions can be obtained analytically by a judicious selection of model structures and parameters. The systems described by Gutkowicz-Krusin (1981) are of that kind. They envisage planar geometry, a barrier of fixed exponential profile, highly extrinsic material of finite thickness, and variable bulk lifetimes, through only two extreme values of the recombination velocity at boundaries (namely zero and infinite). The carrier generation rate is taken as an exponential function of depth. In such circumstances the collection efficiency (defined as the ratio of photocurrent to the generation rate) can be determined by solving the transport equation for holes

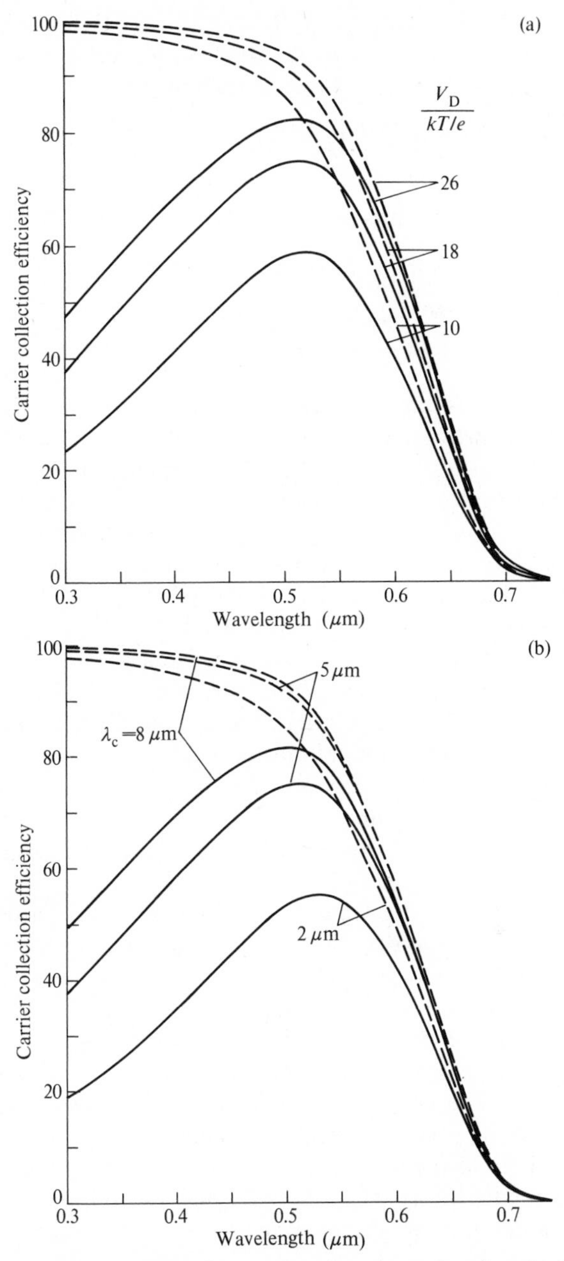

FIG. 5.6. Carrier collection efficiencies as a function of wavelength and other parameters. Solid lines: infinite surface recombination; broken lines: zero surface recombination. Results calculated for contacts on α-SiH_x, under low-level; illumination. (a) For different values of barrier height in units of kT/e. $\lambda_c = 5\ \mu$m. (b) For different values of barrier thickness parameter. λ_c (μm). Specimen thickness, $L = 1\ \mu$m. Diffusion length, $L_p = 0.1\ \mu$m. Band gap, $E_g = 1.7$ eV. After Gutkowicz-Krusin (1981).

alone. This is done, beginning with the continuity relationship eqn (3.2.8) to which, of course, the optical generation term must be added. With the present assumptions, we then have approximately

$$\mu_p \frac{d(pF)}{dx} - \frac{\mu_p kT}{e} \frac{d^2 p}{dx^2} - \frac{p}{\tau_p} + g_L(x) = 0 \tag{5.1.8}$$

where g_L is the optical carrier generation rate, given (at any particular frequency ν) as

$$g_L(\nu, x) = g_0 \cdot \exp\{-a(\nu)x\}. \tag{5.1.9}$$

It is assumed that the field can, with sufficient accuracy, be described by

$$F(x) = F(0)\exp(-x/\lambda_c) \tag{5.1.10}$$

where λ_c could be found by capacitance measurements. Despite these many simplifying provisions, the solution (though complex) has yielded instructive relationships: concentration contours, as well as overall efficiencies as a function of wavelength. Figure 5.6 gives representative results. As one must expect, the outcome depends on the relative magnitude of four characteristic lengths involved here: the specimen thickness L, the diffusion length L_p, the optical absorption length $a(\nu)^{-1}$, and the effective barrier thickness λ_e. Gutkowicz-Krusin applied the method specifically to the analysis of amorphous silicon hydride solar cells. Hydrogenated silicon does not, however, have the sharp band edges assumed by the model. Morel and Moustakas (1981) *inter alia* have shown that the shape of the band tails depends on hydrogen content, and have demonstrated that this has an effect on the open-circuit photovoltage. Phosphorus, fluorine, and boron likewise have such effects (e.g. see Moustakas *et al.* 1982). All this means that unsharp band edges cannot be safely ignored.

An approximate analytic treatment by Dubey and Paranjape (1977) avoids linear superposition but neglects recombination within the barrier, even though it deals with planar systems. We now know from numerical calculations (see Section 3.5.3) that this is not as drastic a simplification as it was at one time believed to be. Dubey and Paranjape also formulate the boundary conditions so as to include the bulk material, excited in varying degree by the penetrating radiation. The analysis is elegant, but the final results must be numerically evaluated, which makes this into a hybrid method. It also raises a question of tactics, in the sense that it might be more satisfactory to apply numerical procedures to the problem as a whole, rather than to the last part of an analytic approximation.

5.1.3 *Illustrative models: power output and fill factor*

In the right circumstances, optical excitation can support electrical power generation. It cannot, of course, do so under open-circuit conditions

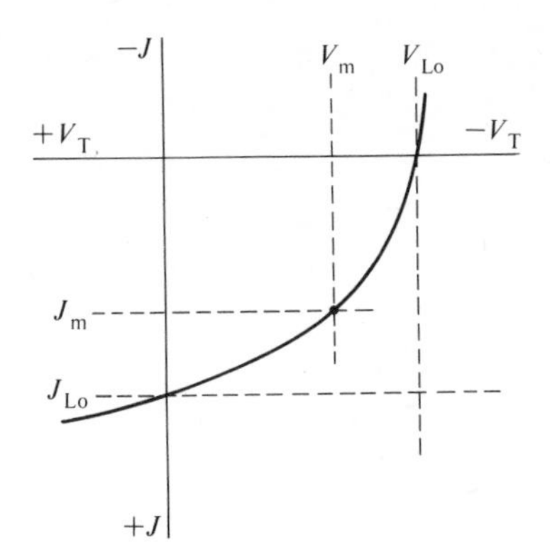

FIG. 5.7. Power generation by a barrier under illumination. Fill factor = (FF) = $J_m V_m / J_{Lo} V_{Lo}$. Reverse currents, positive; photovoltages, negative.

(when $J=0$), nor under short-circuit conditions (when $V_B=0$); power generation is necessarily associated with voltages and currents between those limits. The maximum conceivable power, for given values of V_{Lo} and J_{Lo} would be $V_{Lo}J_{Lo}$ but the actual voltage–current relationship does not pass through this point (Fig. 5.7). How close it comes, depends on its detailed shape. If that shape were correctly represented by eqn (5.1.6), then there would be no choice in the matter: all photocells would behave in the same way (see also Shockley and Queisser 1961). However, eqn (5.1.6) is schematic in many ways, including the facts that it omits all reference to series resistance, and to any departure from 'ideality' which may be associated with the dark-characteristic. To remedy this (albeit still within the bounds of the original key assumptions) eqn (5.1.6) could be rewritten in the form

$$J = J_s\left[1-\exp\left\{-\frac{e(V_T+JR_b)}{\eta kT}\right\}\right]+J_L(V_T) \tag{5.1.11}$$

as suggested and used by Pulfrey (1978), all the cautions of Sections 2.2.5 (concerning η) and 5.1.2 (concerning linear superposition) notwithstanding. In the power-generating regime, V_T will be negative and J positive. R_b features as a current-independent constant, though we know that the series resistance is in fact modulated by light on the one hand, and by injection on the other. The light modulation is especially important. Thus, a higher series resistance (in darkness) may be associated with a wider collection region (thicker barrier), and this beneficial effect may well outweigh other considerations, once photoconduction is taken into account (Wilson and McGill 1978).

On the basis of eqn (5.1.11) a particular current, $J=J_m$, would yield maximum power, and that power would be given by $J_m^2 R_T$, where

R_T is the total resistance of the whole assembly, easily found from eqn (5.1.11). An external voltage V_m corresponds to J_m, and with $V_m = V_T$, eqn (5.1.11) yields

$$\frac{V_m}{J_m} = R_T = R_b + \frac{\eta kT}{eJ_m}\log\left(\frac{J_s}{J_s + J_L - J_m}\right). \tag{5.1.12}$$

This can be shown to simplify to

$$R_T = R_b + \frac{\eta kT/e}{J_s + J_L - J_m}. \tag{5.1.13}$$

The power $J_m^2 R_T$ is numerically maximized for the calculation of J_m, and a *fill factor* (FF) is then conventionally defined as

$$(\mathrm{FF}) = J_m^2 R_T / J_{Lo} V_{Lo}. \tag{5.1.14}$$

(FF) can thus be calculated as a function of η, R_b, J_s, and J_L. Similar calculations have been made by Green (1977), using V_{Lo} as the principal variable. For an alternative formulation of series resistance effects, see Landsberg (1975). R_b depends on the semiconductor thickness, of course, and in solar cell design that must be optimized for the operating conditions envisaged.

The fill factor gives a measure of efficiency, and is widely used, but its simplicity is deceptive, because J_{Lo} and V_{Lo}, as well as the parameters which govern R_T, are not in fact independent variables, nor is (FF) independent of light intensity (since it is a function of J_L). Actual values of (FF) estimated in this way can vary over a wide range and are typically below about 0.8.

Pulfrey (1978) has shown that a high value of the non-ideality factor η, ordinarily regarded as undesirable, can to a small extent compensate for a high value of R_b, other things being equal. An alternative procedure to the use of η is to express non-ideality in terms of a *pseudo-temperature* T_0, as formulated in Section 2.3.3 (see also Padovani 1966). This has been done by Kumar and Sharma (1981), whose results differ from those reviewed above, because T_0 was taken to affect not only the voltage-dependent term of the dark-characteristic, but the saturation current as well. However, to do this is to invest the pseudo-temperature with a deeper meaning than its status as an empirical fitting constant actually warrants.

The series resistance cannot enter into the picture when the current is zero and, accordingly, eqn (5.1.11) yields the same open-circuit voltage as eqn (5.1.1), except for the fact that $\eta kT/e$ has (here) replaced kT/e. The presence of series resistance has a thoroughly undesirable effect on the fill factor, as Fig. 5.8. shows. However, it should still be possible to increase

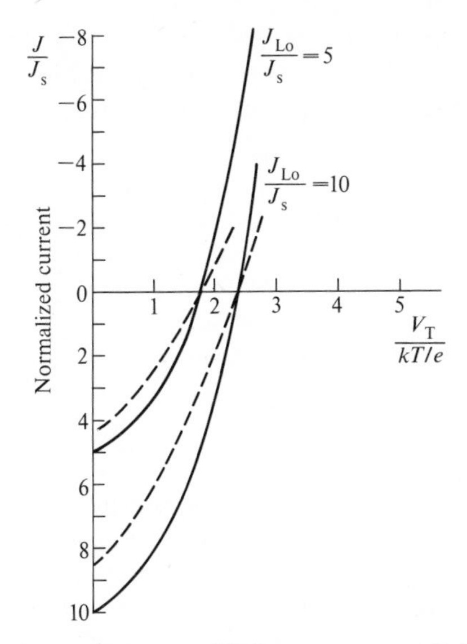

FIG. 5.8. Effect of series resistance on fill factor, as numerically evaluated from eqn (5.1.11), assuming $J_L = J_{Lo}$ = constant, and $\eta = 1$. Two light intensities, assessed by J_{Lo}/J_s. Full lines: $R_b = 0$. Broken lines: $R_b = 0.1$ ohm.

the photovoltaic effect by diminishing J_s; and Card *et al.* (1977) have shown (on Al–n-Si contacts) that this is actually the case.

Attempts have been made (in quasi-analytic terms) to go one step beyond 'linear superposition', by modifying eqn (5.1.11) to

$$J = J_s\left[1 - \exp\left\{-\frac{e(V_T + JR_b)}{\eta kT}\right\}\right] + J_L(V_T) - \left(\frac{V_T + JR_b}{R_{sh}}\right) \quad (5.1.15)$$

(Wilson and McGill 1978). The last term, in which R_{sh} is an effective shunt resistance, is schematically introduced to take care of the interaction between the different carrier fluxes. With V_T negative, and JR_b small and positive, its effect is to augment J_L. However, the real R_{sh} cannot be constant, and although eqn (5.1.15) could provide a better fit to experimental results, the problem of interaction is not thereby solved.

Many systems which are of practical interest have structures more complicated than those shown here, either by involving third phases between metal and semiconductor, or else by involving doping gradients. Thus, for instance, Li (1978) has shown that Au-p–n structures in GaAs can yield high-efficiency solar cells when the parameters (layer thicknesses and doping levels) are properly chosen. Such structures are of special interest, because the barrier heights associated with them can easily approach values equal to the band gap. A full analysis of such systems would have to take non-degeneracy into account. Lee and Pearson (1980) have explored the use of heterocontacts under a semi-transparent gold electrode, and an even more sophisticated method is to use 'graded' heterojunctions, with band gaps monotonically decreasing from high to low, to ensure that all the incoming radiation is absorbed and used for electron–hole pair generation. Calculations of maximum efficiency along those lines have been presented by Pauwels and deVoss (1981). The quoted values (e.g. 86.8 per cent at room temperature) are very high, but structures of the idealized kind to which these values refer have not actually been made. If they could be made, the parameters on which calculations should be based are likely to be very different from those associated with bulk specimens. 'Ordinary' efficiencies, e.g. those calculated for single-gap junction systems, are in the region of 22–25 per cent; those associated with contacts are lower, as one would expect. Values estimated by Pulfrey and McQuat (1974) on the basis of a formulation by Sze (1969) go to about 16 per cent for silicon, depending on the barrier height. Since the maximum conceivable barrier height is E_g, one can arrive at maximum conceivable efficiencies for different material. Pulfrey and McQuat give these as about 12 per cent for Ge, 22 per cent for Si, 25 per cent for GaAs, 18 per cent for GaP, and 16 per cent for CdS. Photon reflection, photon absorption outside the barrier, carrier recombination, and series resistances were neglected in the calculations; not surprisingly, the efficiencies actually achieved with contact systems are considerably lower. An improved model (based in part on numerical solutions for Au–GaAs and Au–Si, but still assuming linear superposition of dark- and photo-currents reflects this, and does indeed yield efficiencies below 10 per cent (McQuat and Pulfrey 1976). In all these instances, the gold electrode must be thin enough (e.g. 50–80 Å) to allow a sufficient amount of light to penetrate and, in the optimization of the system, that thickness must be matched to other structural parameters. (See Wolf (1980) for a comprehensive survey of efficiency estimates for various silicon solar cell structures.)

All the present considerations apply to systems of small area, in which the lateral resistance of the electrodes themselves can be safely neglected. In large systems, the need for optimizing electrode configurations represents an additional problem, in as much as well-conducting electrode materials tend to be non-transparent, and quasi-transparent materials tend to be poorly conductive. This is, of course, a classical network problem (see, for instance, Wyeth 1977, and Rajkanan and Shewchun 1979). Lue and Hong (1978) have pointed out that calculated optimum electrode-thicknesses depend rather sensitively on the expected photo-flux. In a typical example, the optimum thicknes of an Al electrode film might be 50 Å for 20 mW cm^{-2} and 100 Å for 180 mW cm^{-2}.

Of course, not all applications of light-sensitive contacts are governed by considerations of power efficiency. The use of a system for light-detection, as distinct from energy conversion, would depend on high values of V_{Lo} or J_{Lo}, rather than high values of V_{Lo} and J_{Lo}.

5.1.4 *The collection velocity*

The above discussion leaves the actual magnitude of J_L (as a function of V_B) open, but while this is again a strong case for the use of numerical methods, it is interesting to examine approximate solutions and semi-quantitative expectations. J_L can be expressed in terms of a collection velocity S_{cv} by writing (in accordance with the present sign convention)

$$J_L = -j_L = eS_{cv}\{p(\lambda_B) - p_e\} \tag{5.1.16}$$

Under open-circuit conditions, S_{cv} is thus closely related to the interface recombination velocity, which is defined (for extrinsic material only) by an expression of the same form, but with reference to $p(0)$ rather than $p(\lambda_B)$. However, there is a subtle difference between recombination at a free surface, and recombination at a contact in circuit (as Card (1976a) has pointed out). In the former case, a surface state that has captured a hole must capture an electron from the conduction band, and that process is more or less difficult. In the latter case, a surface (interface) state that has captured a hole can capture an electron from the metal, where the supply is plentiful.

A numerical solution could take details of arbitrarily complex barrier structures into account, but in the search for non-numerical approximations, we shall assume a normal Schottky barrier. It is then useful to distinguish between situations in which $V_B \simeq 0$, and situations in which V_B is high and positive (reverse voltages). For the latter, the barrier would constitute a perfect sink. One would therefore expect S_{cv} to approach some saturation value, always well below the thermal velocity of the carriers (i.e. $\sim 2.5 \times 10^7$ cm s^{-1} at room temperature). This expectation should hold, at any rate, until new effects begin to dominate, e.g. image-force barrier lowering or carrier extraction. The latter must, of course, take place at all reverse currents, and its role is to counteract the effect of illumination by reducing $p(\lambda_B)$. Which of these effects will actually 'win' must depend on circumstances: high-level illumination will ordinarily dominate over low-level extraction but, at high reverse current, the extraction rate might actually be greater than the optical generation rate. The boundary concentration $p(\lambda_B)$ should then be low, but J should, for that very reason, be sensitive to optical augmentation (see Section 3.4.2).

For mechanism (a) of Section 5.1.1, extraction is the principle which makes J_L dependent on V_B, although the dependence is not actually steep. When mechanism (b) is active (strongly absorbed light), there is an additional effect: the barrier region widens with increasing V_B, thereby increasing the distance over which electron–hole pairs are separated by the field. This leads to an additional V_B-dependence of J_L.

Under zero applied voltage, the short-circuit current J_{Lo} must consist of two components, of which one is J_L. The other (generally smaller) component arises from augmented majority-carrier diffusion due to the change from n_e to $(n_e + n_L)$. 'Linear superposition' is equivalent to neglecting this component. Along those lines, dealing separately with the electron- and hole-fluxes, one could expect the collection velocity of holes to be of the same order-of-magnitude as the D_p/L_p term already encountered in eqn (3.1.25). At room temperature (e.g. for Ge), this is only of the order of 5×10^3 cm s^{-1}. The present situation and that discussed in Section 3.1.4 are, in fact, quite parallel. In the earlier discussion, a certain non-equilibrium concentration of holes was maintained at the edge of the barrier by means of the contact bias voltage V_B; here, it is maintained by photogeneration. Recombination (on which L_p depends) within the barrier region is expected to be different from that in the bulk, because of the different trap occupancies prevailing there. Panayotatos and Card (1980) have studied this matter as a function of light intensity. Interface recombination also influences the effective L_p, and can even control it, to an extent which depends on contamination, heat treatments, etc. (see Reed and Scott 1975). The actual measurement of an effective L_p close to a surface remains a major problem, and this illustrates the difficulty inherent in detailed comparisons between theory and experiment. Numerical solutions do not suffer from the problem in quite the same way, because the concept of an 'effective L_p' is avoided. Cell-to-cell solutions simply employ a lifetime appropriate to the surface in the first cell, and a bulk lifetime in subsequent cells, allowing the carrier concentrations to adjust themselves accordingly.

It should, of course, be possible to calculate S_{cv} from first principles, but attempts to do so have not yet yielded satisfactory results. Card (1976a) investigated this matter in connection with low-resistance contacts, to ascertain the extent to which such contacts (on solar cells) represent sinks for photo-excited carriers. However, the approximations, which had to be used to achieve analytic solutions, led to S_{cv} values greater than thermal velocities, and were therefore implausible.

5.1.5 Computed photoresponse characteristics

If analytic treatments have their shortcomings, and it is by now abundantly clear that they do, so have numerical computations, but of a very different kind. Insolvable equations become solvable, but only at the expense of generality. An analytic solution shows clearly how the various descriptive parameters of a situation influence the outcome; a numerical solution does not. A whole series of numerical solutions is needed to provide such an indication, because each is highly specific and applies only to the case to which it applies. The loss of generality would be more grievous if there were a choice, but in practice there is none. In an ever increasing number of cases, as the systems under investigation increase in complexity, we are committed to the numerical exploration of relationships. In any event, their shortcomings are balanced by gains: this kind of exploration can, and does, often reveal important interactions which analytic treatments are forced to ignore (see below). The situations discussed in Sections 2.4.3 and 3.3.2 have already served as examples of that. Contact systems under light have many descriptors, and since these have to be varied one at a time, a complete numerical assessment is a laborious exercise and is, accordingly, rare. More often we deal with 'typical examples', as below.

Figure 5.9, for instance, shows computed characteristics in the power-generating regime for a silicon-like material, using the parameters listed (Moreau 1983a). Contacts (of equal height) on n-type and p-type material behave in a remarkably similar way, considering the very different ratios of majority- to minority-carrier mobility. Even for (say) n-type material, the mobility ratio is found to have surprisingly little influence on the carrier concentration profile, as shown in Fig. 5.10. For the high bulk conductivities and small currents involved here, V_a and V_T are closely similar.

For a barrier on n-type material, the electron concentration diminishes monotonically from bulk to interface. The hole concentration must correspondingly increase, but with a much more complex contour. This is

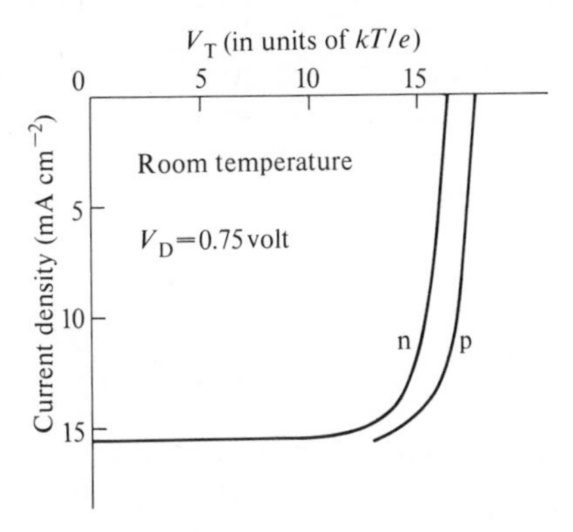

FIG. 5.9. Computed voltage–current characteristics of a contact system on Si-like material in the power-generating regime. Circles: n-type material. Lines: p-type material. Results referred to the barrier voltage V_a. Illumination: 10^{17} photons cm^{-2} s^{-1}. Mobilities: μ_n = 1200 V cm^{-2} s^{-1}. Absorption length: 61.2 μm. Debye length λ_D: 0.07 μm. Specimen length L: 150 μm. Carrier lifetime: 5 μs. Shockley–Read recombination model. After Moreau (1983a).

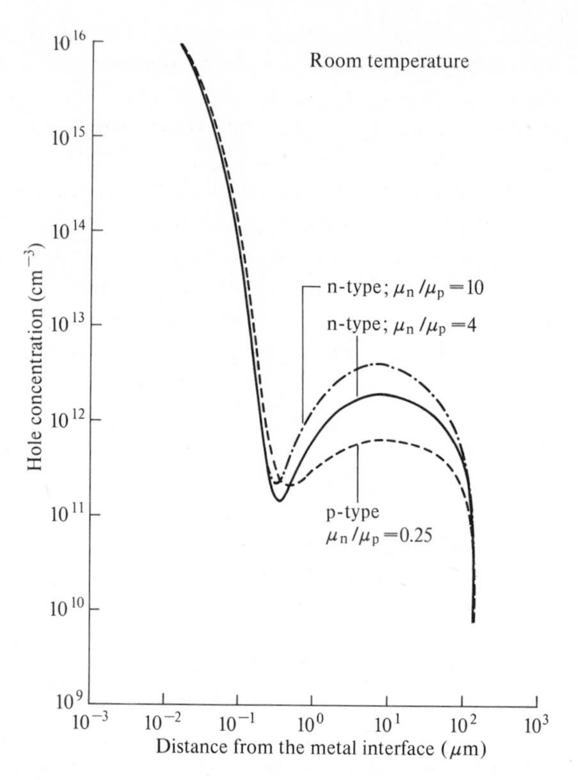

FIG. 5.10. Minority-carrier concentration contours for two mobility ratios. Parameters as specified for Fig. 5.9, except $\mu_n = 1363\ cm^2\ V^{-1}\ s$. $\mu_p = 136\ cm^2\ V^{-1}\ s$. Current density: 6.72×10^{-4} amps cm^{-2}. After Moreau (1983a).

shown in Fig. 5.10, and arises in part from the collecting efficiency of the barrier, and in part from the fact that most minority carriers are generated at distances far in excess of the barrier width. No analytic treatment of this effect is known, and the result is not intuitively obvious.

5.1.6 *Response of potential probes to non-equilibrium situations*

As mentioned in Section 5.1.1, a currentless probe in contact with semiconducting material in which non-equilibrium prevails does not acquire the local lattice potential, but arrives at some value which differs from it by the floating potential. In this case, we are concerned *only* with departures from bulk equilibrium $(n_e + n_L)$ and $(p_e + P_L)$, with $n_L = p_L$, i.e. with mechanism (a) of Section 5.1.1, and not with mechanism (b). In that sense, the behaviour of potential probes used for measurement purposes can differ subtly from that of photo-excited carriers, for which mechanism (b) is ordinarily very important.

The non-equilibrium within the semiconductor on which the probes are situated may be characterized by quasi-Fermi levels ϕ'_n and ϕ'_p. Baron and Mayer (1970) have suggested that the probe reaches a compromise energy given by

$$E_{F'} = \frac{\mu_n \phi'_n + \mu_p \phi'_p}{\mu_n + \mu_p} \tag{5.1.17}$$

and the corresponding floating potential would then be $(E_{F'} - E_F)/e$. However, this formulation cannot be universally correct, because it makes no detailed allowance for the nature of the contact barrier. Thus, the probe response should certainly depend on whether the probe is associated with a Schottky barrier or a flat-band condition or an accumulation layer. Manifacier *et al.* (1984) have computed the probe response for all these conditions, and have also provided analytic estimates of floating potentials, as far as those are feasible.

Figure 5.11 shows a contact system with two small probes in positions x_1 and x_2, and an injecting contact on the left. This main contact could be a metal, or else a junction, and is not the subject of our present concern; we merely assume that the presence of a forward current through it leads to extra (non-equilibrium) holes at x_1 and x_2. In the presence of such holes, floating potentials V_{f_1} and V_{f_2} appear at the probes. When uniform carrier concentrations prevail, $V_{f_1} = V_{f_2}$, and the contact effects cancel, as far as the measured voltage difference between the probes is concerned. In the presence of carrier-concentration gradients (arising naturally in the system envisaged here), we would have $V_{f_1} \neq V_{f_2}$, and the measured value of $V_{x_1} - V_{x_2}$ would be influenced by the difference. It would not, then,

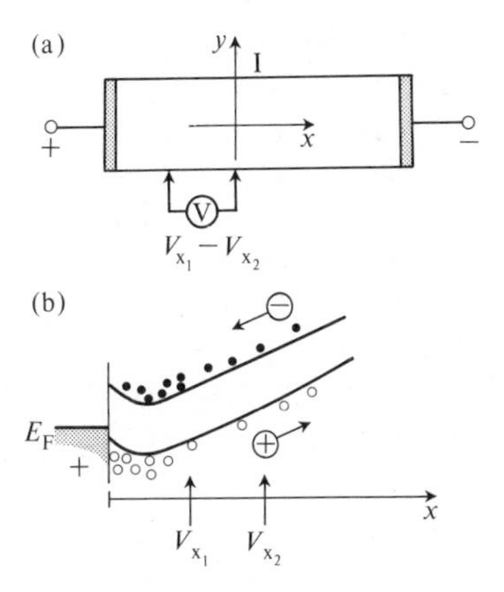

FIG. 5.11. Potential probe system along a one-dimensional structure (x-direction). (a) Schematic probe arrangement. V_{x1} and V_{x2} could alternatively be separately measured with respect to any fixed reference potential. All measurements with high-impedance voltmeters (zero current). (b) Band diagram in the vicinity of the injecting main contact at $x = 0$ with forward current flowing.

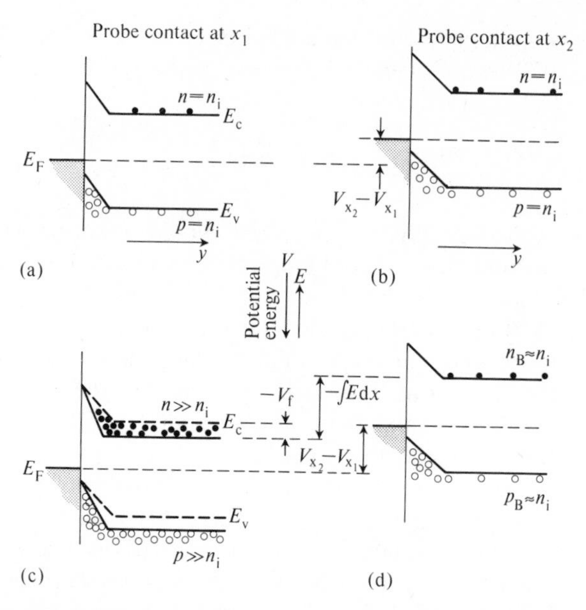

FIG. 5.12. Band diagram in the vicinity of x_1 and x_2, plotted in the currentless y-direction. Intrinsic material. Specimen current in the x-direction. (a) and (b) Under quasi-equilibrium conditions; identical profiles, displaced by the potential drop in the x-direction. (c) and (d) Under non-equilibrium conditions; displacement between (c) and (d) is now due partly to a floating potential at x_1. (b) = (d) because it has been assumed here (for simplicity) that equilibrium prevails at x_2. E_c, E_v denote the *equilibrium* positions of the band edges. E_F constant (here). (From Manifacier *et al.* (1984).)

yield the true potential difference between points x_1 and x_2; and if probe measurements were used for determining bulk resistivity, the floating potentials could lead to substantial errors.

It will be clear that the detailed situation at the probes cannot be represented by a one-dimensional band diagram of the conventional kind. Figure 5.12 therefore shows a pair of such diagrams for each of the contacts (in this case for intrinsic material), plotted in the y-direction, but with a small distance x between them. Figures 5.12(a) and (b) represent the electronic equilibrium situation; Figs. 5.12(c) and (d) the non-equilibrium situation. For simplicity, x_2 is assumed here to be large enough for equilibrium to prevail there, making V_{f_2} zero in

$$V_{x_1} - V_{x_2} = V_{f_2} - V_{f_1} - \int_{x_1}^{x_2} F \, dx \tag{5.1.18}$$

the measured voltage difference. Our concern is then only with $V_{f_1} (= V_f)$.

The above probe contacts are located at $y = 0$, but exactly where the $y = 0$ surface is, in relation to the bulk material, depends on the arrangement envisaged. Manifacier *et al.* (1984) actually examined two plausible

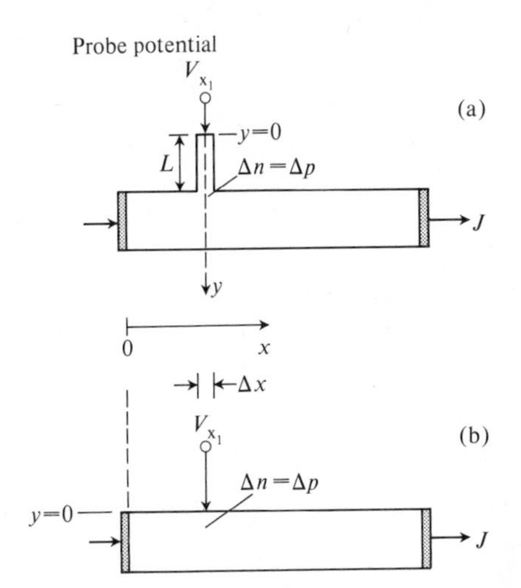

FIG. 5.13. The alternative probe contact systems used for measurement of the local potential (V_{x1}), relative to some fixed reference point.

arrangements, as shown in Fig. 5.13, which demand rather different treatments, but are in excellent agreement as long as L (in Fig. 5.13(a)) is small. In each case, it is supposed that there is a small region Δx over which the carrier concentration can be deemed to be *quasi*-constant in the x-direction. Computer solutions of the transport relationships have been obtained for both configurations under the assumption that the material is a lifetime semiconductor, i.e. that quasi-neutrality prevails ($\Delta n = \Delta p$) even in the regions in which n and p differ drastically from n_e and p_e. At the probe contact itself, equilibrium was assumed ($np = n_i^2$), and the barrier height, as seen from the metal side, was taken as constant. In Fig. 5.13(a), the origin of the disequilibrium at $y = L$ is, of course, the hole flux in the x-direction but, during the analysis of Fig. 5.13(b) (with the x-direction ignored within Δx), additional provision has to be made for the appearance of extra holes. Manifacier *et al.* (1984) did this by assuming a constant bulk generation rate of minority carriers, so adjusted as to yield the desired non-equilibrium values of n and p in bulk, under the prevailing recombination conditions.

Figure 5.14 gives representative results obtained by numerical methods for three different barrier heights, of which $eV_D = 12\,kT$ corresponds to a normal barrier, and $eV_D = -12\,kT$ to a (majority-carrier) accumulation layer. As one would expect, the floating potential is actually greater in the latter case, because electrons have a higher diffusion constant; the discouragement of electron flux therefore needs a higher field than the discouragement of hole flux. Because the carrier mobilities are unequal, and because equilibrium has been assumed to prevail at the interface

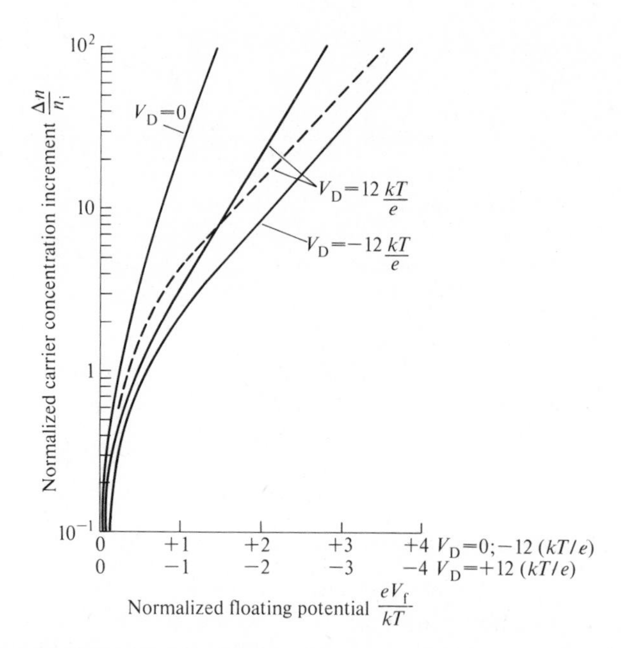

FIG. 5.14. Computed floating potentials V_F for three barrier heights. $V_D=0$ corresponds to the flat-band condition, $V_D>0$ to band bending upward, and $V_D<0$ to band bending downward. Note sign reversal of V_f with V_D. Departure from bulk equilibrium normalized here to n_i, V_f normalized to kT/e. Full lines: intrinsic material; geometry as in Fig. 5.13(a), with $L=50\ \lambda_{Di}$. Broken line: extrinsic material: geometry as in Fig. 5.13(b), with $n_e=10^2$ n_i, $p_e=10^{-2}\ n_i$, $\tau_0=2.6\times10^{-7}$ s. Sign convention for V_f as for V_B.

itself, V_f is not actually zero for the flat-band situation ($V_D=0$). Floating potentials which arise in such (unlikely) circumstances are due to the necessity of satisfying the zero current conditions (in the y-direction). Note the sign reversal of V_f between upturned and downturned barriers. It is sometimes believed that low-resistance contacts are free of floating potentials, but there is in fact no direct or simple relationship between the circuit response of a contact (in darkness) to an externally impressed voltage, and its response to light under open-circuit conditions.

The full lines in Fig. 5.14 correspond to the configuration of Fig. 5.13(a), and the floating potentials computed in this way necessarily tend to zero as L becomes very large. The broken line gives representative results for a system which conforms to Fig. 5.13(b), involving extrinsic material. Manifacier *et al.* (1984) have discussed the detailed circumstances in which probe measurements are 'safe' for the assessment of bulk properties. In practical cases, these circumstances must not be taken for granted, nor can the decision always be made on *a priori* grounds; it should be the subject of experimental checks.

5.1.7 *Illumination of an injection region; interpretation of decay times*

The bulk transport equations examined in Sections 3.2 and 3.3 have also been numerically solved in the presence of uniform volume excitation by light, albeit so far only under limited conditions, namely for unit injection-ratio, $\mu_n = \mu_p$, zero trap-concentration, and small-signal conditions (Popescu and Henisch 1976a). The results are nevertheless instructive, not as regards the precise voltage–current relationships, but as regards the interpretation of the photoconductive decay time.

To illustrate this point, we make use of the p–n relationship prevailing near an injecting contact. In the absence of current, this is at any given point the familiar rectangular hyperbola $\boldsymbol{PN} = \boldsymbol{P}_e \boldsymbol{N}_e = \boldsymbol{N}_i^2$ (Fig. 5.15). In the presence of an injecting current, the relationship is modified, qualitatively as shown. The curves can alternatively be regarded as showing steady-state conditions not simply at one point under different displacements from equilibrium, but at various distances from the injection point under a given initial displacement. Thus, $\boldsymbol{P} = \boldsymbol{P}_e$ obviously prevails at large distances from the point of injection. At smaller distances, $\boldsymbol{P} > \boldsymbol{P}_e$ and at some point, corresponding to $X = 0$, $\boldsymbol{P}$ has its highest value. Every other value of $\boldsymbol{P}$ corresponds to some intermediate value of $\boldsymbol{X}$.

When light ceases, the excess carriers already produced by it must

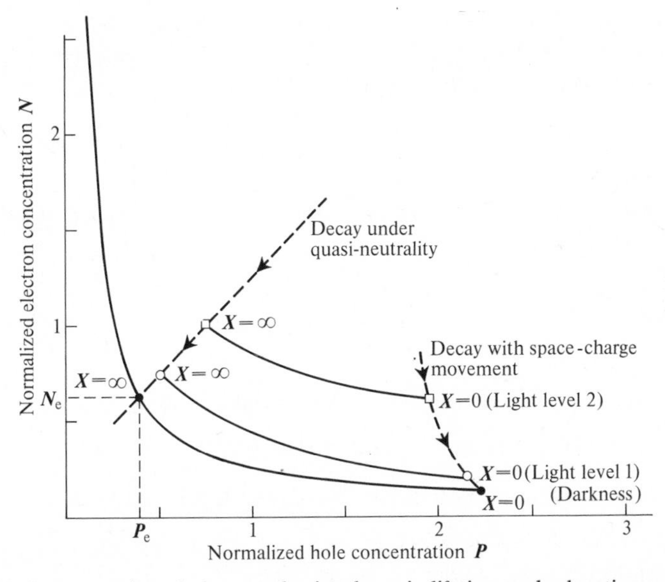

FIG. 5.15. Interpretation of photoconductive decay in lifetime and relaxation semiconductors. $A = \tau_D/\tau_0 = 9.9$, $J = 3.16$ (normalized), $\gamma = 1$, $\mu_n = \mu_p$ assumed. When neutrality prevails, decay (at 45 degrees) measures the lifetime τ_0; decay at some other angle measures a complex mixture of τ_0 and τ_D. Decay 'routes' shown for $x = \infty$ and $x = 0$. Light level 2 corresponds here to $1.4 \times$ light level 1. After Popescu and Henisch (1976b).

recombine. In the bulk of a lifetime semiconductor, they do so under conditions of quasi-neutrality. The photoconductive decay time (in the trap-free case) then measures the lifetime τ_0. The situation is obviously more complicated in a relaxation semiconductor, even when traps are absent. A complete analysis would demand the solution of the transport equations under transient conditions, but this has not yet been achieved. Nevertheless, it is obvious from Fig. 5.15 that the photoconductive decay would still be a measure of τ_0, *if* corresponding points under light and in darkness were situated on lines at 45 degrees on the ***P–N*** chart. The computations show, however, that the points are *not* so situated, and this is what even solutions obtained under restrictive conditions (as here) demonstrate very clearly. Thus, for instance, for a given current density (with $A = \tau_D/\tau_0 = 9.9$, chosen here, $\boldsymbol{X} = 0$ corresponds to the point at $\boldsymbol{P} = 2.2$ (normalized). For the same current density under increasing light excitation (two levels shown), $\boldsymbol{X} = 0$ corresponds to somewhat smaller values which are, however, not on a 45-degree line. Photoconductive decay does not, therefore, measure τ_0 because the decay involves more than recombination; it involves also a movement of space charge, and *that* is governed by the dielectric relaxation time. The photoconductive decay time, as commonly measured, is thus a hybrid of complex origins, and is not easily interpreted for a semi-insulator.

Fig. 5.15 shows, *inter alia*, that contact injection matters less and less with increasing intensity of illumination, and this is entirely what one would expect and, of course, the injection ratio γ would not remain constant.

Further reading

On the detection of α-particles by means of contacts on CdTe: Restelli and Rota 1968.

On the spectral response of photovoltaic cells:
PbS: Schoolar 1970; PbTe: Logothetis *et al.* 1971; MoS_2: Fortin and Sears 1982; Cu_2O: Fortin and Masson 1982.

On the photoresponse of tunnel diodes: Adams and Rosenberg 1969.

On the photosensitive barium titanate diodes: Sawyer 1968.

On infrared detection and mixing in Schottky diodes: Van der Ziel 1976.

On tunnel penetration and its effect on barrier heights measured by photoexcitation: C. A. Anderson *et al.* 1975.

On efficiencies associated with Si-photocell structures: Douglas and D'Aiello 1980, Hall 1981, Minucci *et al.* 1980, Redfield 1980, Ruppel and Würfel 1980.

On amorphous hydrogenated silicon solar cells:
effect of recombination: Moustakas *et al.* 1981, Tidje 1982; general survey: Abeles *et al.* 1982.

On gallium arsenide solar cells: complex structures: Lee and Pearson 1980.

PN photovoltaic devices: Fonash 1980.

On Schottky barriers as electron beam detectors: Namordi and Thompson 1975.

On transient excitation of photo-diodes (p-Si): Janney *et al.* 1976.

On recent work in solar cell device physics: Barnett and Rothwarf 1980, Fahrenbruch and Bube 1983, Fonash 1981.

On the effect of grain boundaries in Al–Si solar cells: Wu *et al.* 1980.

On the modelling of inversion layer solar cells by means of equivalent circuits: Norman and Thomas 1980.

5.2 Photoresponse of composite barrier systems

5.2.1 Function of interface dielectrics; general considerations

By 'complex barrier systems' (here) we shall understand systems of the kind discussed in Section 3.5.5, in which an 'insulating' barrier is interposed between metal and semiconductor. It has already been shown that such a layer can increase the current composition ratio, i.e. that such a layer can increase the current composition ratio, i.e. that it can preferentially support the minority-current component. The function of such layers (when present) in photovoltaic cells is based on precisely this property. The photon-generated minority carriers can flow to the metal with a minimum of impediment, if the thickness of the film is sufficiently small. Thus J_{Lo} tends to be unchanged. In contrast, the flow of the augmented majority carriers is effectively reduced. The supporting arguments are the same as those used in Section 3.5.5, but we have an additional complication here: in Section 3.5.5 it was found that the presence of 'insulating' layers increases the injection ratio, and in Section 5.1.2 it was pointed out that injection is essentially undesirable for solar cell operation. In principle, we are therefore dealing with conflicting needs and processes but, in the present context, the improved collection of photo-generated carriers is (or, at any rate, can be) far more important than the increased injection of thermal minority carriers. Exactly how this delicate balance is achieved is not yet completely clear. When it *is* achieved, it shows itself (by eqn 5.1.7) through increased values of V_{Lo}, and thus ultimately through increased efficiencies. Indeed, efficiencies comparable with those achieved with p–n junctions have been measured. Stirn and Yeh (1975) for instance, have reported 15 per cent for their Au–oxide–GaAs systems, with open-circuit photovoltages as high as 0.7 volts. Shewchun *et al.* (1978) reported 12 per cent for silicon cells with indium–tin-oxide electrodes and SiO_2 interface layers, and estimate that 20 per cent efficiency should be possible after optimization. Analysts have frequently commented on the fact that efficiencies are ultimately limited by the 'back-diffusion of minority carriers', and this is, of course, another way of expressing the fact that the contact is 'injecting'. On those grounds it is often concluded that the ultimate limit of efficiency is given by the performance of homojunctions, but the argument is by no means clear when all the space-charge effects are taken into account.

In practice, there is always a price to pay. For such layers to be transparent to holes (assuming n-type bulk), they have to be very thin, and thin layers are notoriously hard to control (e.g. see Rajkanan *et al.* 1980, and Green 1980). Whether any solar cell made of (say) silicon is ever totally *free* of an interfacial oxide layer is another matter; we are concerned here with systems in which such layers are deliberately introduced.

In the ordinary way, theoretical models treat electron and hole tunnelling currents without reference to any change of carrier mass, on the grounds that the carriers do not spend a long enough time within the insulating layer to interact with it. This is, of course, the colloquial formulation of a very subtle and difficult point. The possibility of such an interaction has actually been considered (e.g. see Temple *et al.* 1974, and Temple and Shewchun 1974), but in practice we do not have sufficient information about the layers and their interfaces that would permit crucial tests of validity to be designed. One of great problems (see below) arises from the fact that the bulk and thin-film parameters of given materials are usually different, and whereas the former may be known, the latter have to be inferred from experiments involving the barrier itself. The danger of circular argumentation is therefore considerable.

Figure 5.16 shows a typical situation, involving communication (as always) between the Fermi-level of the metal and the conduction and valence bands of the semiconductor. However, a new element arises from the presence of interface states between insulator and semiconductor. These states can also 'communicate' with the Fermi level of the metal, and this constitutes a third path, which is effective to an extent governed by the interface state occupancy. That occupancy is ordinarily calculated by a Shockley–Read type model, but that formulation (in its classic form) recognizes only two communication paths, not three. Indeed, there is a fourth path, in as much as transitions can also take place from one

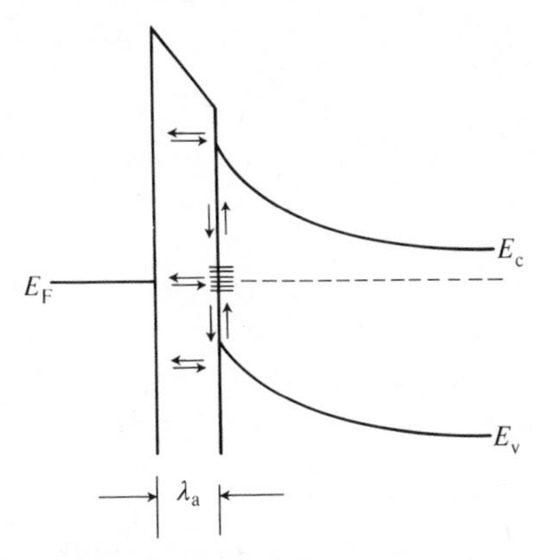

FIG. 5.16. Solar cell structure with 'insulating' interface layer of thickness λ_a; carrier transition paths which maintain the steady state (shown here for zero current). Semi-transparent electrode on the left.

interface state to another. The standard formulation therefore needs to be revised for use in this context. Simplified barrier models also tend to use Boltzmann statistics for the assessment of occupancy when Fermi–Dirac statistics would be appropriate, and sometimes neglect the fact that semiconductor and insulator have different permittivities.

Because the general case is intractable, it is convenient to base calculations on one of three *limiting assumptions*, e.g. as used by Olsen (1977), namely that the Fermi level of the metal remains in equilibrium with (a) the majority carriers in the semiconductor, (b) the minority carriers, or (c) the interface states. Thus, it is clear from Fig. 5.16, for instance, that situation (b) would prevail if the barrier were very high, i.e. comparable with the band gap. In addition to considering these choices, Olsen assumes that interface states are uniformly distributed over energy, and that the interface remains neutral (though observations suggest otherwise; see below). It will be clear that situation (b) offers maximum support for a high injection ratio, since it tends to maintain the hole concentration at the interface in equilibrium, thereby eliminating the hole replacement problem. For solar cell performance, this is not actually the most desirable feature.

The 'insulating' layer can actually affect the system in three ways, as Fonash and Ashok (1981) have pointed out: (i) by suppressing the majority-carrier flow, (ii) by changing ϕ_{ns}, and (iii) by storing charge at the interface, which changes the manner in which applied voltages (including V_{Lo}) are shared between the insulator and the Schottky barrier. Because the three effects are distinct, such complex barrier systems cannot be adequately described in terms of the insulator thickness (λ_a)

alone, as used to be attempted in the earlier days (e.g. see Temple, Green, and Shewchun 1974).

Tunnelling is widely (and in Fig. 5.16) regarded as the principal mechanism of electron- and hole-transmission, but may not actually be the only mechanism, even if thermionic emission *over* the insulating barrier were ruled out. Thin layers of the kind involved here are likely to contain defects, either structural or arising from metal diffusion, which permit some kind of hopping conduction to take place. Lue (1980), for instance, has studied the effect of introducing positive ions into the insulating layer. In any particular case, a combination of mechanisms may be active.

If the 'insulator' thickness in such a structure were very large (e.g. $\lambda_a > 60$ Å) the tunnel current would be negligibly small, and the system would then act as a capacitance of complex structure. However, it does not follow that the adjoining semiconductor would remain in equilibrium under AC excitation, not even in darkness. On the contrary, it should be recalled that injection and other non-equilibrium effects can be induced as transients by capacitive variations of free-surface potential (Section 3.1.1). The distinction which is sometimes made between 'equilibrium diodes' (λ_a thick) and 'non-equilibrium diodes' λ_a thin) is therefore valid only in as much as these terms denote idealized limiting cases. Those limits arise only because an increasing λ_a implies a diminishing deformation of the Schottky barrier by the applied voltage.

5.2.2 *Analytic Models and Observations*

Whatever the mechanisms involved, we may in general terms accept the beneficial effects of interface layers as an empirical fact. For a typical case history, one might go to a set of observations on amorphous silicon cells by McGill *et al.* (1979), who interposed thin TiO_2 layers (of about 20 Å thickness) between the Si and a Ni electrode, using RF glow discharge techniques. As the TiO_2 film thickness increased, the effective barrier height (as assessed by the dark saturation current on the assumption of lateral homogeneity) also increased, suggesting that the band structure of such an oxide is not firmly established while the thickness is very small. Open-circuit photovoltage and short-circuit current changed likewise, in the manner shown in Fig. 5.17. For similar observations of increased V_{Lo} in crystalline silicon cells, see Lillington and Townsend (1976), and Ponpon and Siffert (1976); but increased short-circuit currents were not encountered in these cases, for reasons which may be related to more constant barrier heights. Figure 5.18 shows results for V_{Lo} and J_{Lo} as a function of SiO_2 thickness, with V_{Lo} peaking sharply for $\lambda_a \approx 20$ Å, and this value is in good agreement with an estimate by Card (1977), based on tunnelling. This is also the thickness for which the short-circuit current

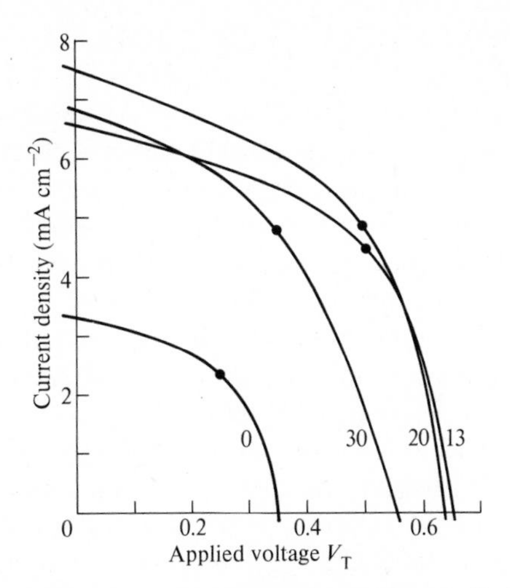

FIG. 5.17. Photoresponse of an Ni–α-Si cell under 60 mW or cm^{-2} white light illumination, as a function of TiO_2 interfacial film thickness. Filled circle-points mark the maximum power points. Thickness in Å units. After McGill, Wilson, and Kinmond (1979).

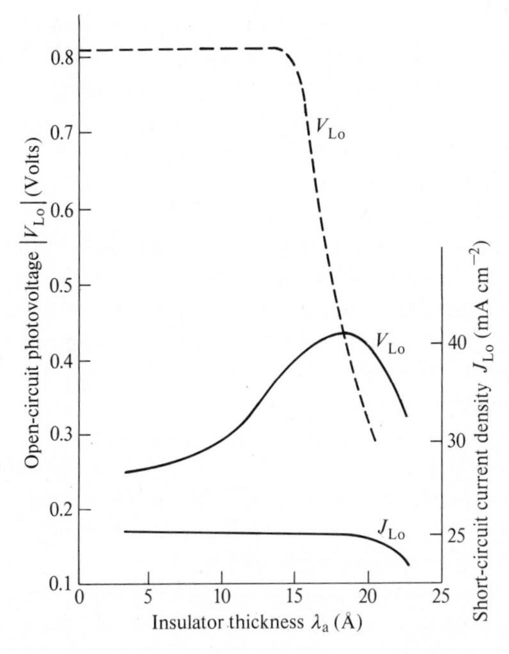

FIG. 5.18. Behaviour of the open-circuit photovoltage V_{Lo} and the short-circuit current J_{Lo} as a function of insulator film thickness λ_a. Full line; Au–n-Si cell, SiO_2 layer. Tungsten lamp illumination of 100 mW cm^{-2}. After Lillington and Townsend (1976) (their Fig. 3). Broken line: computed curve for the ITO–SiO_2–p-Si contact system of Fig. 5.19. After Shewchun and co-workers (1978).

J_{Lo} begins to diminish. A somewhat different pattern of assumptions and parameters used by Srivastava *et al.* (1979) yielded the same results for J_{Lo}, but predicted a V_{Lo} constant for $\lambda_a > 12$ Å. Landsberg and Klimpke (1980) likewise arrived at a critical film thickness of 20 Å, but only for the case in which the recombination coefficient between traps and the valence band is high. When it is low, a less sharply defined critical value of λ_a appears at about 10 Å. There is, thus, a certain measure of agreement on the question of optimum thickness, but the variety of obtainable predictions lends further emphasis to the cautions discussed above.

The descriptive equations of the solar cell system are essentially the same as those given in Section 3.5.5 for the dark case, based on Card and Rhoderick (1971a,b), but whereas those equations were formulated in terms of voltage differentials, the present situation can be thought of as involving (corresponding) differences of quasi-Fermi levels. This leaves the shape of the equations unchanged, but makes them even less accessible to analytic solution, because the quasi-Fermi level contours are not known from any independent evidence. Nor can they be, because they are bound to reflect the carrier distribution as governed by all the prevailing transport relationships and boundary conditions. Accordingly, a pure evaluation of cell properties (i.e. voltage–current relationship under illumination) is impossible along these lines. Of course, the quasi-Fermi level contours can in principle be numerically computed (see Klimpke and Landsberg 1981), but in as much as these levels are merely an analytic convenience, they lose their special role in numerical procedures. It is then just as easy to operate with the carrier concentrations themselves (see also Section 5.2.3).

Model-analysts have been trying to overcome this problem in a variety of ingenious and sometimes brilliant ways, encouraged by the hope (not shared here) that the use of numerical methods might yet be avoided. To this end, simplifying assumptions are frequently made: e.g. $\phi_e = \phi_h$, maximum recombination in a region where $n = p = n_i$, dark current obedient to thermionic emission laws, Boltzmann statistics applicable throughout, no permittivity change at the interface, etc.; but even more accommodation is actually necessary. Thus, models often involve major component parameters (e.g. the actual J_{Lo}) derived from measurement, and then used for the calculation of other cell properties. This is indeed how Card's (1977) estimates were made. In this way, also, Card confirmed that ϕ_{ns} diminishes as λ_a increases; the Schottky barrier then has a diminishing share of the overall voltage drop. It is also generally agreed that ϕ_{ns} depends on illumination; this must mean that the interface states are *not* very tightly bunched, and that the charge in interface states can vary (see Pulfrey 1976). Landsberg and Klimpke (1980), following Olsen

(1977), have indeed assumed a uniform distribution of such states. The nature of that distribution has a critical effect on the calculated properties, since the surface charge distribution governs not only the static value of ϕ_{ns}, but its modification (decrease) due to light (see also Lue 1982).

It often happens that plausible simplifications which yield correct order-of-magnitude results turn out to be quite false when examined in greater detail. Card and Yang (1976), for instance, point out that the ever popular 'linear superposition principle' cannot really apply, and that departures from it have substantial consequences. Nor is agreement with any particular set of observations necessarily a guarantee of correctness. By now it is recognized that a system with so many descriptors can always be brought into agreement with experimental results; as a general rule, all published models are supported by such results! This is one of the reasons why the present treatment presents 'the analytic case' in outline only.

The possibility that the open-circuit photovoltage might increase due to the presence of the 'insulating' layer had already been predicted on the basis of the modified current composition ratio mentioned above (see, for instance, Fonash 1975, Shewchun *et al.* 1974, and Card and Rhoderick 1971a). All details apart, this is the broadly accepted interpretation. However, other interpretations have been proposed.

Thus, it is known that thin films always contain pinholes, and whereas these are generally regarded as a minor complicating feature in the present context, Rothwarf and Pereyra (1981) have pointed out that they are, at any rate in principle, capable of acting as a primary mechanism in the control of photovoltaic cell properties. Their model is based on the suggestions that pinholes, acting in conjunction with the Schottky barrier, can also permit minority carriers to pass preferentially and can, in that sense, act like the tunnelling barriers discussed above, but *without* any tunnelling process. The argument is that minority carriers drift and diffuse towards the interface, whereas majority carriers tend to be reflected by the barrier. If the insulating layer has local pinholes (where only the simple Schottky barrier prevails), holes will exit the semiconductor through those pinholes, either directly, or after lateral diffusion within the plane of the interface. No corresponding mechanism is available to electrons. The total result would be a reduced diode saturation current for the entire system, coupled with a supposedly unimpeded (or only slightly impeded) J_{Lo}. As a consequence, V_{Lo} would increase. Rothwarf and Pereyra neglect all spreading resistance considerations, and do not concern themselves with the all-important lateral diffusion in detail, but arrive at otherwise plausible order-of-magnitude estimates. The pinholes envisaged here would be, of course, in (non-aligned) series with those of the Schottky barrier itself, as discussed in Section 2.3.6.

No detailed model of the pinhole system exists, and the notion that pinholes have an essential role in solar cell behavior has not gone uncontested. Fonash and Ashok (1981) have pointed out that barrier heights (inferred from measurements over a temperature range) are systematically different when an interfacial layer is present, and this certainly precludes any interpretation which ascribes to pinholes the *sole* function of supporting charge transport. However, the possibility of pinholes and intervening areas playing comparable roles is not easily disposed of (nor, of course, is it easily confirmed). In the absence (to date) of more detailed calculations, the pinhole model has failed to receive a great deal of attention; its ultimate status remains to be clarified.

5.2.3 Computations on solar cell models

The cautions recorded in Section 5.1.5 also apply here; accordingly, a *general* representation of results is not feasible, but a few illustrative examples shall be discussed. Thus, for instance, Fig. 5.5 represents schematic voltage–current characteristics in darkness and under light, and it is of interest to compare these with computed results. A set of those was obtained by Shewchun and co-workers (1978) for a complex barrier on p-type Si, involving thin (10–30 Å) SiO_2 layers, and a transparent electrode of (degenerate) indium–tin-oxide (ITO). The model adopted and the starting data of the computation are shown in Fig. 5.19. The same parameters have been also used by Myszkowski *et al.* (1981), who give more details about the way in which the optical absorption was handled. Reflection at the ITO and SiO_2 layers was neglected, and an infinite surface recombination velocity was assumed to prevail at the base contact (not shown in Fig. 5.19), resulting in permanent equilibrium concentrations there. An inversion layer is envisaged, but falling short of degeneracy anywhere. Majority carriers (in this case, holes) were assumed to be completely blocked by the barrier, not only in darkness, but also under light—a frequently made, but not easily supportable, assumption; with increasing light intensity, the majority carrier flow must eventually make itself felt. It should, indeed, play at least some role even in darkness. Tunnelling to and from interface states was regarded as the dominant mode of charge transport through the insulating layer, and the Fermi level was assumed to be pinned in a high position so as to give the electrons (here, minority carriers) good communication with the conduction band of ITO, irrespective of applied voltage. ϕ_{ns} was thus kept constant. The tunneling relationships used for the computation were those developed by Harrison (1961), and the detailed description of the model includes an assumed (double-peaked) distribution of interface states. Tunnel and bulk currents are matched by the computation, as outlined in section 3.5.5.

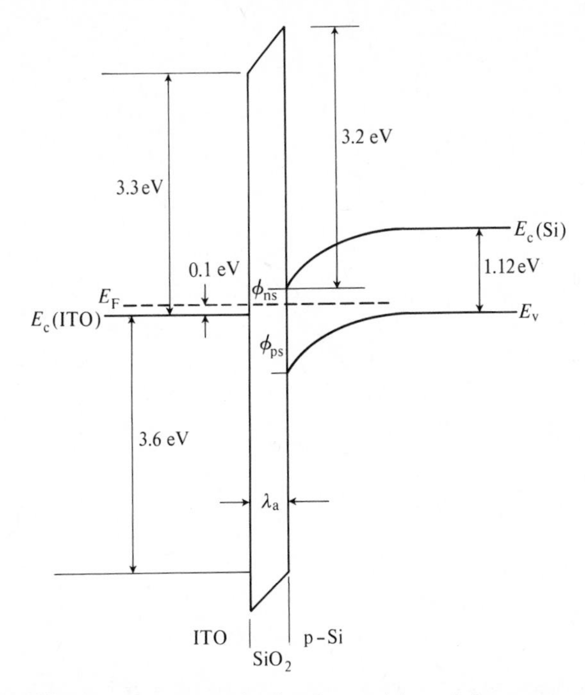

FIG. 5.19. Specification of a composite barrier model used for computation purposes; indium–tin oxide (ITO) electrode on p-type Si, with an intermediate layer of SiO_2 (gap width 8.0 eV). Carrier lifetime in bulk: 1.79×10^{-5} s; dielectric constant of Si: 11.7; dielectric constant of SiO_2: 3.82; total semiconductor thickness: 2.5×10^{-2} cm; insulator thickness: λ_a. After Shewchun *et al.* (1978).

Under such conditions, Shewchun and co-workers obtained a set of voltage–current characteristics in darkness as a function of λ_a (Fig. 5.20) and under light (as shown in Fig. 5.21 which can now be compared with Fig. 5.15). In this idealized dase, J_{Lo} and $J_L(V_{Lo})$ differ by 20 per cent, but the saturation current is (by implication) negligibly small, and J_L almost independent of voltage in the low range. The result is a high fill-factor (0.84), and a very high (computed) efficiency of 19.9 per cent. Needless to say, other assumptions and other initial data would lead to different results, though not necessarily in all respects. The great advantage of using ITO as an electrode derives, of course, from the fact that it achieves a very desirable combination of transparency and conductivity. The difficulties are in processing.

Shewchun *et al.* have also provided information on how various other cell properties vary with λ_a and temperature. The efficiency, for instance, is virtually constant up to $\lambda_a = 14$ Å, and then diminishes drastically to practically zero for $\lambda_a = 21$ Å. It depends, moreover, on the intensity of illumination. Myszkowski and co-workers (1981) have pointed out that

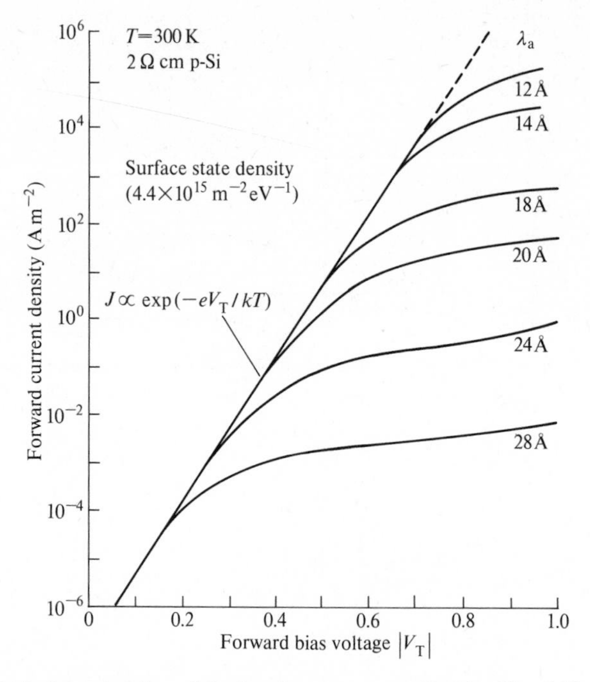

FIG. 5.20. Dark-characteristics of the barrier system specified by Fig. 5.19, computed on the basis of the assumptions explained in the text. Note that 'ideality' is approached under these conditions as λ_a tends to zero. After Shewchun and co-workers (1978).

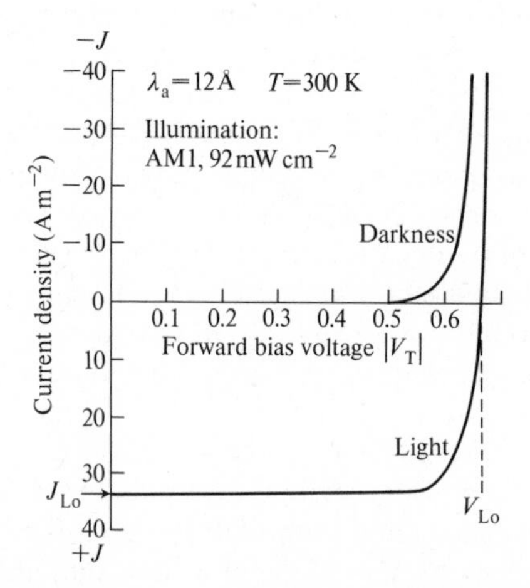

FIG. 5.21. Computed characteristics of the barrier system specified by Fig. 5.19 under illumination (AM1; 92 mW cm^{-2}). After Shewchun and co-workers (1978).

the higher intensities (e.g. 100 AM1) could in principle lead to higher efficiencies (up to 22.5 per cent), but do so only when the insulator is very thin, e.g. λ_a smaller than about 10 Å. In such circumstances, the use of bulk constants for calculations relating to barrier characteristics is quite unsafe. Moreover, at such thicknesses, uniformity is difficult to achieve. Thus, although sunlight could be artifically concentrated upon a solar cell, it is not necessarily easy to take advantage of the extra light intensity. At one AM1 illumination, the short-circuit current J_{Lo} remains surprisingly constant up to about $\lambda_a = 18$ Å, and diminishes drastically after that. The open-circuit voltage V_{Lo}, on the other hand, is least sensitive to variations of λ_a, and remains constant throughout the range explored (up to 20 Å). The fact that the computed efficiency is independent of λ_a in the lower range could be deemed to suggest that one might just as well do away with the insulating layer altogether, but, of course, this would not be correct. At smallest values of λ_a, the simplifying assumptions made by the computational model would become completely untenable. In particular, majority-carrier currents would have to flow, and diminish the efficiency very sharply. The insulator thickness is therefore bound to have an optimum value. Olsen's (1977) analytic approximations reflect this, yielding optimum values between 6 and 20 Å for various configurations. There is, as one would guess, also an optimum doping level for bulk silicon. Myszkowski *et al.* (1981) give it as 2×10^{23} (acceptors) m^{-3} and that figure, fortunately, appears to be independent of illumination level. Olsen has estimated the optimum band gap as close to 1.5 eV. In general, all the parameters have to be optimized together, a task not yet fully achieved.

Further reading

On the increase of photovoltage due to 'insulating' interface layers: Anderson *et al.* 1974, Charlson and Lien 1975, Fabre 1976, Fonash 1976, Green and Godfrey 1976, Landsberg and Klimpke 1977.

On the efficiency of crystalline metal-oxide–silicon systems: W. A. Anderson *et al.* 1976.

On design guidelines and optimization: Dalal 1980, Daw *et al.* 1982, Singh *et al.* 1980.

On trapping processes within interfacial oxide layers: Van Meirhaeghe *et al.* 1982.

On amorphous silicon solar cells: Carlson 1977, Wilson and McGill 1978 (many other references given there), Wronski 1977.

On the photovoltaic response of contacts between metals and amorphous chalcogenide alloys: Wey and Fritzsche 1972.

On the photovoltaic response of interfaces between amorphous selenium and organic polymers: Okumura 1974.

On MOS solar cells using InP substrates: Pande 1980.

On semiconductor–insulator–semiconductor solar cells: Ashok *et al.* 1980, Shewchun *et al.* 1980.

On titanium nitride layers as diffusion barriers: Seefeld *et al.* 1980.

6

SPECIAL TOPICS AND PROBLEMS

6.1 Lateral conduction; field effect systems

THE density and distribution of interface states enter into most contact considerations, and one of the best ways of assessing these parameters is by measurement of lateral conductivity and its field-modulation.

6.1.1 *Modulation of lateral (interface) conductance*

Everywhere else in this book, we deal with conduction processes at right-angles to the plane of some contact interface. However, there are also highly interesting aspects of the situation which involve conduction parallel to the contact plane. Such phenomena cannot be explored when the contacting member is a metal, because lateral conduction in that metal would predominate over all other effects. However, in the presence of an insulating film separating metal and semiconductor, lateral conduction processes can be useful and informative. Interest in these matters began in the 1950s, in the context of attempts to come to grips with the reality and importance of surface and interface states. By now field effects which modulate lateral conductance are not only a scientific tool, but the basis of the manufacturing industry concerned with field-effect transistors. Accordingly, the subject now has a very substantial literature of its own, too large to be summarized and appraised here. The present section gives the merest outline of these matters, partly for the sake of nominal completeness, and partly as a guide to further reading. The fundamentals have already received sophisticated coverage in books by Many *et al.* (1965) and Frankl (1967), where many other references to earlier work can be found (see also Zemel (1961) and the earlier volume edited by Kingston (1957)). A cross-section of more recent work appears in the proceedings of a conference edited by Quinn and Stiles (1976).

The basic configuration has already been encountered in Section 2.4.3 We have an artificial barrier layer, across which some of the barrier field appears, and an adjoining space charge layer, which may or may not involve minority carriers. The field at the semiconductor surface would be continuous, were it not for interface states on the one hand, and for differences of permittivity on the other (Fig. 1.19). In one respect, the picture is now simplified. If the artificial barrier is high enough and thick enough to prevent all current flow (as is assumed here), then the Fermi level will certainly be flat throughout the space charge layer. Overall neutrality must prevail, and the positive charge Q_T within the Schottky barrier is ordinarily balanced by a charge Q_{ss} (per unit area) in interface states, and the charge on the metal surface. This applies not only under static conditions, but also in the presence of any external voltage V_E between the metal and the bulk semiconductor. The static Q_{ss} is evidently negative for $V_E = 0$ under the conditions here envisaged. In the course of experimentation with large applied voltages, one is concerned with Q_{ss} and Q_T themselves, but many measurements involve only small ripple voltages about a static bias, and thus involve ΔQ_{ss} and ΔQ_T, which may be of either sign. If the screening by interface states were perfect, the application of external voltages would leave the space-charge barrier entirely unchanged. Since it is not perfect, V_E causes the space-charge barrier to be modified in height and thickness. These modifications show themselves in turn as changes of lateral conductance, because they involve changes in the total number of free carriers.

By monitoring these changes, the barrier height itself can be assessed, and thereby also, the density and distribution of the interface states. Indeed, this is one purpose of such measurements. Another is to derive information on the energetic position of trapping states. Such evaluations are possible, as long as a unique relationship can be established between lateral conductance and the height of the space-charge barrier. If one could succeed in carrying out experiments of this kind with vacuum as the 'insulating' layer, the results would, of course, refer to surface- rather than interface-states.

A schematic measurement circuit is shown in Fig. 6.1 but somewhat more complex arrangements tend in practice to provide greater versatility and accuracy. Frankl (1967) has described several of those in some detail. Measurements can be performed with DC or AC or pulses (Harnik *et al.* 1960). A complete analysis would, of course, entail consideration of the time-dependent elements involved, but for illustrative purposes it will be sufficient here to review only the static large-signal case. It covers low-frequency situations also, though these are sometimes complicated by

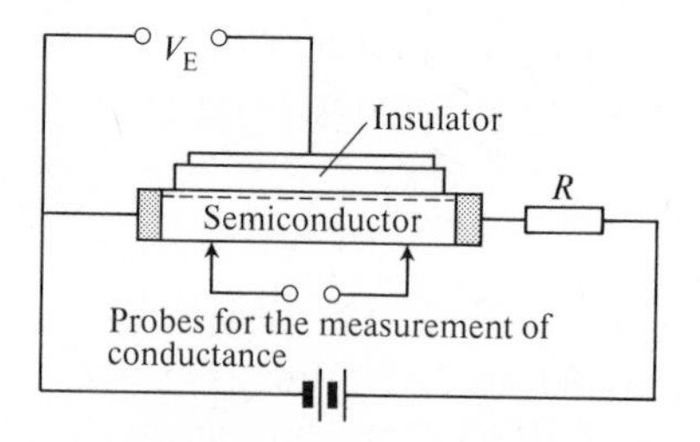

FIG. 6.1. Basic circuit for modulation of interface conductance measurements. Vertical dimensions not to scale; insulator layer is thin, but its resistance is high enough to ensure practically zero current through V_E. The series resistance R is high, to ensure a practically constant lateral specimen current.

the presence of oxide layers which give rise to 'slow' surface states. These aspects are ignored here (see Many *et al.* 1965, Statz *et al.* 1956, and Terman 1962).

A given barrier can be distorted by voltages of either sign, applied to the field electrode relative to the bulk semiconductor. It is convenient to begin a qualitative assessment with the 'flat-band' condition, i.e. the condition under which there is neither a barrier nor an accumulation layer at the surface. It will be recalled that this condition is unattainable for a metal–semiconductor contact as such, but in the presence of an insulating layer, i.e. with zero current in the x-direction, it is entirely accessible. For metal–semiconductor contacts, we have regarded ϕ_{ns} as a constant (except for the considerations of Section 2.3.5), and all barrier distortion was discussed in terms of $V_D + V_B$. It is here necessary to do the opposite; because the flat Fermi-level varies in position, ϕ_{ns} is now a variable. Nor are we now limited to positive space-charge regions within the semiconductor, since large voltage-swings applied to the external field electrode generate space-charge regions within the semiconductor which may be of either sign. In any event, the changes which arise from a variable ϕ_{ns} should not be confused with those due to variable voltages V_B and a fixed V_D in other parts of the book. The latter are always associated with *currents*, and thus with quasi-Fermi levels which have non-zero gradients. In contrast, we are concerned here (at any rate in the x-direction) with a purely electrostatic situation. Control fields of the order of 10^6 V cm^{-1} can be achieved in practice, before dielectric breakdown sets in. There is one other difference: in the analysis of metal–semiconductor contacts, it was most convenient for the Fermi-level of the metal to be kept constant, as a reference level ($V = 0$); but it is more appropriate here (Fig. 6.2) to

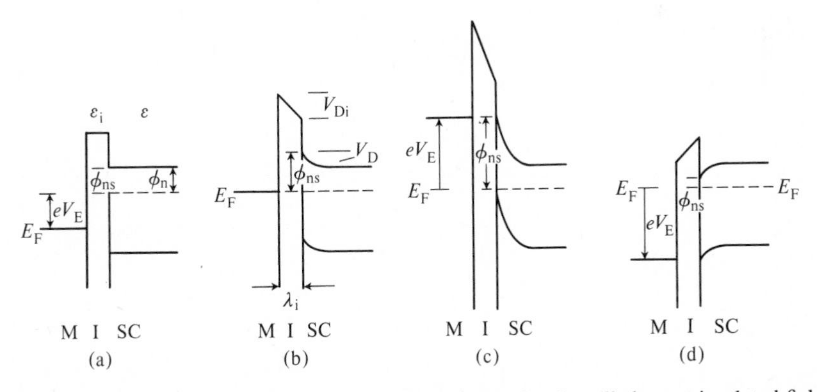

FIG. 6.2. Modulation of lateral interface conduction by signal applied to an insulated field electrode. (a) Flat-band condition. (b) Quiescent condition. (c) Inversion layer formed. (d) Accumulation layer formed. $\epsilon_i = \epsilon$ assumed here.

attach the designation $V=0$ to the Fermi level in the semiconductor (far from the metal), and to allow the Fermi level of the metal to vary.

We begin with the flat-band condition, characterized by $\phi_{ns}=\phi_n$ (Fig. 6.2(a)). When external voltages are applied which make $\phi_{ns}-\phi_n>0$ (Fig. 6.2(b)), the lateral specimen conductance G_s must first decrease, because majority carriers are driven from the newly created barrier region. In due course, as $\phi_{ns}-\phi_n$ increases further, the *minority* carriers will make themselves felt. A high-conductivity inversion-layer will be formed (Fig. 6.2(c)), which will cause G_s to increase. We therefore expect a conductance minimum G_{sm} for some positive value of $\phi_{ns}-\phi_n$. Conversely, when external voltages are applied which make $\phi_{ns}-\phi_n$ negative, majority carriers will form an accumulation layer (Fig. 6.2(d)), and G_s will therefore increase.

The general result is a relationship of the kind shown in Fig. 6.3, where the lateral conductance *change* $\Delta G_s=G_s-G_{sm}$ is plotted schematically against ϕ_{ns}. ΔG_s depends on the external voltage V_E in a similar manner. The space-charge system is (here) being distorted in a way very different from that encountered in Chapter 4 and, as a result, the *capacitive* behaviour of surface layers modulated in this way differs significantly from that of Schottky barriers (see Pfann and Garrett 1959, and Frankl 1961). As ϕ_{ns} increases from the flat-band position, the capacitance goes through a pronounced minimum, just as ΔG_s does.

The practical aspect of all this is that the lateral conductance can be sensitively controlled by external voltages, applied via an insulating layer,

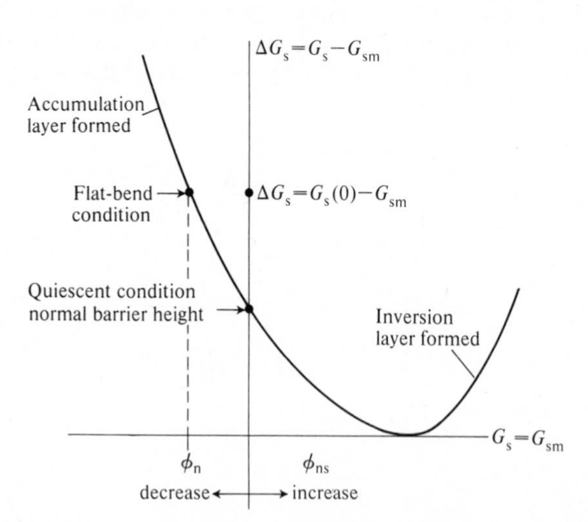

FIG. 6.3. Change of lateral conductance versus barrier height ϕ_{ns}; 'Schrieffer Curve'. The minimum conductance G_{sm} is taken here as the (zero) reference level for comparison purposes. Under flat-band conditions, $\phi_{ns}=\phi_n$.

and therefore not associated with continuous currents. In the present context, this control is a research tool, but it is also the basis of an important set of devices: field-effect transistors. For a survey, see Chapter 10 of Sze (1969) and specialized transistor literature.

Calculations of the lateral specimen conductance demand a knowledge of the carrier mobilities μ_{ns} and μ_{ps} in the space charge region. Because of additional scattering processes (see below), μ_{ns} and μ_{ps} are lower than the corresponding bulk values μ_n and ρ_n. We then have

$$G_s - G_s(0) = e\left\{\mu_{ns}\int_0^{\infty}(n(x)-n_e)\,dx + \mu_{ps}\int_0^{\infty}(p(x)-p_e)\,dx\right\}. \tag{6.1.1}$$

The integration should, strictly speaking, be taken only as far as the specimen thickness L_s but, since all the changes take place close to the surface ($x = 0$), the difference is unimportant. $G_s(0)$ refers to the *flat-band* condition; it is simply the lateral conductance of a barrier-less specimen. $G_s(0)$ thus depends on the specimen thickness, but $\{G_s - G_s(0)\}$ does not, and neither does ΔG_s. G_{sm} is, of course, more readily identified in experimental plots than $G_s(0)$, and this makes the use of ΔG_s for further interpretation very convenient (Section 6.1.2). Frankl (1967) has given tabulated values of the above integrals (see also Kingston and Neustadter 1955, and Young 1961). However, the evaluation of the integrals calls for a knowledge of the potential contour of the space-charge region, and that can differ from case to case, depending on the assumptions made concerning the concentration and energetic distribution of traps. From each model, the contour must be derived, analytically or numerically, from Poisson's equation, using the assumed barrier height as a boundary condition. This means that the curve in Fig. 6.3 is not unique, but is model-dependent. Many *et al.* (1965) have paid special attention to this matter (see also Macdonald (1958) and Rose (1956) for treatments of discrete and continuous trap distributions respectively). Of course, the most frequently considered relationship is that for the trap-free case, mainly because it is the simplest. For that case, every measured value of ΔG_s can be uniquely identified with a definite value of ϕ_{ns}; e.g. by means of (a quantitative version of) Fig. 6.3.

The original work on surface mobilities was done by Schrieffer (1955), and the complexities have continued to be explored ever since. There are actually two classes of problems:

(a) Problems related to the additional scattering of carriers which drift laterally in the z- and y-directions but collide in the x-direction with the walls of any potential well in which they may be contained.
(b) Problems of quantization and band modification.

Both complications are important for holes when we are dealing with lateral conduction in an inversion layer, and for electrons when an accumulation layer is present. Analysis of (a) yields via the Boltzmann equation, new values for μ_{ns} and μ_{ps} in terms of the bulk values and the shape of the potential well. The exact form of the results depends on the simplifying assumptions made. The simplest version regards the carrier collision time as constant (independent of energy), the field $F(x) = F(0)$ as uniform and the carrier masses as scalar. Under such conditions, and for very high fields, Schrieffer found

$$\frac{\mu_s}{\mu} \approx \frac{2}{\sqrt{\pi}} \cdot \frac{(2m_{np}kT)^{1/2}}{\tau e|F(0)|}, \tag{6.1.2}$$

where m_{np} is the effective mass of holes or electrons, and τ the mean collision time. Many refinements have since become available (e.g. see Greene *et al.* 1960, Many *et al.* 1960, and Grover *et al.* 1961). In any event, it is clear that μ_s must depend on ϕ_{ns} and thus on V_E. Greene *et al.* showed that high negative values of ϕ_{ns} can reduce the mobility to something like one-tenth of its bulk value. Although there is some surface scattering under all conditions, $\mu_s = \mu$ is a very good approximation for the flat-band condition and small excursions from it, and that is why the bulk values are used elsewhere in this book.

The quantization problem arises likewise from the presence of a high field, but in a very different way. When the space-charge field becomes high enough to be comparable with the crystal field, then it can no longer be simply superimposed upon the band structure derived by solving the wave equation for the bulk crystal. Instead, the wave equation must be solved *ab initio* with both fields present. Since the space-charge field itself depends on the band structure and its local variation, a computation that yields a self-consistent solution is required. This results in a more pronounced quantization of states within the potential cusp; e.g. see Eger *et al.* (1976) and Uemura (1976), as well as many other references listed in Quinn and Stiles (1976). The effect is, of course, negligible, except at the highest fields. The fact that it can actually occur in conductance modulation experiments has been demonstrated by Hess *et al.* (1975) on silicon inversion-layers, though only at very low temperatures. The effect is otherwise unlikely to be identified as a disturbance in ordinary metal–semiconductor work. When it does occur, it affects m_{np} in eqn (6.1.2), of course.

6.1.2 *Evaluation of the interface charge and trap densities*

By measuring changes of interface conductance, it is possible to arrive at the charge density stored in surface states, and that process is one of the principal applications of the field-effect method in research. The problem can be analysed in a number of ways; one of the simplest is as follows;

(a) Assume that there are no interface states, i.e. that we have field continuity at the insulator–semiconductor interface, except for the difference in permittivities. Thus $\epsilon F(0) = \epsilon_i F_i$, where the subscript i refers to the insulator.

(b) We obtain an experimental relationship between the external voltage V_E and the lateral conductance change ΔG_s, e.g. by applying a low-frequency large-amplitude signal to the field electrode. By Fig. 6.3, every value of ΔG_s then corresponds to a particular value of ϕ_{ns}. It is also true that every ϕ_{ns} corresponds to a particular value of the boundary field $F(0)$ within the semiconductor. For every value of V_E, ϕ_{ns}, and ΔG_s, we can thus find

$$V_i = \lambda_a F(0) = \lambda_2 F_i \tag{6.1.3}$$

the potential drop across the insulating layer.

(c) In a similar way, $(\phi_{ns} - \phi_n)/e$ is the potential difference across the space-charge region. Ignoring the difference of thermionic work-functions, the total potential drop across the system is thus

$$V'_E = V_i + (\phi_{ns} - \phi_n)/e. \tag{6.1.4}$$

However, not all this drop is *measurable*, because it includes two built-in diffusion components, V_{Di} across the insulating layer, and V_D across the space-charge layer (Fig. 6.2(b)). These components must first be subtracted, as shown under (d) below, to provide a basis for comparison between the model and experiment.

(d) One particular value of ΔG_s, as measured under (b), corresponds to $V_E = 0$, i.e. to the quiescent condition. The potential drop V'_E, as calculated under (c) above, then equals $V'_E(0)$, the sum of the two diffusion voltages V_{Di} and V_D, since V_i is then V_{Di}. $V'_E(0)$ represents the unmeasurable component of the total voltage drop across the system. Accordingly, $\{V'_E - V'_E(0)\}$ represents the measurable component under all other conditions, ready for comparison with the actual V_E applied. In general, the values will not agree, because the barrier is partially screened by surface states. For a given applied voltage, the conductance changes will thus be smaller than expected on the basis of an interface-state-free model.

(e) For every measured ΔG_s, we can ascertain a value of

$$\Delta V = V_E - \{V'_E - V'_E(0)\} \tag{6.1.5}$$

which represents the departure from the expectations inherent in the interface-state-free model. $\Delta V \neq 0$ is a system of the fact that $\epsilon F(0) \neq \epsilon_i F_i$

(compare (a) above). Since the conditions inside the space-charge layer are already identified by ΔG_s, the discrepancy must be associated with a modified field in the insulating layer. Accordingly, $\Delta V \neq 0$ implies an interface charge density Q_{ss} given by

$$Q_{ss} = \epsilon F(0) - \epsilon_i(\Delta V/\lambda_a). \tag{6.1.6}$$

In this way, Q_{ss} can be evaluated as a function of the externally measured voltage V_E on the field electrode and, much more significantly, as a function of ϕ_{ns}. When screening is highly effective, the method becomes insensitive, since ΔG_s becomes very small. However, V_E can be large, which means that the screening power of interface states must eventually cease; it does so, of course, when all the states have been emptied or filled.

Trapping states within the Schottky barrier complicate these relationships, because they destroy the uniqueness of Fig. 6.3. However, the problem becomes tractable again *if* we can assume that interface states are *absent* (or else present only in negligible concentrations). That has the simplifying effect of ensuring field continuity at the interface (apart from the ϵ/ϵ_i factor). Thus, Malhotra and Neudeck (1975) have successfully analysed the lateral conductance modulation observed on layers of amorphous germanium. As a model, they used a hypothetically exponential trap density profile which could be characterized by two simple constants. The analysis itself is then entirely straightforward, though the final equation for ΔG_s has to be numerically solved (see also Neudeck and Malhotra 1975a,b).

As an alternative to the assumption of a particular trap distribution (e.g. exponential, as here), the experimental results may be empirically fitted to sets of calculations made with different adjustable parameters. To make this process less arduous, Goodman, Fritzsche, and Ozaki (1979) have (in outline) described a computer program that finds the trap density profiles which reproduce observed values of the surface conductance as a function of V_E.

By methods of this kind, trap state density profiles have been obtained for amorphous Si by Madan, LeComber, and Spear (1976) over an energy range of up to 0.5 eV. It was shown that the states have a well-defined minimum concentration close to the centre of the mobility gap. After annealing, the density of states was smaller by one or two orders-of-magnitude. However, when a situation is characterized by an unknown distribution of interface states and an unknown distribution of traps, the conductivity modulation results cannot be unambiguously resolved.

6.2 Electrothermal effects

6.2.1 *Self-heating; thermal turnover; temperature gradients*

In the analyses given in previous chapters, it was tacitly assumed that rectifying systems remain isothermal at all parts of their voltage–current relationship. In practice, it cannot be so, except in the range of very small continuous power dissipations, or else when measurements are performed with very short current pulses. As voltage and current increase, the power dissipated in the barrier may lead to significant self-heating, to an extent which depends on the prevailing thermal links to the outside. Self-heating, in turn, modifies the contact characteristic, particularly in the reverse direction of current flow. For point contacts, where power dissipation is concentrated in a very small volume, this effect is particularly serious; it is equally true that the effect can be entirely negligible, e.g. in contacts of large area with good heat sinks. When self-heating is prominent, it can lead to a voltage turnover, as shown on Fig. 6.4.

To calculate the properties of self-heating systems, any of the characteristics deduced in Section 2.2, for instance, may be used as a starting point. For simplicity, here we shall take (say) eqn (2.2.39a), assume complete ionization of all donors (σ almost temperature-independent), and write:

$$J = J_c \exp(-eV_D/kT_c)\{1 - \exp(-eV_B/kT_c\} \tag{6.2.1}$$

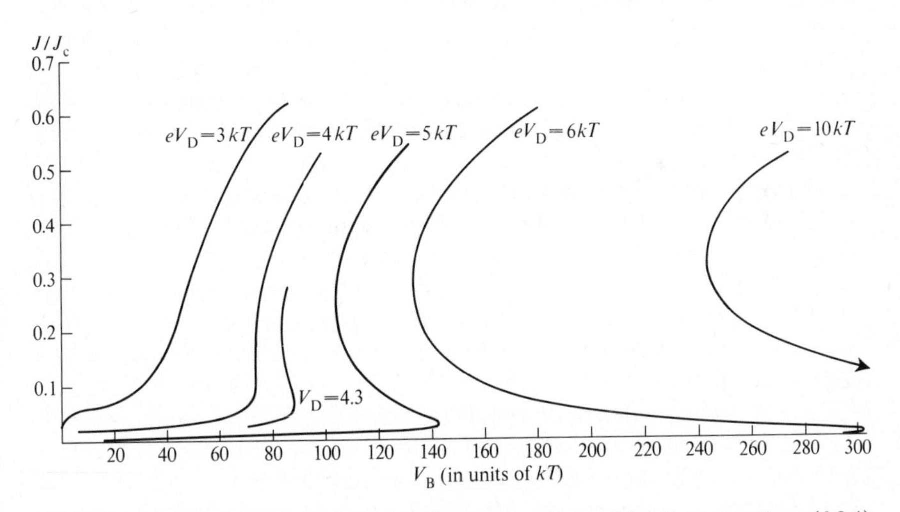

FIG. 6.4. Calculated voltage–current characteristics under self-heating; equations (6.2.1) and (6.2.2), with $H_{TC} = h_{TC}J_c/T_a = 0.1$. Curves for different barrier heights. Note that turnover ceases to occur somewhere between $V_D = 4kT/e$ and $3kT/e$. Here $T = T_a$.

J_c then is a constant, and T_c the actual contact temperature, given by

$$T_c - T_a = h_{TC} V_B J \tag{6.2.2}$$

where T_a is the ambient temperature, and h_{TC} a thermal coupling constant. This formulation implies a temperature-independent thermal conductivity (Hunter 1951); other plausible assumptions could certainly be made (e.g. see Burgess 1955) but they do not alter the general course of events in any drastic way, only in matters of detail.

For a constant thermal conductivity K_{th} (and no other heat losses considered), the thermal coupling constant can easily be shown to be given by

$$h_{TC} = 1/2\pi K_{th} r_0 \tag{6.2.3}$$

where r_0 is the spherical contact radius. h may then be considered a *thermal spreading resistance.*

For a given voltage V_B, eqn (6.2.1) may have more than one solution, but for a given current only one value of V_B satisfies it, although that has to be obtained by numerical methods. The results of such calculations are shown on Fig. 6.4, in which the currents are normalized to J_c. The appropriately normalized (new) thermal coupling constant H_{TC} is given by $h_{TC}J_c/T_a$ and is, for purposes of this particular calculation, fixed at 0.1. For low barrier heights, reverse currents are relatively high, but the temperature dependence of the contact is small; hence no turnover. As the barrier height increases, the turnover characteristic becomes increasingly prominent, and soon dominates the whole picture. It is, of course, a direct consequence of internal positive feedback. At some stage, the temperature may indeed become high enough to permit intrinsic conduction, and that would accelerate the thermal breakdown process enormously.

The negative resistance characteristics which follow thermal breakdown have been discussed by Burgess (1955), with special reference to their behaviour under dynamic conditions, and their equivalent circuits. As long as Newton's law of cooling holds, the system can be described in terms of a single time constant, and this leads to a choice of (only two) simple equivalent circuits (with temperature-independent components) as shown in Fig. 6.5.

Actual measurements of point-contact temperatures were made by Tipple and Henisch (1953), using a thermoelectric method. For a tungsten point contact on (extrinsic) 6 ohm-cm germanium (n-type), the actual temperature at the turnover point was typically about 135 °C, independent of ambient temperature and presumably due to the onset of intrinsic conduction. The peak back-*currents* were approximately constant, and the peak back-power was a linear function of ambient temperature. It extrapolated to zero at $T_a = 135\,°C$, for the particular specimen of germanium under test. Of course, the peak back-power can be greatly exceeded for short periods of time, e.g. under pulse conditions. This was originally demonstrated by Lempicki and Wood (1954); and, when sufficiently short pulses are applied, turnover can be avoided altogether, as one must expect (see Bennett and Hunter 1951), and Nasledov and Yashukova 1960).

In the above formulation, the presence of temperature *gradients* as such has been neglected, but there is good reason for believing that this is not realistic in the high reverse direction. Indeed, an effect due to gradients which, at a point contact, may easily reach $10^5\,°C\,cm^{-1}$, is already implicit in the results quoted above. The existence of such enormous gradients makes the difference between 'point' and 'area' contacts more subtle than is generally believed. A temperature difference of (say) $100\,°C$ could exist across a spreading region of 10^{-3} cm, smaller than the diffusion length; and this could not fail to have serious consequences. Thus, a material might well be within the intrinsic conduction range at the metal–semiconductor interface, but highly extrinsic within significant regions of the spreading resistance. With contact temperatures constant at turnover, different values of T_a imply different thermal gradients at the contact interface, and this is, indeed, what makes one such turnover situation different from another (Henisch and Tipple 1954). For a given contact temperature, the total resistance increases with increasing gradient, as one would expect, not only in Ge, but in GaP (Bhargava 1969). The whole matter is no longer as important as it used to be because modern device technology achieves substantially better thermal environments than the 'cat's whisker' point contact of old.

Gundjian and Habert (1970) have used a transmission-line analysis similar to that outlined in Section 1.1.4 (but including capacitive elements) to calculate the response of point contact spreading resistances to pulse excitation in the self-heating mode. For a discussion of other matters relating to non-isothermal voltage–current relationships, see Numata (1959).

Ankerman and Erler (1977) have drawn attention to the effect of temperature gradients in a different sense, i.e. to regions in which the lattice temperature is uniform but the electron temperature is a local variable (as, of course, it is in Section 2.2.7). Even at low-resistance contacts, such gradients can give rise to non-linear characteristics and can thereby enable such devices to function as infrared mixers.

6.2.2 *Thermal inertia and self-oscillations*

The present discussion is concerned with the apparent impedance of a temperature-dependent circuit element, such as a rectifying contact, when

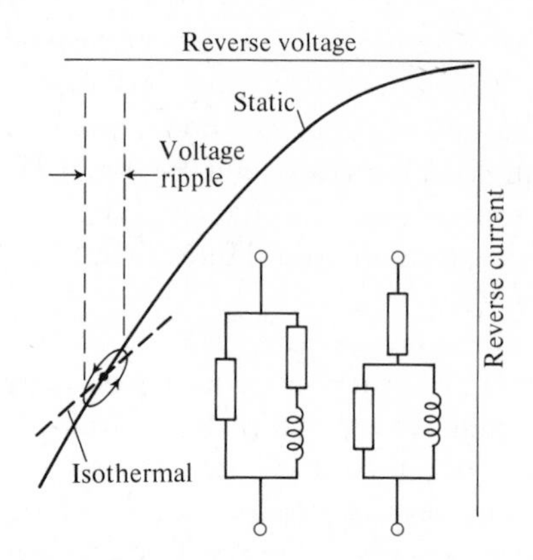

FIG. 6.5. Thermal inertia of contact characteristics, resulting in inductive behaviour. Choice of equivalent circuits. Broken line: isothermal characteristic through the operating point. Full line: static characteristic. After Burgess (1955).

a small ripple voltage is superimposed on a steady bias voltage, e.g. in the reverse direction of current flow. If the frequency of the ripple voltage is very low, the corresponding current fluctuations will follow the static voltage–current relationship. If, on the other hand, the frequency is very high, the current fluctuations will be governed by the isothermal relation. In intermediate cases, the temperature of the contact will lag behind the power dissipation and this, in conjunction with the usual temperature dependence of the barrier resistance, simulates an electrical inductance. The current fluctuations then follow an ellipse, as shown in Fig. 6.5.

A great deal of sophisticated work has been done on electrothermal systems, but one of the simplest analyses is due to Burgess (1955), who assumed that all circuit elements are linear and that the resistance is completely determined by a single temperature. The effective resistance and reactance which such a system presents to the alternating ripple can be calculated if the mechanism of heat dissipation is known. In the simplest case, Newton's law of cooling may be assumed to hold, and the thermal behaviour of the device can then be described in terms of a single thermal time constant. This leads to simple inductive equivalent circuits with components which are themselves independent of frequency. The magnitudes of their components depend, of course, on the operating point. If the temperature dependence of the isothermal voltage–current relationship and the prevailing power dissipation are sufficiently great, the resistive component of the impedance is negative for the lower frequencies, though positive when a critical frequency is exceeded. If sufficient capacitance is present in parallel with such a circuit element, the negative resistance and inductive reactance give rise to oscillations in the low-frequency range. This possibility is not limited to rectifying contacts, but exists for any circuit element with temperature-sensitive characteristics. The self-capacitance of the contact rectifier itself is generally insufficient to produce oscillations, but stray circuit capacitances may do so.

When Newton's law of cooling does not apply, the rate of heat dissipation becomes a function of frequency, and the system can then no longer be described in terms of a single thermal time constant. Components of an equivalent circuit can still be evaluated, but they are then frequency dependent.

6.3 Contacts with semi-insulators

6.3.1 *Semi-infinite systems; classical Mott and Gurney case*

The subject of metal–insulator contacts remained almost totally neglected until Mott and Gurney (1940) analysed the prevailing circumstances for semi-infinite systems. If $(\phi_m - \chi)$ were not too large for a given pair of materials, some electrons could be emitted from the metal and could penetrate some way into the insulator. The insulator was assumed to have no free carriers of its own; all its $n(x)$ carriers at a distance x from the metal interface would thus contribute to a space charge, and hence to band curvature, as shown in Fig. 6.6. In these simple circumstances, the energy profile resulting from the electron influx can be readily derived. Mott and Gurney's classical solution is given here for the sake of completeness. The band gap of the material concerned is assumed to be large, making minority carriers irrelevant. Traps, which are the key to real insulator behaviour, are also ignored, which means that the model

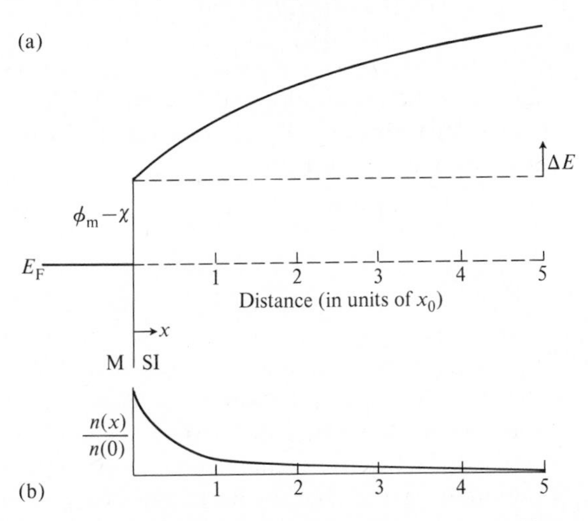

FIG. 6.6. Metal contact on semi-insulators without interface states. (a) Energy profile; eqn (6.3.4). (b) Carrier concentration profile; eqn (6.3.6).

aims at metal contacts with something intermediate between a normal semiconductor and a full insulator. Such carrier transport as the material can support is governed by eqn (2.2.7) which, for zero current (i.e. in equilibrium) implies the Boltzmann relationship. We can therefore write

$$n(x) = n(0)\exp(-\Delta E/kT) \tag{6.3.1}$$

which, with Poisson's equation, gives

$$\frac{d^2E}{dx^2} = -\frac{e^2 n(0)}{\epsilon}\exp\left(-\frac{\Delta E}{kT}\right). \tag{6.3.2}$$

After one stage of integration, we have

$$\frac{1}{2}\left(\frac{dE}{dx}\right)^2 = \frac{e^2 n(0) kT}{\epsilon}\exp\left(-\frac{\Delta E}{kT}\right)$$

or

$$\frac{dE}{dx} = \frac{d\Delta E}{dx} = \left\{\frac{2e^2 n(0) kT}{\epsilon}\right\}^{1/2}\exp\left(-\frac{\Delta E}{2kT}\right). \tag{6.3.3}$$

The second stage of integration yields

$$\Delta E(x) = 2kT \log\left(\frac{x}{x_{mi}} + 1\right) \tag{6.3.4}$$

if one takes into account that ΔE must go to zero as x goes to zero. The constant x_{mi} is a characteristic distance given by

$$x_{mi} = \left\{\frac{2\epsilon kT}{e^2 n(0)}\right\}^{1/2} \tag{6.3.5}$$

which equals $\sqrt{(2)}\,\lambda'_D$ in the terminology of Section 2.4.2, where λ'_D is a Debye length defined in terms of $n(0)$. By combining eqns (6.3.4) and (6.3.1), Mott and Gurney also obtained

$$\frac{n(x)}{n(0)} = \left(\frac{x_0}{x_{mi} + x}\right)^2. \tag{6.3.6}$$

Equation (6.3.4) is already 'contained' in Fig. 6.6(a) and eqn (6.3.6) is plotted in Fig. 6.6(b).

The calculated energy profile in Fig. 6.6(a) is not easily verified by experiment, but it nevertheless incorporates a profound truth about insulators, and that is its great didactic virtue: insulators *insulate* not merely because they have no free charge carriers of their own, but because they accumulate space charge (here, only in the form of free carriers, but ordinarily in traps) which discourages the entry of further charges and thereby stops the current.

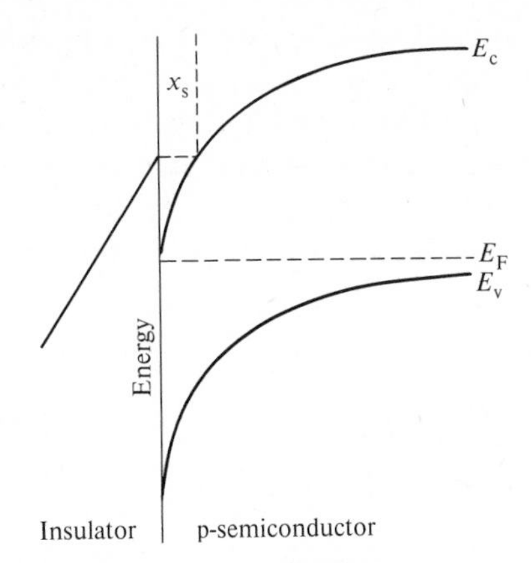

FIG. 6.7. Insulator–p-semiconductor contact, with 'forward' voltage applied, leading to electron injection into the insulator. x_s denotes critical scattering distance. Schematically after Kolk and Heasell (1980a,b).

Contacts between insulators (e.g. Al_2O_3) and semiconductors (e.g. p-Si) have also been studied, e.g. by Bosselar (1973), Verwey (1973), and by Kolk and Heasell (1980a,b). Such contacts involve a model as shown in Fig. 6.7. Incoming electrons from the right will experience the spike-like barrier if they are scattered within the distance x_s. Otherwise, they will not experience it, and an analysis of the voltage–current relationship can be based on this fact for any given magnitude of the mean-free-path. The existence of the spike-like barrier represents a classic problem in hetero-junction work (Milnes and Feucht 1972, Kajiyama *et al.* 1979). For a comprehensive analytical treatment of space-charge conduction in thick systems with traps, see Lampert (1956), and Lampert and Mark. (1970). In the models discussed there, the contact serves only to provide the majority carriers (of which it is assumed to be an inexhaustible source) and plays no other role, but this is not necessarily a realistic assumption (see O'Reilly and DeLucia 1975, and Rosenthal and Sapar 1974).

6.3.2 Thick metal–'insulator'–metal systems

In the context of electrical conduction in near-insulators, the semi-infinite model is not actually very useful. In the present context, we again shall consider 'thick' to mean impermeable to tunnelling electrons, but thin enough to permit some current to flow by drift and diffusion processes, for which the electrodes (whether identical or not) provide the carrier reservoirs.

The linear part of the energy profile presented in Fig. 1.19 is a symptom of the fact that space charges have been neglected within the λ_a region. However, if there were *no* free carriers (resident) in the insulator, then strict linearity of the contour would be impossible, even at zero current, while the insulator is in contact with conducting materials. If the Boltzmann distribution were to prevail in the layer *after* contact, it could be so only on the basis of carriers gained from the electrodes, and these carriers would constitute a space charge. The profile would thus have to have a convex curvature (see below). The problem is analytically com-

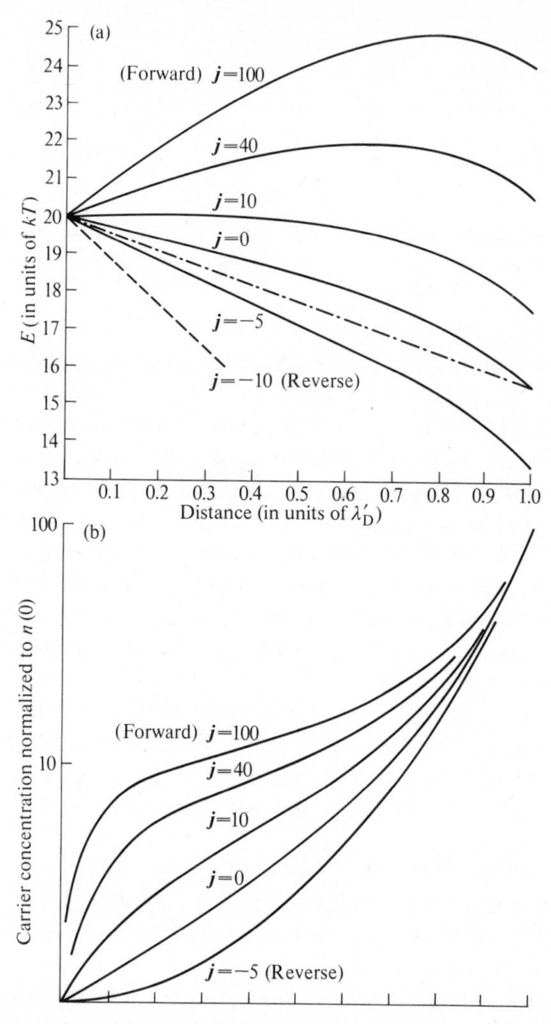

FIG. 6.8. Numerically calculated characteristics of an artificial barrier layer with free-carrier space charges. Results for a specimen thickness $L = \lambda'_D$ (Debye length). Because $n(0)$ is small, this corresponds to a layer of substantial thickness. Layer undoped and free of traps. λ'_D referred to $n(0)$. Contact with two different metals is taken here to involve $\mathbf{N}(L)/\mathbf{N}(0) = 100$, and hence $V_{Di} = 4.61kT/e$. (a) Energy profiles for different (normalized) currents, forward and reverse. (b) Corresponding concentration contours. After Henisch *et al.* (1980).

plex. Fan (1948) has provided a solution in terms of tabulated Airy integrals, but these are not convenient to use, and the same is true for the implicit solutions derived by Skinner (1955a,b) in terms of ordinary and modified Bessel functions of fractional order. (Many references to earlier work in this field will be found in these papers, which also deal with the subject of contact rectification.)

It is simpler to proceed on the basis of numerical solutions, along the lines described in Section 2.4.1. (Henisch *et al.* 1980). Typical results are given in Fig. 6.8(a) in terms of the energy contour, and in Fig. 6.8(b) in terms of the concentration profile. Both are instructive. Figure 6.8(b) shows that, for zero current, the concentration does indeed depart from the simple exponential dependence on distance, precisely because a space charge is present even under those conditions. As expected, the departure is smallest where the electron gas is most dilute, i.e. towards $\boldsymbol{X}=0$. In Fig. 6.8(a), an energy maximum appears only for substantial forward currents, but if $\boldsymbol{N}(L)/\boldsymbol{N}(0)$ had been taken to be considerably smaller (than 100, as used here) the maximum would have appeared for much lower forward currents, and even for zero current.

The application of reverse voltages, as always, augments the field and diminishes the role of diffusion. Accordingly, it tends to make the profile more linear. It can easily be shown that the presence of space charges also diminishes rectification.

In the presence of carrier traps (neglected above and also by Skinner (1955a)) all the curvatures would, of course, be increased. That situation has been analysed by Suits (1957), but the treatment yielded only approximate results, limited to the case of low trap occupancy, and then only in terms of Bessel functions of fractional order. Within the framework of the present arguments, a more general trap correction is easily introduced. If, in normalized terms, there are $\boldsymbol{M}_0$ traps per unit volume, all originally neutral, and if these capture $\boldsymbol{N}_\mathrm{T}$ electrons, then we can put

$$(\boldsymbol{M}_0-\boldsymbol{N}_\mathrm{T})\boldsymbol{N}=K \tag{6.3.7}$$

where K is an equilibrium constant, proportional to the trap concentration. When the total space-charge density (now proportional to $\boldsymbol{N}+\boldsymbol{N}_\mathrm{T}$ is introduced into Poisson's equation, the analysis yields in normalized terms:

$$\frac{\mathrm{d}^2\boldsymbol{N}}{\mathrm{d}\boldsymbol{X}^2}=\boldsymbol{N}^2+\boldsymbol{N}\boldsymbol{N}_\mathrm{T}-\frac{\mathrm{d}\boldsymbol{N}}{\mathrm{d}\boldsymbol{X}}\left(\boldsymbol{j}-\frac{\mathrm{d}\boldsymbol{N}}{\mathrm{d}\boldsymbol{X}}\right)\frac{1}{\boldsymbol{N}} \tag{6.3.8}$$

which contains $\boldsymbol{M}_0$ and K implicitly. The current density $\boldsymbol{j}$ is a constant parameter here. The numerical solutions are the carrier concentration contours. Figure 6.9 gives typical results, Fig. 6.9(a) in the absence of

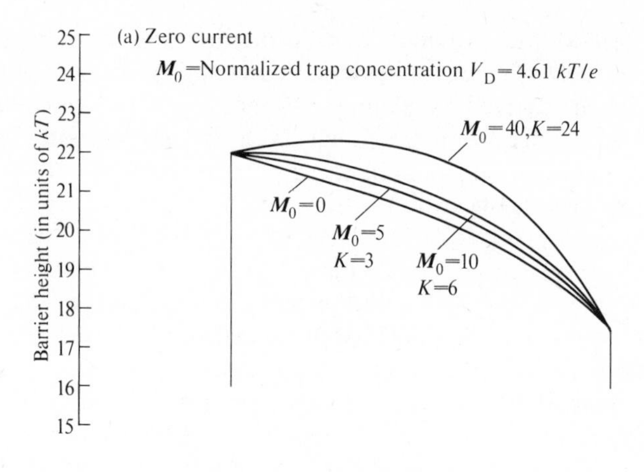

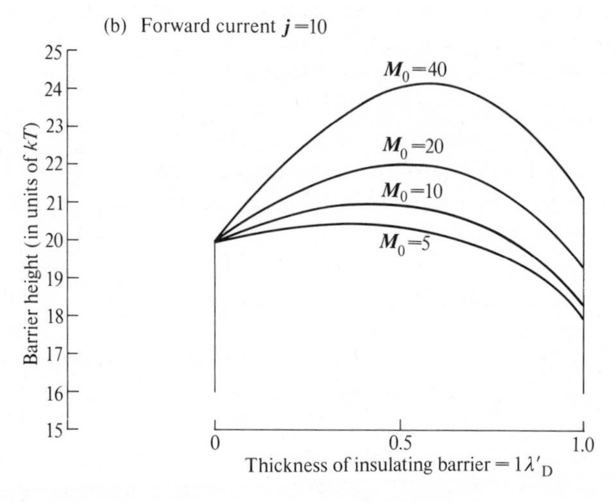

FIG. 6.9. Numerically calculated characteristics of an artificial barrier layer, parameters as in Fig. 2.45, but with variable trap density $\boldsymbol{M}_0$ (normalized to $n(0)$). (a) Energy contours at zero current. (b) Energy contours for a forward current of $\boldsymbol{j} = 10$. After Henisch *et al.* (1980).

current, and Fig. 6.9(b) for $\boldsymbol{j} = 10$. From (a) we see that, as the trap concentration $\boldsymbol{M}_0$ (here again normalized to $n(0)$) increases, a stage is reached in which the resident field is cancelled at $\mathbf{X} = 0$ and even exceeded. The curve for $\boldsymbol{M}_0 = 40$ reflects a situation in which the artificial barrier layer has trapped so much charge that its zero-current shape is generically different from that of the trap-free case. With increasing current, traps augment the distortion, as seen from (b). In Fig. 6.8 (zero trapping) a forward current of $\boldsymbol{j} = 10$ was just sufficient to establish the flat-band condition at $\mathbf{X} = 0$. Now $\boldsymbol{j} = 10$ leads to a substantial maximum,

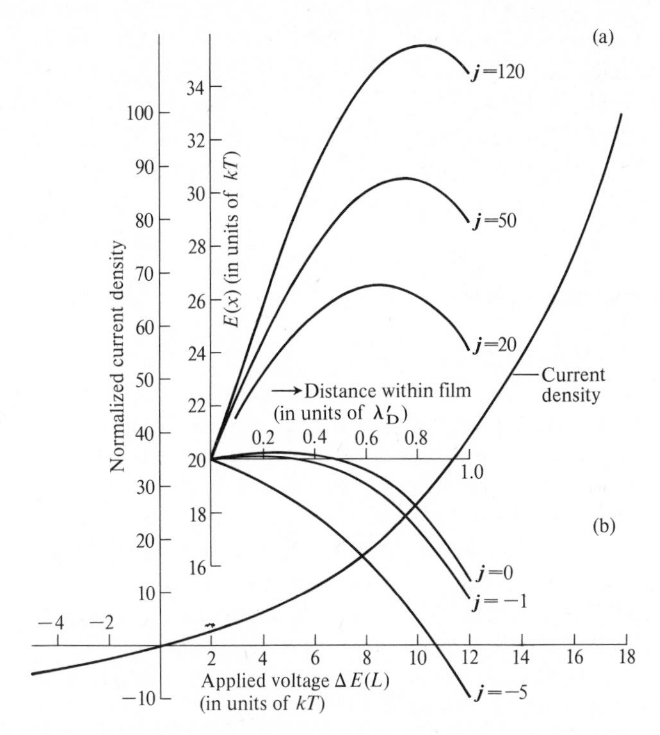

FIG. 6.10. Numerically calculated characteristics of an artificial barrier layer; trap density $\boldsymbol{M}_0 = 40$, $K = 24$, $V_D = 4.61kT/e$, $L = \lambda'_D$. (a) Energy contours for varying currents. (b) Voltage–current relationship. After Henisch *et al.* (1980). $\mathbf{N}(L)/\mathbf{N}(0) = 100$.

even for a trap concentration as low as $\boldsymbol{M}_0 = 5$, and, of course, to a higher maximum for higher trap contents. At the same time, the rectification ratio further deteriorates. As forward current begins to flow, the effective barrier height increases, and this counteracts the current itself (Fig. 6.10). As the applied voltage (represented here in terms of $E(L)$) increases, the height of the maximum likewise increases. This goes a long way towards explaining why 'insulating layers' between different metal electrodes do not in practice make good rectifiers; all other complications apart, they contain too many traps. In practice, the 'insulating' layer is often an oxide, and because oxide layers grown in different ways can contain different trap concentrations, they can give rise to different barrier heights (Bhattacharya and Yeaman 1981). On the basis of the above analysis, the total barrier height is expected to increase with insulator thickness, and Lewicki and Mead (1966) found this to be true for Al–AlN–Mg structures, with AlN layers varying between 40 and 90 Å. Over that range, the barrier height increased from about 1.6 to 2.6 eV.

It should be noted that, in such systems, the fields at the two electrodes are not the same, which means that the capacitance does not conform to classical definitions (Callarotti 1981). The general shape of the voltage–current characteristic, and the existence of rectification as such, have been well confirmed (Sullivan and Card 1974).

Actual situations are not always as simple as those described here, but the model gives a good deal of insight into what happens and why. Moreover, though the immediate context has been that of artificial barrier layers, the same considerations apply to any space-charge-controlled conduction in bulk material, as long as the assumed boundary conditions prevail.

Image-force corrections have been ignored in the above discussions, but they should in fact enter in an interesting way. A charge located anywhere between the metal electrodes will have an image in each electrode, and each image will have a secondary image, and so on. The total effect is similar to that of an object between parallel mirrors. Geppert (1963) has calculated its influence on an initially rectangular barrier contour. The effect is, of course, to round that contour, and to lower the electron energy (in varying degree) everywhere (Fig. 6.11). Thin barriers experience the greatest change; ΔE_s is inversely proportional to the barrier thickness. (The multiple image feature has recently been the subject of interesting notes on summation procedures by Newcomb (1982) and Anon. (1982).)

As far as transport is concerned, we have here considered only drift-diffusion, but barriers of this kind can also be looked upon in another light, namely as limiters of Schottky emission from the negative contact electrode. Soukup (1975) has analysed such systems, and has calculated the associated voltage–current characteristics, which are of course sharply non-linear.

Of course, the above computations can also be carried out for symmetrical contact systems. The voltage–current characteristics are then likewise symmetrical, the role of diffusion currents is diminished, that of field currents is enhanced, but the results are not otherwise very different.

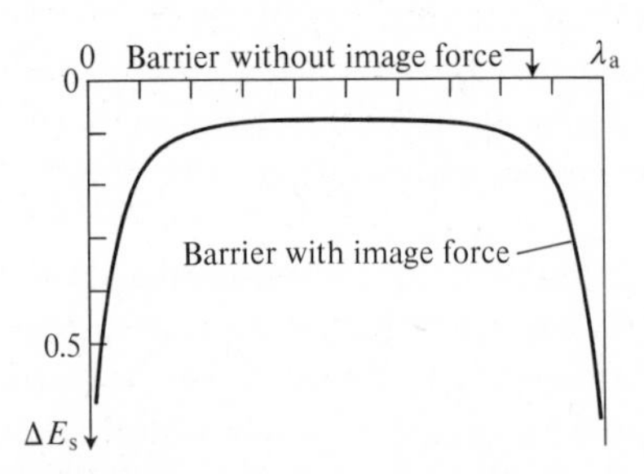

FIG. 6.11. Barrier lowering by multiple image-force effects in a thin insulating layer. Numerical for $K = 4$, and barrier thickness (λ_a) = 25 Å. (See Geppert 1963.)

At higher applied voltages, one or more of several other effects are expected to come into play: image forces, tunnelling, hot-carrier generation and field ionization (if the mean-free-path is sufficiently large), dielectric breakdown, and self-heating (see also Suits 1957). If one were to consider semi-insulating barriers of diminishing thickness, the onset of tunnelling would be very sudden (between 30 and 10 Å), and all other transport mechanisms would quickly cease to be competitive. A comprehensive treatment of thermally assisted tunnelling through *thin* 'insulating' films has been provided by Simmons (1963).

Further reading

On contacts:
with CdS: Courtens and Chernow 1966, Fuchs and Heime 1966, Learn, *et al.* 1966; with CdF_2: Kessler 1981; with ZnS: Edwards and Jones 1978.

On contacts with ceramics, molecular crystals, and polymers: Wintle 1975.

On interfaces between an insulator and a semiconductor: Williams 1974.

On double injection into anthracene: Schwob and Williams 1974.

On the evaluation of trapping parameters in metal–insulator–metal systems: Gupta and Van Overstraeten 1976.

On the properties of contacts with organic materials:
Al/Mg phthalocyanine: Ghosh *et al.* 1974; tetracene: Ghosh and Feng 1973.

6.4 Semiconductor–semiconductor contacts

6.4.1 *Single contacts between semiconductors of the same type*

We have already seen that the surfaces of non-polar semiconductors tend to be characterized by a relatively high density of surface states, often densely bunched at some energy region within the forbidden band. When a contact is made with a metal, these states are modified but their presence (now as interface states) continues to make itself felt. A charge exchange takes place between them and the metal (Section 1.3.2) and if the density of those states is high enough, the space-charge region within the semiconductor is completely screened by them. We would then expect contact properties which are completely independent of the thermionic work-function of the contacting metal. In practice, the screening is always imperfect, but we still expect contact properties to be much less dependent on work functions than they would be in the absence of interface states.

When two equally doped semiconductors of the same phase come into contact, the identity of their Fermi levels makes a charge exchange unnecessary. On the other hand, what the resulting interface state density really is remains a moot point. One might simplistically consider that each contacting crystal contributes its own surface states, giving twice the number of interface states after contact. Their distribution might well be unaltered, and the intergranular barrier would thus be of the same height as that of a metal–semiconductor barrier on the same material. The result would be the symmetrical barrier profile shown in Fig. 6.12(a). There is, however, some evidence which suggests that matters are not quite so simple. Other things being equal, intergranular barriers (on Ge, for instance) tend to be distinctly lower than metal–semiconductor barriers, typically half as high, namely ≈ 0.25 eV (Rieder 1979). Stratton (1956b) found $V_D = 0.28$ volt from his measurements. Such values lead to the tentative conclusion that there are actually *fewer* interface- than surface-states in such a case, in which the interpenetrating wave-functions are essentially the same. A total disappearance of such states cannot be contemplated, because of the expected interface disorder, in varying degree, but it is not implausible to believe that the mismatch is much smaller when identical lattices are involved on the two sides of the interface. The notion that the nature of interface states is bound to depend on both of the contacting materials has already been discussed by Andrews and Lepselter (1970), and though these authors were specifically concerned with metal–semiconductor contacts, the conclusions should have a more general validity. Experimental situations are usually complicated by geometrical factors and by the difficulty of specifying the nature of surface contaminations, which is why the evidence relating to semiconductor–semiconductor contacts is still not totally conclusive. A model of intergranular contacts based on the presence of interface states was first proposed for such situations by Mitchell and Sillars (1949), in connection with the properties of microcrystalline SiC. The classic papers on Ge–Ge contacts are those of Benzer (1947), and of Taylor, Odell, and Fan (1952).

Let us now consider that we have a fixed density of interface states, distributed in some manner over the energy range of the forbidden band, irrespective of how that density and distribution might vary with the nature of the contacting members. The question still remains how the interface-state *occupancy* may be expected to depend of voltages applied and currents flowing. Since that occupancy governs the barrier height, this is an important matter, and two situations may be envisaged:

(a) that the interface states behave like a metal which can be charged or discharged virtually without limit, or

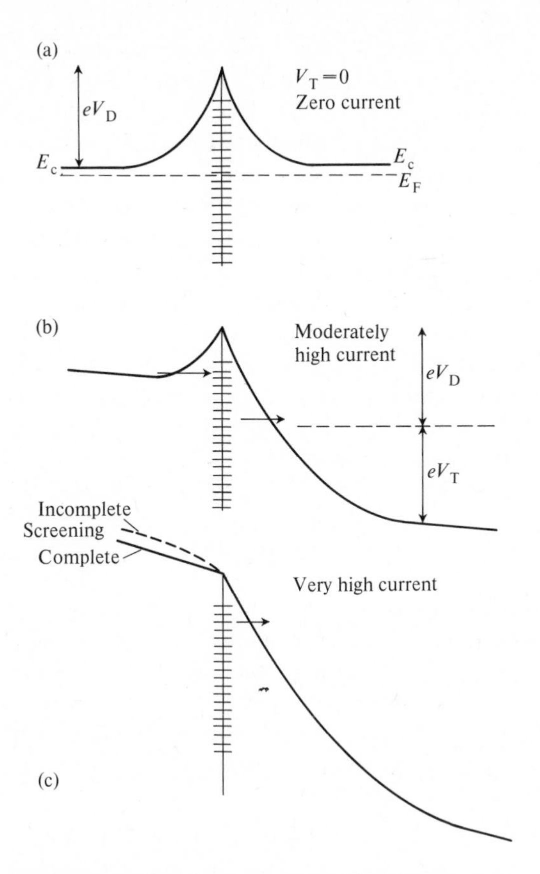

FIG. 6.12. Semiconductor–semiconductor contact. (a) At zero current; high density of interface states. (b) With moderately high voltage applied; high density of interface states. (c) At very high current; V_T no longer defined. Full line: complete screening. Broken line: incomplete screening; negative space charge; interface states completely filled. The horizontal arrows denote tunnelling possibilities.

(b) that the interface states are essentially isolated, and characterized by a fixed charge which remains independent of external voltages and the resulting current flows.

Neither assumption is realistic, but as limiting cases, they have their uses. Meanwhile, let us note that the cusp-like double-barrier (Fig. 6.12(a)) is expected, on all the grounds discussed here, to behave in a manner very different from the back-to-back barrier systems considered in Sections 2.2.9 and 4.1.5.

The case of unlimited interface charge When a reverse voltage is applied to a metal–semiconductor contacts, the barrier widens, the positive charge within it increases. Correspondingly, the negative surface-charge on the metal also increases, and the system as a whole is neutral. Thus, if interface states were to behave like a layer of metal, in the sense that their charge can be freely changed, we would be dealing with two non-interacting barriers, non-interacting because completely screened from one another. The only considerations which can be usefully applied to such a system are then circuit considerations. One would have a forward voltage across it, the other side a reverse voltage, as shown in Fig. 6.12(b). However, the current is clearly common, which means immediately that this situation can prevail only over a very small range of currents (i.e. within the realm of reverse currents). The relationship between the forward and reverse voltage components (V_{BF} and V_{BR}, *both regarded here as inherently positive*) would be governed by

$$\{1-\exp(-eV_{BR}/kT)\}=-\{1-\exp(eV_{BF}/kT)\} \tag{6.4.1}$$

in a single-carrier system, as long as there is not tunnelling. Also, we have $V_{BR}+V_{BF}=V_T$, this being the total voltage. This assumes that the contact exhibits total reverse saturation, and in such circumstances, V_{BF} can never exceed $0.7(kT/e)$, a very small value. As a result, the forward-biased barrier changes very little, as the other barrier increases in width. The overall change of barrier width $\lambda_{BR}+\lambda_{BF}$ should be reflected by the behaviour of the intergranular capacitance. If we denote the barrier capacitance (per unit area) for zero bias by C_{BO}, then C/C_{BO} is given by eqn (4.1.4), in conjunction with the fact that we are dealing with the series combination of the two barrier components. Thus,

$$\frac{C}{C_{BO}}=\frac{2V_D^{1/2}}{(V_D+V_{BR})^{1/2}+(V_D-V_{BF})^{1/2}} \tag{6.4.2}$$

and this relationship is illustrated in Fig. 6.13 for three values of V_D. Such gently S-shaped curves have been observed by Lou (1980) and by Emtage (1977) for ZnO aggregates, i.e. for systems governed by multiple grain boundary capacitances in complicated series and parallel connection. These measurements may be regarded as a qualitative confirmation of eqn (6.4.2), but the relationships in a multigrain assembly are, of course, greatly complicated by geometry and by the presence of voltage-dependent series and shunt resistances. When such contacts enter the breakdown region, there is a drastic increase of capacitance (e.g. see Mahan *et al.* 1979, and Vanadamme and Brugman 1980) which is not shown within the voltage range covered by Fig. 6.13.

When high voltages are applied, current saturation cannot be expected to be maintained. The forward-biased barrier will tend to be wiped out, and the reverse-biased barrier will then behave more or less as a single

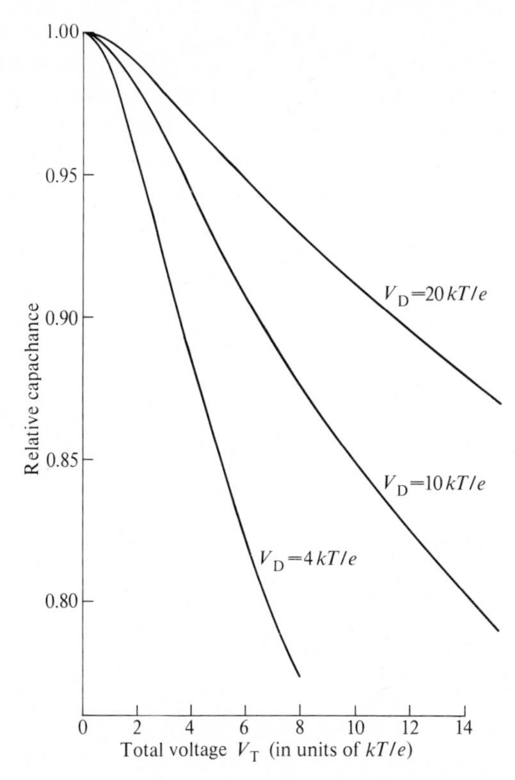

FIG. 6.13. Capacitance of a semiconductor–semiconductor contact with a high density of interface states, as a function of total applied voltage V_T for various barrier heights V_D (eqn 6.4.2).

entity. The ultimate result should be an energy profile as given in Fig. 6.12(c).

The case of limited interface charge This case is more general and more realistic but has suffered from neglect in the literature. Under extreme conditions, the interface charge would be completely fixed and, in the resulting absence of all screening, the current-dependent charge distribution in one barrier would inevitably affect the charge in the other. The basic transport equation (2.2.7) would then have to be solved numerically for appropriate boundary conditions but, as far as is known, no such calculations are on record to date, not even for a single-carrier system. For generality, partial screening would have to be envisaged, by including a relationship between the interface charge and the excess free-carrier concentration Δn at $x = 0$. A Shockley–Read recombination model would serve for this purpose, as it does in many similar contexts.

Taylor and co-workers (1952) did, however, begin a treatment of grain-boundary properties by considering the interface charge as totally constant for small currents (*zero* screening). They neglected free carriers and, in such circumstances, overall neutrality would have to be maintained entirely by an adjustment of λ_{BR} and λ_{BF}, both defined along standard (Schottky) lines. In terms of the present symbols, the essence of their argument can be represented as follows.

The total charge per unit area of a *single* barrier is given by

$$Q_B = eN_d\lambda_B \tag{6.4.3}$$

if we neglect the free-carrier contribution, then by eqn (2.2.1b) this can be written as

$$Q_B = \{2N_d\epsilon e(V_D + V_B)\}^{1/2}. \tag{6.4.4}$$

In the absence of current, the charge density due to each barrier is clearly

$$Q_{ss} = 2\{2N_d\epsilon eV_D\}^{1/2}. \tag{6.4.5}$$

When current is flowing, while Q_{ss} is kept constant, we must have

$$Q_{ss} = \{2N_d\epsilon e(V_D + V_{BR})\}^{1/2} + \{2N_d\epsilon e(V_D + V_{BF})\}^{1/2} \tag{6.4.6}$$

with, again, $V_T = V_{BR} - V_{BF}$, where V_{BR} is positive and V_{BF} negative, as usual. This suggests a voltage-sharing process very different from that represented by eqn (6.4.1), with which it is in conflict because of the different underlying assumptions. Indeed, without some special provisions (see below), it would be in conflict with the requirement of current continuity. *With* such provisions, it is an acceptable approximation in the present (limited) context, and we see that $V_{BF} = -V_D$ when the forward-biased barrier is wiped out.

Thus, from the last two equations,

$$2\{2N_d\epsilon eV_D\}^{1/2} = \{2N_d\epsilon e(V_D + V_{BR})\}^{1/2} \tag{6.4.7}$$

or

$$V_{BR} = 3V_D \quad \text{and thus} \quad V_T = 4V_D.$$

This gives the limiting value of the total voltage for which electrostatic voltage-sharing could possibly be maintained, and even then one would have to assume that there are sufficient surface states present not to impose any additional (earlier) limitation. In contrast, eqn (6.4.1) yields the maximum $V_T \approx 6(kT/c)$ (for about 99 per cent saturation), and this value is generally much smaller. When V_T is increased beyond the limit set by eqn (6.4.7), Q_{ss} cannot remain constant; it must likewise increase, always assuming that the interface-state concentration is high enough to allow this. When the available interface states are filled, further negative charge must accumulate in the adjoining bulk region, as shown schematically by the broken line in Fig. 6.12(c).

Taylor *et al.* (1952) calculated the barrier conductance (at zero applied voltage) also, and did so by using the diffusion relationship in the form of eqn (2.2.8), while adjusting the carrier concentrations at $x = 0$ so as to maintain current continuity at that boundary. (In contrast, eqn (6.4.1) is based on the assumption that $n(0)$ is fixed at its equilibrium value.) They did so, however, for electron- and hole-currents independently, without considering injection or any cross-modulation by space-charge effects. The importance of injection is shown by the fact that semiconductor–semiconductor contacts are capable of displaying electroluminescence (e.g. see Lossew 1940, and Henisch 1962). Voltage–current relationships of a particularly simple injection system have been analysed by Board and Darwish (1982). That model postulates a plane of acceptors, separating two regions of n-type material and, of course, generating a cusp-like barrier indistinguishable from that generated by acceptor-like interface states. Board and Darwish derive the effective barrier heights as 'seen' from two sides on the basis of the Schottky depletion approximation, and calculated the voltage–current relationship, using diode theory for electrons and drift-diffusion for holes. By implication, electronic equilibrium is assumed to prevail at the interface. A more precise formulation would have to be solved numerically, taking account of the fact that the interface has a dual role: it is the seat of trapped charge and it also acts as a recombination sink. As one would expect, such barriers rectify when the doping on the two sides is unequal. See also Baccarani *et al.* (1978) for an approximate treatment of intergranular barriers in polycrystalline silicon.

Another interesting treatment of a simplified contact situation is due to Stratton (1956a), envisaging only single-carrier participation on the basis of a fixed lifetime in interface states. Such a model can approach validity for highly doped material of substantial band gap (see also Card and Yang 1977).

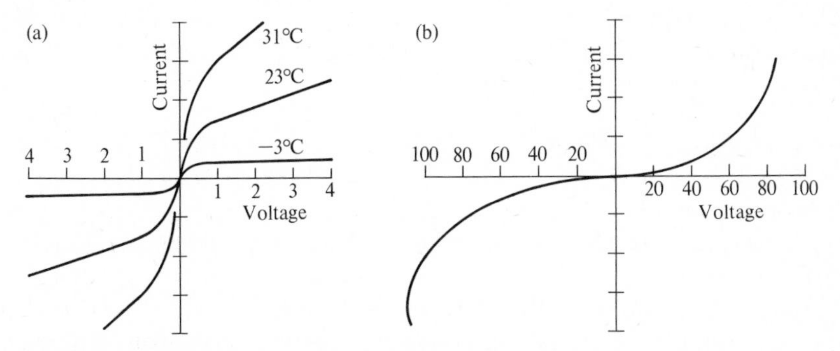

FIG. 6.14. Characteristics of contacts between specimens of n-type germanium. (a) Low-voltage range. (b) High-voltage range. Note small degree of asymmetry, probably due to slightly different donor contents. Measurements by pulse methods to avoid heating. Curves corrected for voltage drop in bulk material. Schematically after Taylor *et al.* (1952).

Figure 6.14 shows results for contacts between specimens of n-type germanium, obtained by means of fractional multisecond pulses, to avoid heating. As expected, the curves resemble Fig. 1.4(c) because they are essentially 'reverse'-characteristics, a fact which is also confirmed by their temperature-dependence. For germanium, Taylor *et al.* also report that grain-boundary barriers 'disappear' upon neutron bombardment, which makes the specimens p-type. This is true only in the sense that the boundary layers become less and less resistive (see Section 1.3.2 for a more detailed description of the behaviour).

6.4.2 Heterocontacts; heterojunctions

'Heterojunction' is the more common term, but it is useful to maintain a distinction between transitions from one material to another which are known to be gradual (e.g. like those of p–n junctions in a single semiconducting phase), and those that are for practical purposes abrupt. The latter are designated as 'contacts', rather than 'junctions', and the term 'hetero-' refers to the fact that two phases are involved. We have already seen that the abruptness is actually in doubt even for the simplest metal–semiconductor system, which is undoubtedly the most common form of heterocontact. Here we are concerned with systems in which the materials on both sides of the interface are semiconductors. Such systems have far more descriptive parameters than metal–semiconductor contacts, in the sense that either side can be n-type or p-type, and either side can have the greater band gap, electron affinity, mobility ratio, etc. On those grounds, such structures are in principle much more 'designable', hence their device-interest. In practice, the 'designability' often encounters severe limitations, but many opportunities are still open in this promising field. Gradual ('graded') transitions from one phase to another are, besides, of great interest in the conceptual realm, since their unharmonically periodic crystal fields involve some profound wave-mechanical considerations (e.g. see Mehely 1968), and Marshak and van Vliet 1978a,b). In the ordinary way (i.e. in homojunctions or uniform bulk material), the slope of the band edges represents the local field, but since *field* is a uniquely defined and univalued function of space, the band edge slopes of a graded heterojunction (e.g. see Fig. 6.15) cannot *both* represent *the* field. Moreover, the parts of the energy contour that have *gradients* (in the absence of current) are not true band edges at all, but *envelopes* of attenuated states.

The matter can be fundamentally and precisely resolved only by recourse to the wave equation, taking structural changes from cell to cell into account. However, as long as the electron gases are non-degenerate, a widely used shortcut method is available. We first recognize that eqn (2.2.7) is no longer adequate as the starting point of a transport analysis,

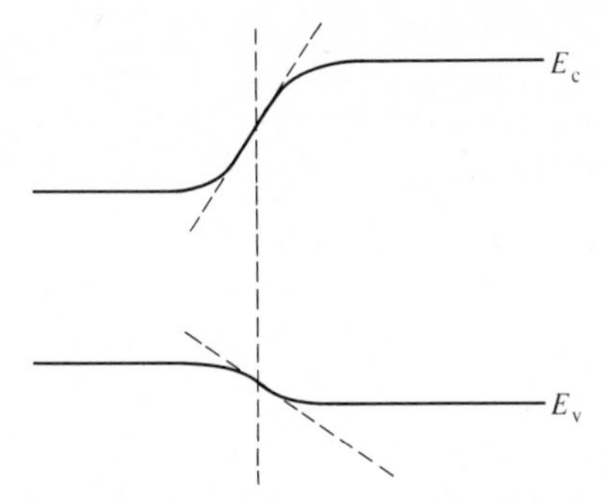

FIG. 6.15. Schematic energy contour for a heterojunction without trapping state.

because allowance has to be made for the rate of change not only of n, but of χ (the electron affinity), E_g (the band gap), and $\mathcal{N}_c$, $\mathcal{N}_v$ the effective densities of states in the conduction and valence bands. When the movement of charge carriers in such an environment is calculated, it turns out that electrons behave *as if* the gradient of E_c were the effective field acting upon them, and holes behave as if the gradient of E_v fulfilled the same function for them. Thus, whereas there can only be one (electrostatically defined) field, we have *two effective fields* which serve the calculations simply and well. Many treatments found in the literature actually *assume* this conclusion and use it without further justification. For a successful application of these principles to various GaAs-based heterojunctions, see Sutherland and Hauser (1977). The situation is analogous to that encountered in connection with quasi-Fermi levels. There can be only one true field, which must, of course, be common to electrons and holes. However, when diffusion is taken into account in a material of uniform band structure, the gradients of E_{Fn} and E_{Fp} are known to act as effective fields for electrons and holes respectively. In the same way, the gradients of E_c and E_v in a continuously graded heterojunction act as effective fields as long as the carrier concentrations are in equilibrium. When they are not, quasi-Fermi levels can be again defined which fulfil this role.

A detailed discussion of graded junctions (and, indeed, all junctions as such) is actually outside the present scope, which is limited to abrupt or, at any rate, quasi-abrupt transitions. Although this subject has also a very large literature by now, no integrating works seem to have appeared since those of Milnes and Feucht (1972) and Sharma and Purohit (1974). A brief outline of the situation will have to suffice here.

We have already seen that, in the presence of interface states between the contacting materials, the transport properties of the assembly as a whole depend on the extent to which these states provide electrical screening and recombination facilities. If the screening were perfect, we would be dealing with the series connection of two quasi-metal semiconductor contacts in opposition. In practice the screening is never perfect, which means that the space charges on each side of the interface interact with one another in varying degree.

By way of example, let us first consider a heterocontact of GaAs and Ge, materials which have electron affinities of 4.07 and 4.13 eV, and band gaps of about 1.45 and 0.7 eV respectively. Because they also have very similar lattice constants (5.654 Å for GaAs and 5.658 Å for Ge), these crystals provide an excellent illustrative case of contacts which can be made with a minimum of lattice mismatch and thus, one might hope, with a minimum number of interface states per unit contact area. Depending on which of the materials is n-type and which p-type, substantially different band configurations can be obtained, as Fig. 6.16 shows. In Figs 6.16(a) and (c), the situation is represented as it exists before the establishment of thermodynamic equilibrium; in Figs 6.16(b) and (d) as it is afterwards. The spike-like discontinuity in the conduction band of Fig. 6.16(b) could be dismissed as minor, but that in the valence band of Fig. 6.16(d) is highly prominent. The contours on Figs 6.16(b) and (d) are shown rounded near the interface, to signify space charges. No interface states are envisaged by this diagram (because of the lattice match in this

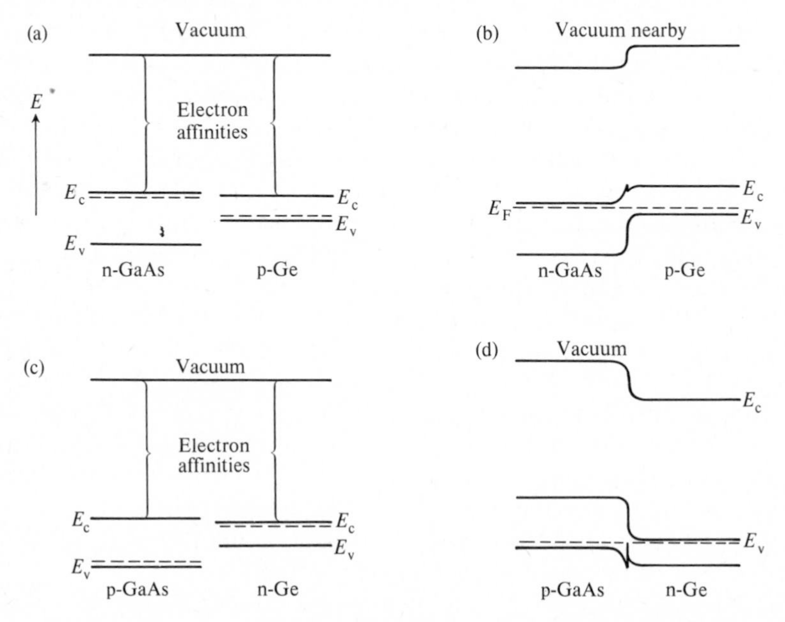

FIG. 6.16. Heterocontact configurations for GaAs–Ge, in the absence of interface states. (a), (c) Material apart. (b), (d) Materials in contact and in equilibrium.

instance), which means that the total space charges on the left and the right must be equal. Each material can be said to contain a Schottky-like barrier, and the two diffusion voltages must evidently add up to the original difference of the Fermi levels in Figs 6.16(a) and (c). Of course, both materials could be n-type, or p-type in varying degree, or else intrinsic; and many different models can therefore be constructed. Milnes and Feucht (1972) give extensive lists of data on which such constructions can be based for a variety of semiconductor pairs. Sharma and Purohit (1974) provide a great deal of empirical information on heterojunctions made and measured. Many references to the earlier literature will be found in both works.

Interface states (which depend in practice on how the heterocontact is made) change the situation drastically, as shown in Fig. 6.17 (Oldham and Milnes 1964). The resulting problems have not yet been solved with any high degree of generality, but *ad hoc* analyses are available of rectifying

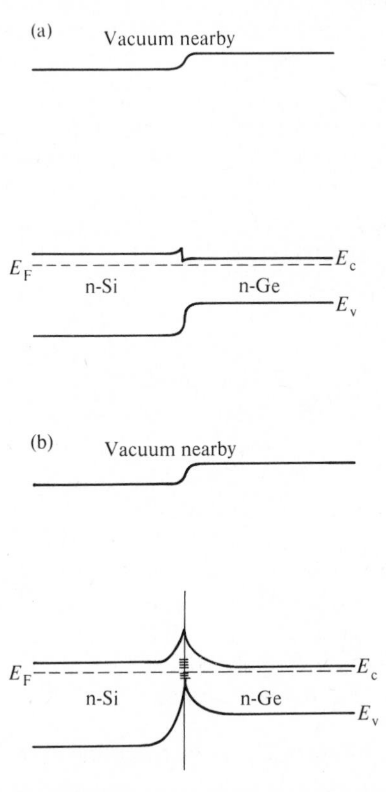

FIG. 6.17. Heterocontact configuration for n-Si–n-Ge. Schematically after Oldham and Milnes (1964). Electron affinity difference of 0.2 eV assumed. (a) Without interface states. (b) With interface states.

and photo-sensitive behaviour; e.g. see investigations by Gale *et al.* (1975). Even switching properties are occasionally observed, as for instance in p-Ge–n-CdS systems (Duncan *et al.* 1973).

In Fig. 6.17 it is arbitrarily assumed that the two barriers have a comparable importance, but one or other of them could in fact be made very thin, and thus transparent, by additional doping. The precise distribution of surface states is another potent factor in determining the barrier profile (Fig. 6.17(b)), and so is the ratio of dielectric constants. Not many systems have been analysed in detailed from this point of view, but a set of self-consistent calculations for Ge–GaAs has been provided by Pickett *et al.* (1977, 1978). Quite separate from the problem of interface-state distribution is the question of how these states are charged and recharged when external voltages are applied. *Inter alia* this effects the interpretation of practical observations, as Chandra and Eastman (1980) have shown in the course of their studies on n-GaAs–n-$Al_xGa_{1-x}As$. See also Wu and Yang (1979) for a detailed consideration of tunnelling problems, and Pellegrini (1975b) for a general treatment of capacitance relationships, applicable to many kinds of transitions from one material to another.

The case in which one of the semiconductors of a heterocontact is amorphous is of special interest, since the unsharp band edges offer additional opportunities for tunnelling. One would on general grounds expect an amorphous member to play a role similar to that played by the metal in metal–semiconductor contacts. Riben and Feucht (1966) have suggested that multiple recombination centres (with corresponding band gap states) could actually cause even a crystalline semiconductor to behave in this way. The many amorphous–crystalline contacts studied include α-Ge–Si (England and Hammer 1971, Norde and Tove 1977) and α-Si–Si (Döhler and Brodsky 1974, Tove *et al* 1979).

It is, of course, perfectly possible for both components to be amorphous. The sputtered ZnO–Bi_2O_3 system studied by Lou (1979) may well come under this heading, and is of special interest because of its relevance to Varistor devices (see Section 6.4.3). The practical potential of all-amorphous heterocontact systems deserves careful examination, considering how easily thin-film structures of this kind could be prepared for device purposes. Of course, *very* thin films tend to some complications, in as much as end-contact barriers may overlap with the heterocontact space charge (e.g. see Lee and Pearson 1980).

6.4.3 *Intergranular contacts in microcrystalline systems*

It is clear that the behaviour of intergranular contacts determines the properties of 'Varistor'-type devices, and because this field is technologically important it has a substantial literature of its own. (For surveys see,

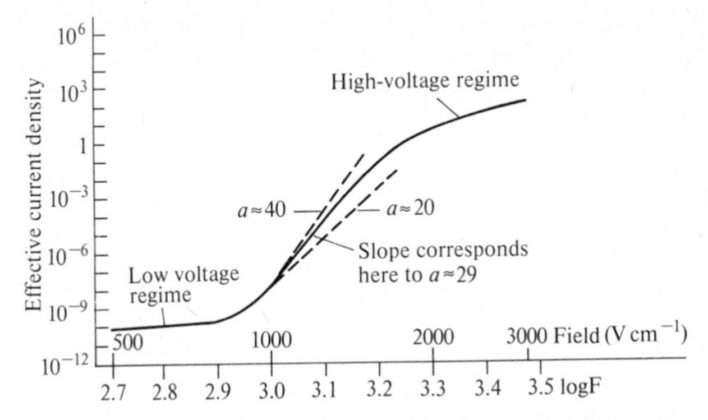

FIG. 6.18. Typical Varistor characteristics. From information by Emtage (1977).

for instance, Bernasconi *et al.* (1976), Philipp and Levinson (1977) and Mahan, Levinson and Philipp (1979); many other references will be found in these papers.) A comprehensive review of the engineering implications is outside the present scope, but the physical principles involved will be discussed. All such systems are characterized by a pronounced and, indeed almost catastrophic, superlinearity in the voltage–current characteristic, and this is, indeed, the primary reason for our interest. After a certain ('breakdown') voltage is reached, the current increases very rapidly by something like four orders-of-magnitude, as shown in Fig. 6.18 (see also Levine 1975). Such steep current increases have already been encountered (e.g. the curves for low barrier heights in Fig. 6.4) but only in the context of self-heating and thermal activation. There is, however, overwhelming evidence that self-heating is not fundamentally involved in Varistor systems, and that new current-flow mechanisms play a role there, especially at low temperatures, e.g. see Matsuoka (1971), Emtage (1977), and Lou (1979). The reactive behaviour could in principle provide further clues, but a combination of factors causes the capacitance of multigrain systems to be frequency-dependent, and to unravel the structure of such assemblies from measured characteristics is not a simple matter (see also Levinson and Philipp 1976).

As a rule, intergranular characteristics are also pressure-dependent, presumably because the surface-state distribution is affected by deformation. Experiments which demonstrate this have important theoretical implications, but are in practice difficult to perform under sufficiently controlled conditions.

The voltage–current characteristic of such systems consists of a prebreakdown region, believed to be controlled by field-assisted thermionic emission (barrier lowering by the Schottky effect; see Section 2.1.5), followed by a high-current region which can usually be represented by a power law of the form $I \propto V_T^2$. This kind of behaviour is ascribed to a tunnelling process. In SiC conglomerates, the exponent a rarely exceeds 7, but it can be as high as 60 in ZnO-based Varistors. The relationship then looks like a Zener characteristic (Einzinger 1975), but Varistors are capable of handling much more current and power than conventional Zener diodes.

The above interpretation of the pre-breakdown region is well supported by observations of Levinson and Philipp (1974, 1975), according to which log I varies linearly with $F_{av}^{1/2}$ over a substantial current range, where F_{av} is the average field. On the other hand, Vanadamme and Brugman (1980) found that a more drastic form of barrier-lowering had to be invoked to account for their observations. In view of the uncertainties which govern the distribution of $F(0)$ values, and the significance of F_{av}, such discrepancies are not unexpected and are not sufficient to change the basic conclusion. In the regime that follows, the steep power law is also not obeyed with total accuracy, and especially not at highest currents, for which the gradient diminishes sharply (Fig. 6.18). Nevertheless, tunnelling represents the best available interpretation of that regime. For ZnO ceramic micro-conglomerates, Levinson and Philipp have evaluated the quiescent barrier height V_D as about 0.8 volt, from the exponential temperature dependence of the voltage–current characteristic at zero applied voltage. Amongst other things, this high value explains why ZnO is so much better for Varistor purposes than (say) germanium.

The analysis of commercially produced devices is complicated by the presence of additives which are necessary for the optimization of device properties. Thus a disordered layer of other oxides generally surrounds the ZnO grains in a ZnO Varistor (Clarke 1978), and this layer may be more conductive than the grains themselves (Emtage 1977). Whether it is invariably so, remains uncertain but it is clear that high values of a depend very much on the presence of such oxide phases. For commercial material of high-exponent, a typical composition (in mole per cent) is ZnO:97, Sb_2O_3:1, and Bi_2O_3, CoO, MnO, $CrCO_3$:0.5 each (see also Wong, Rao, and Koch 1975). The *average* thickness of the intergranular layer is about 0.1 μm, but the average value is not necessarily a significant parameter. There is, for instance, no certainty that the intergranular layers are continuous and, accordingly, it is problematical whether they affect all micro-contacts equally. Clarke (1978), for instance, found that the intergranular material accumulates mostly at 3-grain and 4-grain meeting points, and that *most* of the grains are *not* surrounded by it. These are major uncertainties, but the electrical importance of the intergranular regions seems clear in broad principle; it is believed to arise from their ability to trap charge in one way or another. Since ZnO as such is believed to be virtually free of surface states (see Section 2.1.6), the very existence of intergranular barriers calls for the presence of some other intergranular material. The detailed manner in which these intermediate phases discharge their function is still not fully resolved (e.g. see Emtage 1977, Matsuoka 1971, and Levine 1975), and this is in large part due to the difficulty of devising independent tests for the different features of a model. On the other hand, there is in general agreement about barrier heights in ZnO Varistors, which are in harmony with those derived from direct Schottky barrier experiments (Neville and Mead 1970). Nor is there any doubt about the basic role of intergranular contacts as such, considering that the *effective* dielectric constant of a ZnO conglomerate is of the order of 1000; only a distributed barrier capacitance can explain such a value. (One of the remarkable things is that this distributed capacitance behaves as a simple one, in as much as it is directly proportional to the reciprocal of the specimen thickness (Levinson and Philipp 1976).)

The problem of model making is to find a process which will yield not only a sufficiently high value of the exponent a, but one which can prevail at high current densities. The slight temperature dependence of the current in the exponential regime makes the tunnelling hypothesis highly plausible. Avalanche breakdown is ruled out by the fact that it leads to the expectation of a breakdown voltage with a positive temperature coefficient, which is contrary to observations. The carriers which are supposed to do the tunnelling are those in the thin intergranular flux region, no matter whether that region is conductive or 'insulating'. Such carriers have been called 'trapped', but a variety of concepts has in fact been accommodated under that heading.

One of the interpretational difficulties arises quite generally from the smallness of the voltage which can appear across any intergranular contact in a conglomerate, considering that the grain size is only about 10–20 μm on average, and that this average is the outcome of a substantial grain size distribution (Wong 1976). There is also uncertainty as to what constitutes the most effective conduction path between the two device electrodes, and this can lead to quite different estimates of the number of intergranular contacts actually involved. Something of the order of 30–60 contacts appears to be a plausible number for a specimen on 1 mm thickness but, as far as is known, no percolation analysis has ever been attempted. Geometry is another problem. While it is always tempting to represent a conglomerate of randomly shaped grains as a simple cubic matrix (Fig. 6.19), that is obviously a serious oversimplification. Yet we have no choice, amongst other things because there is as yet almost no understanding of how barriers react to voltages applied in directions other than perpendicular to the interface.

In a remarkable series of experiments on thinly ground specimens of

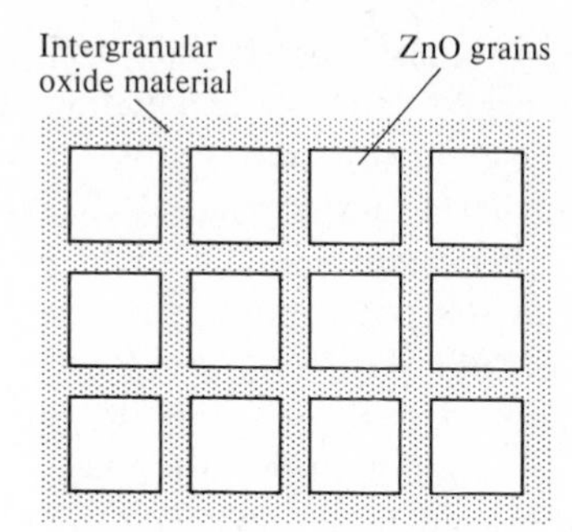

FIG. 6.19. Highly idealized model of grain structure. The intergranular material is assumed here to be much less conductive than the ZnO grains.

about 1 μm thickness, using microprobes, Einzinger (1975) was able to confirm that the ZnO grains themselves have an ohmic n-type resistivity, typically 0.1–1.0 Ω cm. In the same way, Einzinger proved that the characteristic of single intergranular contacts are indeed a significant index for the overall behaviour, in agreement with findings by Wong (1976). Only a negligible fraction of the field appears within the ZnO grains. This also explains the linear dependence of the breakdown voltage on the thickness of macro-specimens. Observations of photoconductivity and its sensitive dependence on the applied voltage lead to the same conclusion (Philipp and Levinson 1975). Another basic result by Einzinger was the demonstration that the characteristic of an individual micro-contact is almost independent of the thickness of the intergranular material. Along those lines, the Varistor effect would have to result from a barrier which surrounds each ZnO grain and which, by implication, resides within it, corresponding to Fig. 6.20(a). This, in turn, is electrically similar to the barrier system shown in Fig. 6.12(b). Detailed conduction mechanisms were not considered, nor the grave problem of preventing the ZnO grains (barriers and all) from being simply short-circuited by highly conductive interface material.

In sharp contrast, a model proposed by Levinson and Philipp (1975) is shown in Fig. 6.20(b). Here the intergranular material is regarded as substantially 'insulating', but with field-ionizable trapping centres. Both models thus depend on tunnelling, but quite different roles are ascribed to the oxide additives. Here, certainly, the resulting properties would be highly dependent on the thickness of the interface layers, contrary to Einzinger's findings. On the other hand, the short-circuiting problem mentioned above is altogether avoided.

We are confronted by conflicting models, but in view of the fact that different researchers tend to work on specimens of different structure, this

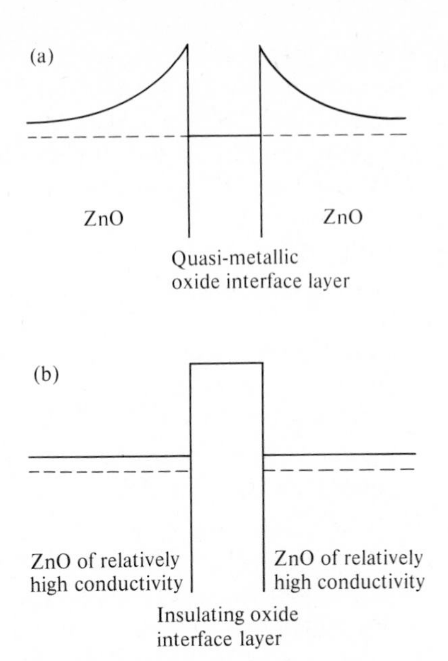

FIG. 6.20. Intergranular barrier models I. (a) Schematically after Einzinger (1975) (b) Schematically after Levinson and Philipp (1975)

is not altogether surprising. A structure that is in some ways a compromise is shown in Fig. 6.21 (Mahan *et al.* 1979). It envisages an intergranular oxide film which may be tunnelled through, wholly or in part, but which is no longer believed to represent the essential barrier. The breakdown process is here considered to start with the generation of holes in an inversion layer, formed when a sufficiently large voltage (estimated on the

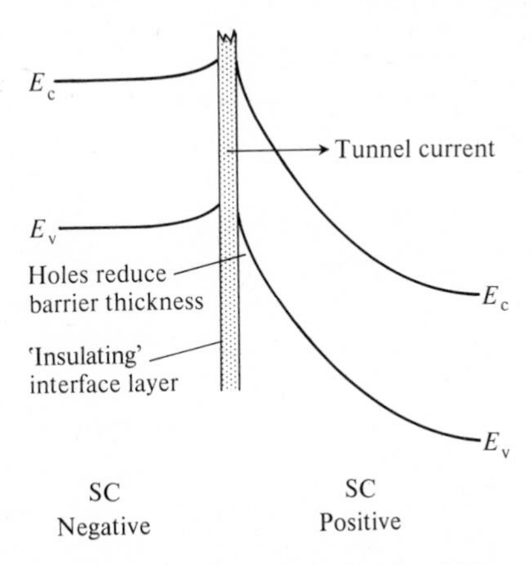

FIG. 6.21. Integranular barrier models II. After Mahan *et al.* (1979).

basis of equilibrium statistics to be about 2.4 volts per barrier) is applied. The contention is that the holes, with their positive charge, would drastically thin the barrier, and thereby make it more penetrable to electrons by tunnelling. Indeed, they would, but the hole concentration which provokes this cannot be reliably estimated by recourse to equilibrium statistics; and the dynamic situation is highly complex. As far as is known, it has not yet been theoretically surveyed. Moreover, the above threshold voltage estimate seems implausibly high, and even the role played by the space charge of holes is still problematical; Vanadamme and Brugman (1980) managed to envisage high values of the exponent a without it (on the basis of Zener breakdown). A great deal of work remains to be done, in the face of a disappointing truth: the commercial specimens which are of greatest practical interest are the least well suited to interpretation by simple models.

We have been concerned here mainly with structures which exhibit highly prominent barriers, but the importance of the basic considerations is by no means limited to such cases. Even low barriers in a microcrystalline aggregate of small grain size can be responsible for substantial contributions to the overall resistivity and to the (effective) field-dependence of all electrical properties.

6.4.4 *Intergranular contacts under illumination*

Since intergranular contacts are characterized by barriers, they are also photoconductive. The circumstances are very similar to those described in Chapter 5, except that we have nothing here corresponding to a constant ϕ_{ns}. Not only is the carrier concentration within the material modulated by light, but the barrier height is itself illumination dependent. This arises qualitatively in the following way.

Figure 6.22(a) shows the quiescent barrier profile in darkness. Additional carriers Δn_L and Δp_L are produced by the incoming radiation.

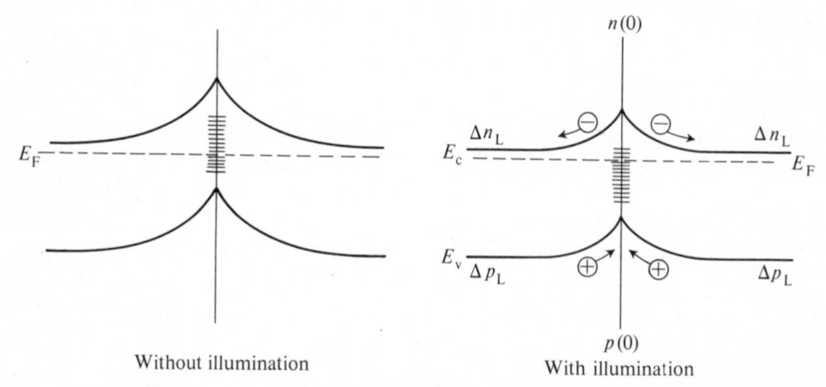

FIG. 6.22. Integranular barrier under illumination. Continuous distribution of interface states assumed. Barrier-lowering increases $n(0)$ and thereby promotes recombination.

Electrons tend to move into the bulk material under the influence of the barrier field; holes move into the barrier for the same reason. There they constitute a positive space charge (as long as they have a non-zero lifetime). This charge resides partly in the interface states themselves, and partly in the adjoining cusp of the barrier profile. The barrier height is thereby lowered (Fig. 6.22(b)), which means that $n(0)$ is increased, as it *must* be, to attract electrons with which the inflowing holes can recombine. The band bending is thereby reduced, and this helps to keep the hole flux under control. Effectively, it is as if a forward voltage had been applied to both sides of the double barrier. Card and Yang (1977) have pointed out that the barrier profile adjusts itself so as to permit the maximum amount of recombination to take place. If the capture cross-sections of the interface states for holes and electrons were equal, then $\Delta n(0)$ and $\Delta p(0)$ would likewise be equal after the barrier adjustment.

The arguments of Shockley–Read (1952) recombination statistics can be applied to this system (as long as there is no tunnelling to and from interface states), and can be used to calculate ΔQ_{ss}, the change of surface charge resulting from the extra carrier concentrations Δn_L and Δp_L. This is, indeed, what Card and Yang have done. How Q_{ss} affects the barrier height must depend on the interface distribution over energy. Tightly bunched states will give rise to very small effects; uniformly spaced states (e.g. 10^{10}–10^{14} cm^{-2} eV^{-1}) will support high light sensitivity. As a result of barrier lowering, the conductance of the system will be (almost exponentially) decreased.

Grain boundaries are contacts of this kind, and the fact that they are associated with double barrier cusps allows them to gather-in minority carriers very efficiently. That is why grain boundaries are as potent as they are in reducing the effective bulk lifetime.

The quantitative estimates obtained by Card and Yang apply specifically to silicon; see Philipp and Levinson (1975) for a discussion of thermally activated tunnelling by photo-excited carriers in metal oxide. Varistors.

Further reading

On SiC–SiC contacts: Henisch and Roy (eds) 1969, O'Connor and Smiltens (eds) 1969 (many other references given there).

On the effective resistance of microcrystalline systems: Card and Hwang 1980, Kuznicki 1976.

6.5 Low-resistance contacts

6.5.1 General principles; carrier degeneracy

Our detailed understanding of low-resistance contacts has been greatly impeded by the lack of crisp experimental data. The determination of interpretable voltage–current characteristics over an appreciable voltage range demands very high currents, and those could be envisaged (without disturbing thermal side-effects) only under pulse conditions. Moreover, the separation of bulk and contact effects, never totally easy in any circumstances, becomes more and more difficult as barrier resistances decrease. All these reasons are responsible for the fact that we know much less about low-resistance contacts than about contacts of high resistance on semiconductors.

When a metal contact is applied to a semiconductor, the resulting energy contour may correspond to a *barrier*, as is assumed in most of the previous discussions, or else to an *accumulation layer*, as described in Section 2.1.2. Between these two situations is the quiescent *flat-band condition*, a very special case which we do not expect to encounter in practice. An accumulation layer represents a reservoir of free carriers, and its presence contributes less to the total resistance than equal thickness of bulk semiconductor. It is therefore the best type of low-resistance contact that can be hoped for, but its practical achievement is not always simple and, indeed, not always possible, depending, as it does, on a combination of the right interface-state distribution and thermionic work-functions. On most of the technologically familiar materials we thus expect to find barriers. In order to minimize their contribution when low resistances are desired, it is common practice to dope a thin boundary layer of semiconductor material with extra donors (n^+), or (when dealing with p-type material) with extra acceptors (p^+). To some extent, this procedure simulates an accumulation layer (Fig. 6.23(a)). Many recipes for achieving such structures, and for assessing their level of success, have been given by Schwartz (1969). (The book has an excellent index, and provides a supplementary bibliography, in addition to the hundreds of works cited by contributing authors.) For a recent overview of low-resistance contacts on III–V compounds, see Piotrowska *et al.* (1983).

Figure 6.23(a) assumes that the n^+ region (x_{n^+}) is much thicker than the barrier itself, but other models have also been examined. Thus, Shannon (1976) has considered the case of x_{n^+} being very small, and Popović (1978) the cases of x_{n^+} just equal to the value of λ_0 in that region, and just greater (Fig. 6.23(b)). The latter case involves a narrow degenerate region but, as far as tunnel currents are concerned, only electrons above the bulk E_c take part in the quantum-mechanical exchange with the metal (if the mean-free-path is long enough). As far as is known, this intriguing structure has never been exhaustively analysed.

In the presence of high doping levels, the carrier gas becomes degenerate, and equations based on Boltzmann statistics (which means most of the transport equations discussed in previous chapters) cease to be applicable. Contrary to popular belief, no simple algebraic adjustment can

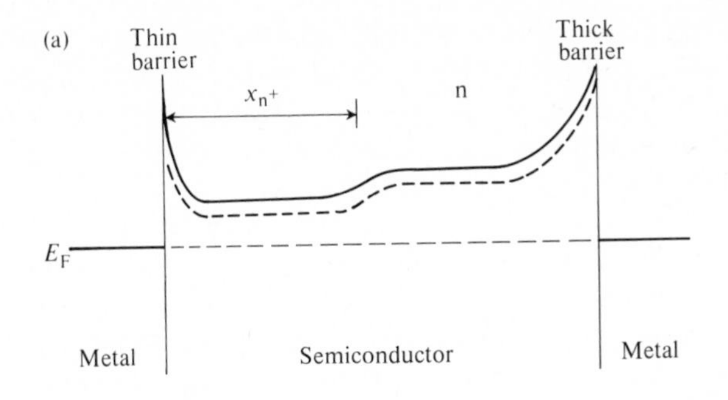

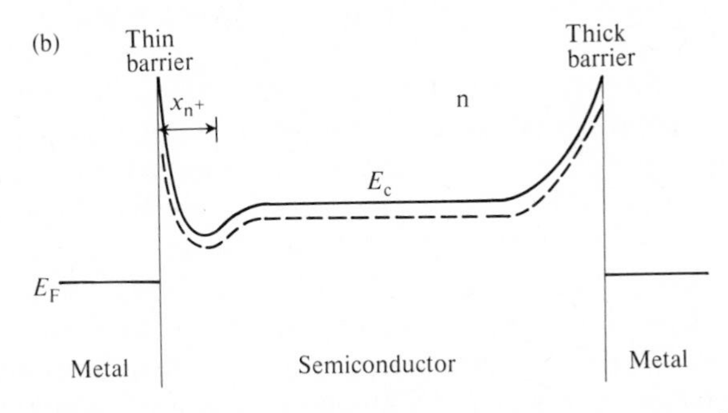

FIG. 6.23. Preparation of a low-resistance contact by additional doping. (a) $x_{n+} \gg \lambda_0$. (b) $x_{n+} \approx \lambda_0$.

take care of this situation. Let us, for instance, consider the free-carrier concentration, which would no longer be given by

$$n_e = \mathcal{N}_c \exp(\xi) \tag{6.5.1}$$

but would have to be calculated from

$$n_e = \frac{4\mathcal{N}_c}{\pi^{1/2}} \mathcal{F}_{1/2}(\xi) \tag{6.5.2}$$

and

$$\mathcal{F}_{1/2}(\xi) = \int_0^\infty \frac{x^{1/2}\,dx}{1+\exp(x-\xi)} \tag{6.5.3}$$

with $\xi = \phi_n/kT = (E_c - E_F)/kT$. ξ itself is, of course governed by the neutrality condition, and can be evaluated either graphically, as shown by Shockley (1950), or else by numerical methods. For a given ξ, $\mathcal{F}_{1/2}(\xi)$ can then be calculated numerically, or looked up in available tables: McDougall and Stoner (1938) or Rhodes (1950). Yet another possibility for finding n_e from a known ξ is to make use of algebraic approximations (of the above integral), e.g. as first introduced by Ehrenberg (1950). A comprehensive overview of such approximations has recently been provided by Blakemore (1982) (see also Smith 1969). However, as far as is known, all previous work assumes sharp band edges, whereas in a semiconductor which has a degenerate electron gas as a result of heavy doping this condition is not likely to be satisfied. The matter remains to be analysed in detail, even as far as bulk relationships are concerned. Until this happens, we cannot hope for precise quantitative agreement between contact theory and experiment.

The free-carrier concentration is only one example of an affected term; Boltzmann relationships abound in conventional contact theory, and none of those could be allowed to stand. However, a straightforward replacement of Boltzmann terms by Fermi–Dirac terms would not be meaningful. Thus, for instance, a space-charge barrier in highly doped material is not expected to conform to any Schottky model, not even to a schematically modified one. Donor clustering, lattice distortion, discontinuous charge distribution, tunnelling, hopping, and image-force effects augmented by enormous local fields are some of the factors which invalidate the conventional assumptions. They cease to be small perturbations and become principal features of the system, causing contact barriers to behave in complex and not straightforwardly predictable ways (see also Section 6.5.2).

In practice, high-level doping is often successful in achieving low contact resistances, but it is not a universally potent formula. There are indeed materials which do not tolerate high dopant contents, e.g. zinc sulfide, zinc selenide, boron arsenide, aluminium nitride, etc. Aven and Swank (1969) have discussed such cases, and have given other suggestions for improving 'contactability'. One of these is to form a mixed crystal of suitable band gap, in lieu of forming an accumulation layer. Another method is to increase the real contact interface by making the contacts non-planar (e.g. multiple spikes).

Many instances are on record in which specific procedures have yielded low-resistance contacts on ordinarily 'difficult' materials, for reasons which remain unclear. Thus, Gu and co-workers (1975) prepared excellent low-resistance contacts on CdTe by evaporating gold in high vacuum, and firing in a hydrogen atmosphere. To this day, the number of empirically successful recipes greatly exceeds the number of systems understood in detail. Mysteriously, a sintered Au–Zn–Au sandwich layer has proved its worth as a low-resistance contact on InP (Schiavone and Pritchard 1975), and many other such examples can be found in the literature.

When, for one reason or another, local doping is not feasible, there is still the possibility (particularly for materials of wide band gap) of choosing a metal with an appropriate work function, low for n-type and high for p-type material. Aven and Swank (1969) have described successful uses of this method on a variety of semiconductors.

It will be clear from Chapter 3 that the term 'ohmic' contact is not associated with any implication concerning the mechanism and composition of the currents flowing. The 'low-resistance' contacts discussed here may or may not be 'ohmic' for small current densities; for high current densities, departures from Ohm's law are inevitable, though they may be in practice, quite unimportant.

6.5.2 *Thin barriers*

In general terms, barriers can have a low resistance for one of two reasons: because they are thin, or else because they are low. That tunnelling lowers the effective height of a barrier by making its top transparent to electrons has already been discussed in Sections 2.1.5 and 2.2.8. This aspect was then most directly relevant to the analysis of reverse characteristics, but the question is now whether it can can have other consequences, and whether it might ultimately eliminate the barrier height altogether for practical purposes.

All the models described in Section 2.2 are characterized by barrier conductances proportional to N_d. Increased doping thus reduced contact resistance in absolute terms, but leaves it unchanged in relation to the bulk resistance. To reduce the contact resistance also in relative terms, the tunnelling mechanism must be brought into play; processes which depend on thermionic emission cannot do it, as McColl *et al.* (1971) have pointed out. Moreover, tunnelling which entered into the discussions of Sections 2.1.6, 2.2.8, and 2.3.3 mainly as a correction term for the reverse characteristic becomes equally important for the forward characteristic when a barrier is very thin.

If the entire current were carried by thermally excited barrier tunnelling, and none by thermionic emission *over* the barrier, then one would expect the current for complete donor ionization to be given by

$$j = j_0' \exp\left\{-\frac{4\pi}{h}\left(\frac{\epsilon m_n}{N_d}\right)^{1/2} V_B\right\}. \qquad (6.5.4)$$

(See Padovani and Stratton 1966a, Crowell and Rideout 1969b, and Mead 1969.) A plot of $\log|j|$ versus V_B would then have a slope very different from (and smaller than) e/kT, and Mead has shown this to be true. As N_d goes up, the total current increases, but the slope of the relationship decreases, being proportional to $N_d^{-1/2}$. Here ϵ is the static permittivity. One of the characteristics of a tunnelling current is its temperature independence and, on the basis of this criterion, such a current can easily be distinguished from one due to thermally activated carriers. Millea *et al.* (1969) have shown, at any rate for Schottky barriers on heavily doped GaAs, that tunnelling dominates at or below 77 K, and thermionic emission at room temperature. The same has been demonstrated for contacts on InSb (McColl and Millea 1976). For totally degenerate material, tunnelling is always important, even at room temperature.

The effect of tunnelling can also be assessed by reference to the barrier resistance R_B at zero applied voltage, and Mead (1969) has indeed shown for gold contacts on GaAs that log R_B is a linear function of $N_d^{-1/2}$. The measured contact resistance varied by more than four orders-of-magnitude as N_d varied from about $10^{18}\,cm^{-3}$ to $10^{20}\,cm^{-3}$. One would expect something like this from the considerations of Simmons (1963) who concerned himself with tunnelling through initially rectangular barriers, distorted by the application of external voltages across them. For the very low voltage range, Simmons arrived at an ohmic expression which, in the present terms, may be given as

$$j = j_0' \bar{E}_B^{1/2} V_B \exp(-A_T \bar{E}_B^{1/2}) \tag{6.5.5}$$

where

$$j_0' = \{(2m_n\epsilon)^{1/2}/\lambda_a\}(e/h)^2 \tag{6.5.6}$$

and

$$A_T = (4\pi\lambda_a\beta/h)(2m_n\epsilon)^{1/2}. \tag{6.5.7}$$

Here λ_a is the barrier width, β a small correction term of the order of unity, and $\bar{E}_B$ the average barrier height, defined at zero bias by

$$\bar{E}_B = \frac{1}{\lambda_a}\int_0^{\lambda_a} \{E(x) - \phi_n\}\,dx. \tag{6.5.8}$$

The use of this concept gives the treatment a certain degree of generality, in as much as various other barrier profiles might be averaged, within the framework of approximation here involved. According to eqn (2.1.6), V_D is proportional to $\lambda_0^2 N_d$ and it is a very simple matter to show by means of eqn (6.5.8) that $\bar{E}_B = eV_D/3$ for a Schottky barrier. Using eqn (2.1.8b), $(\bar{E}_B)^{1/2}$ will be seen to be proportional to λ_a/λ_D, which means that it is independent of N_d. This makes $A_T(\bar{E}_B)^{1/2}$ proportional to λ_a and hence, for a given value of V_D, proportional to $N_d^{-1/2}$, as actually found. On the other hand, the approximations involved are not totally compelling, and conflicting results are also on record. Thus, Heiblum *et al.* (1982) report finding only an astonishingly weak relationship between contact resistance and N_d (in the range $N_d = 10^{11}$–$10^{19}\,cm^{-3}$) for AuGeNi contacts on GaAs. Such matters are now the subject of many investigations, and the time for general conclusions has not yet come.

A quantitative analysis by Andrews (1974) based on these considerations also draws attention to the fact that the tunnelling phenomenon, e.g. as expressed by eqn (6.5.4), is sensitive to the effective mass. The pre-exponential term j_0' turns out to be sensitive to the total barrier height and, together, the two parameters lead to variations of contact resistance (calculated, in this case, at the absolute zero of temperature) varying over 32 orders-of-magnitude, while the barrier width varies between 30 and 500 Å. The possibilities for control are thus enormous, which accounts for the central fact that additional doping is highly effective in reducing contact resistances (see also Lepselter and Andrews 1969). A more detailed (but mathematically non-transparent) model by Pellegrini and Salardi (1975) aims at the optimization of the dopant content for otherwise fixed contact parameters.

Some of the complications associated with high-level doping (and thus with thin barriers) have already been mentioned in Section 6.5.1, but there are in fact several others, e.g.

(a) Interaction between donors and interface-states.
(b) Donor segregation at the interface.
(c) Interaction between the contacting metal and donors.
(d) Interaction between the contacting metal and other (non-donor) defects of the semiconductor created by Strain (see Bachrach 1978).
(e) The likelihood of a position depending band structure arising; see Section 6.4.2 and Lundstrom *et al.* (1981).
(f) The diffusion of contact metal ions into the semiconductor, and their possible role as donors or acceptors, as envisaged by Kröger *et al.* (1956).

All these tend to represent more serious disturbances in thin, rather than in thick barriers. However, it is also true that features which appear as difficulties in one context, appear as opportunities in another. Thus, for instance, Sebestyén (1976) has suggested the design of an entirely new class of low-resistance contacts, based on the band-structure consequences of deliberately arranged crystalline–amorphous transitions close to the metallic contact (see also Pruniaux 1971).

Further reading

On low-resistance contacts:

with miscellaneous III–V compounds: Pounds *et al.* 1974, Rideout 1975; With GaAs: Edwards *et al.* 1972. Gopen and Yu 1971, Grosvenor 1981, Iliadis and Singer 1983, Jaros and Hartnagel 1975, Nathan and Heiblum 1982, Ramachandran and Santosuosso 1966, Tantraporn 1970; with GaP: Itoh *et al.* 1980, Pfeifer 1976; with InP: Barnes and Williams 1981, Becker 1973, Kuphal 1981, Valois and Robinson 1982; with CdTe: Triboulet and Rodot 1968; with CdS: Scholten 1966; with Si: Card 1976b, Rolland *et al.* 1977, Saxena 1971.

On contact resistances in the absence of barriers: Gossick 1970, 1971.

On the theory of very low barriers: Gossick 1969a.

On low-resistance electrolyte–insulator contacts: Elsharkawi and Kao 1976.

On carrier recombination at low-resistance interfaces: Heasell 1979.

On ion-implantation techniques for the preparation of low-resistance contacts: Feuerstein and Kalbitzer 1973, Yamaguchi *et al.* 1981.

On low-resistance contacts applied to graded heterojunctions: Sebestyén 1982.

6.6 Electrical breakdown

6.6.1 *Avalanche and Zener breakdown; general principles*

Avalanche and Zener breakdown both show themselves as a very rapid ('catastrophic') increase of current at a particular reverse voltage $V_B = V_{BD}$, sometimes to the accompaniment of permanent damage, but often in the form of a non-destructive reversible process. The mechanisms and dynamics are complex, but a few basic facts are easily reviewed.

Both processes are supported by high electric fields, and for Schottky contact barriers, the highest occurring field F_M is obviously $F(0)$, i.e. the field at the interface, as long as the image force is neglected. When the image force is taken into account, the maximum field occurs at some distance $x_M > x_m$ (Fig. 6.24). To calculate x_n, we need only consider that the field $F(x)$ is made up of two components, one due to the image force, and one due to the space charge. Accordingly, from eqns (2.1.23) and (2.1.29), we have in the presence of an applied voltage:

$$F(x) = \frac{e}{16\pi\epsilon_\infty x^2} + \frac{N_d e}{\epsilon}(x - \lambda_B) \tag{6.6.1}$$

which has a maximum where $x^3 = x_M^3 = \epsilon/8\pi N_d \epsilon_\infty$. The maximum field

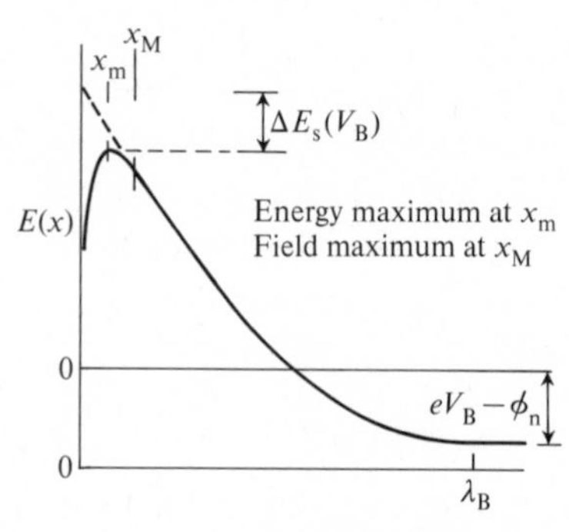

FIG. 6.24. Maximum field at $x = x_M$ of a Schottky barrier with image-force correction. Reverse voltage V_B applied.

itself is

$$F(x_M) = (N_d e/\epsilon)(x_M - \lambda_B) \tag{6.6.2}$$

and this is the field with which we are concerned here. Since $|F(0)| = N_d e \lambda_B/\epsilon$, $|F(x_M)|$ is obviously smaller, but only slightly so, because x_M is generally much smaller than λ_B. For practical purposes, therefore, $F(0)$ can be used in place of $F(x_M)$ without incurring any major error. In any event, the maximum field of a Schottky barrier occurs only near the metal interface; $F(x)$ diminishes linearly throughout the barrier, until $F(\lambda_B) = 0$. In contrast, as we have seen (Section 2.1.4) the field of a Mott-type barrier is constant. In each case, the maximum field increases with applied voltage V_B when $V_B > 0$ (reverse direction). For an avalanche to be sustained, the prevailing field must exceed the ionizing field not only at $x = 0$, but over a range of distances into the barrier.

In any high field, carriers can become 'hot', in the sense that their energy becomes comparable with (and, indeed, greater than) their average thermal energy $3kT/2$. To what extent the carriers are actually heated depends on the mean-free-path, i.e. on the scattering mechanism which predominates at the temperature concerned. At low fields, the energy lost on impact with a scattering centre is easily dissipated in the form of phonons, but at high fields this would demand a multiple phonon collision, and is therefore unlikely. Accordingly, the average energy ('temperature') of carriers continues to increase, until a new and inelastic collision process is encountered (Shockley 1951). Impact ionization is such a process, schematically illustrated in Fig. 6.25. Increasing electron temperatures are graphically represented by electron paths which fail to follow the band edge contour. However, since momentum tends to be conserved in ionizing electron collisions with the lattice, the quantitative features of such diagrams may be misleading; the ionization energy (corresponding, as it does, to a direct transition) may be substantially greater than the band gap. When at least one (in principle) accelerated electron has enough energy to cause impact ionization, the ionization products, an electron and a hole, both find themselves in the same electric field, and can themselves participate in the process. They do so with different ionization energies when the transitions are indirect and with identical probabilities when they are direct. Thus, after a single collision, there may be three carriers in place of one. The build-up of free charge carriers can therefore be very rapid, but it always *opposed* by two other processes: recombination, and carrier loss by transport out of the high field area. The extent to which these balancing mechanisms are effective governs the 'hardness' of the breakdown relationship (Fig. 6.26). Efficient loss processes 'postpone' breakdown, and make it 'soft'. Eventually, with increasing

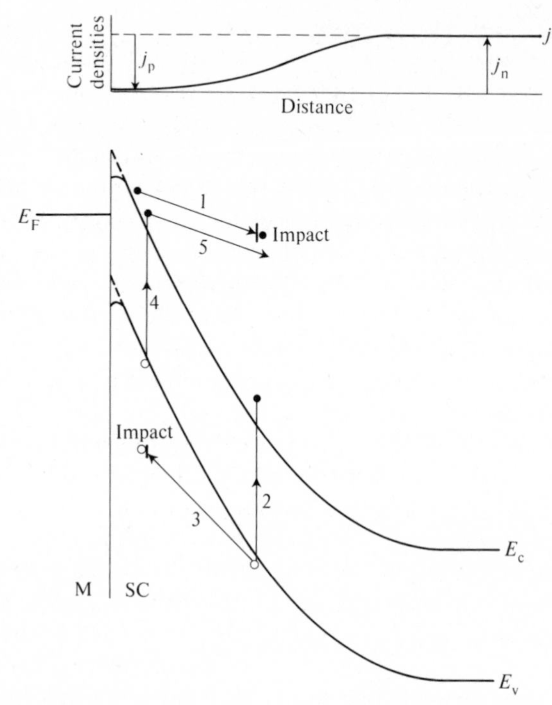

FIG. 6.25. Mechanism of impact ionization; schematic order of events. (1) Electron acceleration to impact. (2) Creation of electron–hole pair. (3) Hole acceleration to impact. (4) Creation of new electron pair. (5) Acceleration of new electron pair.

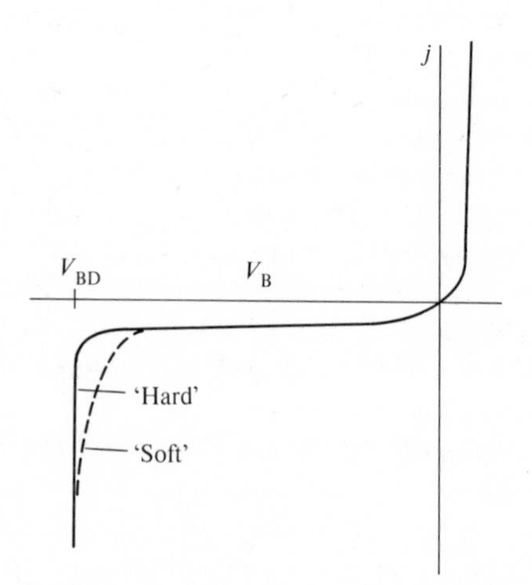

FIG. 6.26. 'Hard' and 'soft' reverse breakdown; qualitative classification, irrespective of breakdown mechanism.

applied voltage, the current increases so rapidly as to amount to catastrophic breakdown. Negative resistance characteristics cannot be obtained, except in the presence of some additional positive feedback mechanism. Because the various parameters (e.g. the scattering length l_s) are sensitively temperature-dependent, avalanche breakdown likewise is very temperature-dependent, as distinct from Zener breakdown (see below) which is not. Higher temperatures correspond to higher values of the breakdown voltage V_{BD}.

The ionization probability α_i is, of course, strongly field-dependent. In simplest terms, one can argue that the probability of any electron achieving a collision-free path of length l is given by

$$n(l) = n_0 \exp(-l/l_s) \tag{6.6.3}$$

where l_s is the *mean-free-path*. If l were in the direction of the field, then such a carrier would acquire an energy $\Delta E = eFl$ in that process, which means

$$n(l) = n_0 \exp(-\Delta E/Fel_s) \tag{6.6.4}$$

When $\Delta E = E_i$, the ionization energy (where by $E_i \geqslant E_g$), and bearing in mind that not all free paths will be in the direction of the field, we can write

$$\alpha_i(x) = \alpha_{i0} \exp[-E_i/b_{ls}F(x)] \tag{6.6.5}$$

where b_{ls} is a constant, proportional to l_s. This is valid only as long as the mean electron energy is much less than ΔE_i. One can show that $\alpha_i(x)$ depends on $\exp(-E_i/b_{ls}F_{(x)}^{-2})$ when the opposite applies (Allen 1973).

When an electron travels a distance dx, the carrier concentration will be augmented by

$$\mathrm{d}n(x) = (\alpha_{in} + \alpha_{ip})n(x)\,\mathrm{d}x \tag{6.6.6}$$

where α_{in} and α_{ip} are x-dependent. For a rigorous calculation of α_i, see Baraff (1962). Of course, hot electrons can also ionize impurities (as distinct from the lattice), and Allen has presented an analysis of that problem (see also below).

A coarse but, for present purposes, useful estimate of E_i can be obtained as outlined by Carroll (1970). Neglecting band structure subtleties (anisotropy, etc), we can consider that the incoming electron of mass m_n has a thermal velocity v_{Tb} before its ionizing collision, and v_{Ta} thereafter. The electron–hole pair generated by the impact has masses and velocities m_n, m_p, and v_{Tn}, v_{Tp}. Then, conservation of momentum and energy yield

$$\begin{gathered} m_n v_{Tb} = m_n v_{Ta} + m_n v_{Tn} + m_p V_{Tp} \\ \frac{m_n v_{Tb}^2}{2} = E_g + \frac{m_n v_{Ta}^2}{2} + \frac{m_n v_{Tn}^2}{2} + \frac{m_p v_{Tp}^2}{2}. \end{gathered} \tag{6.6.7}$$

We look for the minimum incident energy that will achieve the transfer, and this is given by $v_{Ta} = v_{Tn} = v_{Tp}$. When this is the case, eqns (6.6.7) can be manipulated into the form

$$\frac{m_n v_{Tb}^2}{2} = E_i = E_g\left(\frac{2m_n + m_p}{m_n + m_p}\right) \tag{6.6.8}$$

For $m_n = m_p$, this yields $E_i = 3E_g/2$, which gives the order of magnitude. A more thorough treatment would have to take account of the fact that m_n may be different for the incident and impact-generated electrons.

Calculations which aim at the carrier concentrations as a function of time and distance through the barrier are highly complex; see, for instance, Baraff (1962, 1964), Chynoweth and McKay (1957), and Chynoweth (1965). There are, however, simplified treatments (e.g. Sze 1969) which assume (explicitly or implicitly): (i) single-carrier conduction in bulk, (ii) the absence of recombination (presumably because carriers move very quickly through the high-field regions) (iii) absence of hot-carrier loss to the metal before ionizing impact, (iv) absence of space-charge and diffusion effects, and (v) steady-state conditions. In spite of these restrictions, some useful insights can be gained in this way.

Adapting Sze's (1969) analysis of breakdown in p–n junctions, we shall assume that an electron current of density $j_n(0)$ enters the barrier (on n-type material) in the presence of a high reverse voltage. Through impact ionization, that current will increase from $x = 0$ to $x = \lambda_B$. At the edge of the barrier, the electron current must constitute the entire current, since we are dealing with a system sustained by single-carrier conduction in bulk. Thus

$$j_n(\lambda_B) = j = \mathcal{M}_n j_n(0) \tag{6.6.9}$$

where $\mathcal{M}_n$ is a current multiplication factor for electrons. As electrons are accelerated in the field and make impact with the lattice, they generate not only extra electrons but also holes. Those are accelerated towards the left and, in turn, create new electron–hole pairs by impact (Fig. 6.25). Thus the hole-current density, which is zero at $x = \lambda_B$, increases towards the left. As always, we must have $j = j_n + j_p = \text{constant}$, and if diffusion effects are negligible (as they are likely to be in the high reverse direction, except at the immediate edge of the barrier), the component currents are proportional to the respective carrier concentrations. Thus

$$d(j_n/e) = (j_n/e)(\alpha_{in}\, dx) + (j_p/e)(\alpha_{ip}\, dx) \tag{6.6.10}$$

because the change of electron current arises partly from the ionizing function of the electron current itself, and partly from that of the hole current. It follows that

$$dj_n/dx - (\alpha_{in} - \alpha_{ip})j_n = \alpha_{ip} j. \tag{6.6.11}$$

This is a standard equation which can be solved (e.g. see Davis 1962) as

$$j_n(x) = j\left\{\frac{1}{\mathcal{M}_n} + \int_0^x \alpha_{ip} \exp\left\{-\int_0^x (\alpha_{in} - \alpha_{ip})\,dx'\right\} dx\right\} \Big/ \exp\left\{-\int_0^x (\alpha_{in} - \alpha_{ip})\,dx\right\}. \tag{6.6.12}$$

Integrating from $x = 0$ to $x = \lambda_B$, this gives

$$1 - \frac{1}{\mathcal{M}_n} = \int_0^{\lambda_B} \alpha_{ip} \exp\left\{-\int_0^x (\alpha_{in} - \alpha_{ip})\,dx'\right\} dx \Big/ \exp\left\{-\int_0^{\lambda_B} (\alpha_{in} - \alpha_{ip})\,dx\right\}. \tag{6.6.13}$$

Multiplication occurs, of course, at lower voltages, but it becomes catastrophic when $\mathcal{M}_n = \infty$, which means that the right-hand side of eqn (6.6.13) must be unity. For direct-transition material (only), we have $\alpha_{in} = \alpha_{ip} = \alpha_i$, and this gives

$$\int_0^{\lambda_B} \alpha_i\,dx = 1 \tag{6.6.14}$$

as the breakdown condition, where α_i is derived from eqn (6.6.5). The field for which eqn (6.6.14), with (6.6.5), is satisfied *is* the breakdown field F_n. In a Schottky barrier, this corresponds (via eqns 2.1.29 and 2.2.1b) to a breakdown voltage

$$V_{BD} + V_D \approx V_{BD} \approx -F_M \lambda_B/2 \approx F_M^2 \epsilon/2eN_a \tag{6.6.15}$$

whereby $F_M = F(0)$ has been assumed. In view of the approximations made, eqn (6.6.15) is expected to represent an underestimate. The fact that the above zero-loss condition (condition (iii) above) may not be satisfied can be seen, *inter alia*, from measurements by Card (1974) and Verwey (1972) who demonstrated that hot-carrier injection *from* the silicon into an adjoining oxide or nitride layer. Another piece of evidence for losses is, of course, the radiation discussed in Section 6.6.2. However, proceeding along present lines, it is clear that F_M (in eqn 6.6.15) will not depend very sensitively on doping, which means that V_{BD} is expected to be almost inversely proportional to N_d. Sze and Gibbons (1966) have calculated these relationships for Ge, Si, GaAs, and GaP, taking values of α_{io} from the literature, and solving the above equations by numerical integration. The reciprocal relationship is approximately fulfilled, and though the calculations refer explicitly to p–n junctions, the abrupt barrier case which is relevant here should not be very different. For germanium, for example, the breakdown field is of the order of 2×10^{15} V cm^{-1}. Sze (1969) has compiled data on ionization rates for several materials. For Ge and Si, eqn (6.6.5) is reasonably well obeyed; for GaAs and GaP, the results are better represented by an exponential dependence on $F(x)^{-2}$. The breakdown voltages calculated by Sze and Gibbons

(1966) take this modified dependence into account. Values for V_{BD} itself vary from about 1 volt (Ge, $N_d = 10^{18}$ cm^{-3}) to about 1000 volts (Ge, $N_d = 10^{14}$ cm^{-3}). Other things being equal, V_{BD}-values for silicon are higher by a factor of about 3, and in GaP by a factor of about 6. Measurements should be performed by pulse methods to avoid the superposition of thermal effects (see also Baraff 1962).

Crowell and Sze (1966a) have calculated the temperature-dependence of V_{BD} on the basis of the temperature dependence of l_s, when l_s is governed by optical-phonon scattering. V_{BD} increases with increasing temperature (see also above), at a rate which depends on N_d. High values of N_d reduce the temperature sensitivity. The time required for avalanche build-up is, of course, directly related to $\mathcal{M}_n$, and has been shown to be of the order of 5×10^{-13} s (see Kaneda *et al.* 1976).

At very low temperatures, not only the lattice but the impurity centres themselves can be impact-ionized, because they are then mostly full in the bulk. However, they are empty within the barrier, which means that this particular ionization process can occur only close to $x = \lambda_B$. As far as is known, the matter has never been systematically investigated. In ultra-thin systems, the additional phenomena discussed in Section 3.5.2 come into play (see also Kumar *et al.* 1975). In *any* system, reliable breakdown measurements call for the careful elimination of edge effects, e.g. by means of some of the structures shown in Fig. 1.9. The importance of this precaution has been repeatedly demonstrated, *inter alia*, by Yu and Snow (1968), Yu (1970), Zettler and Cowley (1969), and by Saltich and Clark (1970).

Whereas avalanche formation is the most important breakdown process in relatively thick barriers, *very* thin barriers (corresponding to highly doped material) support Zener breakdown, i.e. tunnel penetration of the forbidden band gap. When this happens, eqn (6.6.15) ceases to be valid,

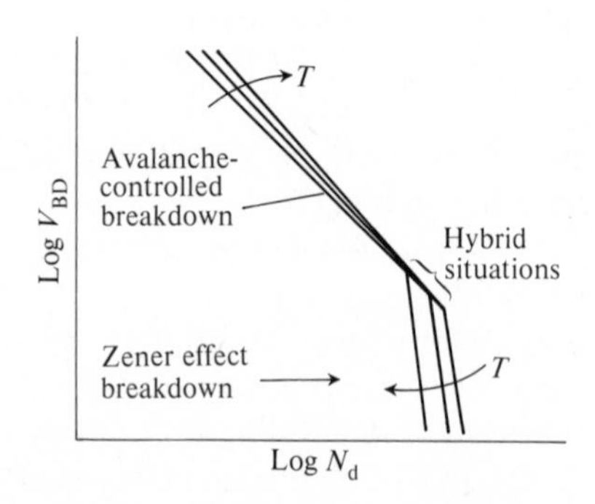

FIG. 6.27. Avalanche and Zener breakdown, as a function of donor concentration and temperature. V_{BD} = breakdown voltage. From information by Chang and Sze (1970).

as is schematically shown in Fig. 6.27. Chang and Sze (1970) have calculated the (very slight) temperature-dependence of the Zener effect (defining an arbitrary 'breakdown current' density of 5000 (j_s), and have shown that the Zener breakdown voltage corresponding to that fixed current is temperature-dependent in the opposite direction from that of V_{BD} governed by avalanche formation, i.e. it decreases with increasing temperature. The matter is ultimately controlled by the temperature-dependence of E_g.

6.6.2 *Avalanche-supported filament formation; light emission*

All space-charge effects have been neglected in the discussions of Section 6.6.1, but this is not always permissible, as Rose (1957), Gunn (1957), and Hoefflinger (1966) have pointed out. A more complete theory predicts a negative resistance characteristic in the breakdown region, and there are general thermodynamic arguments (e.g. see Ridley 1963) which envisage instability and filament formation in such circumstances. Muller (1968) has presented the commonsense argument for filament formation as such, along the following lines.

Let us assume that the system is not totally uniform in the y–z direction, whether as a result of structural defects or temporary fluctuations. If, then, the applied voltage exceeds V_{BD}, the *extra* voltage (sometimes, and particularly in the context of switching, called the *overvoltage*) will be greatest in the region where the current density is greatest. Hence, the growth rate of the carrier density will also be greatest there, and filament formation is an immediate consequence of that. However, that tendency will be increasingly counterbalanced by the build-up of lateral diffusion gradients. Eventually, a steady state is reached, when the carriers lost to the filament by radial diffusion just balance those gained by impact ionization within it. On the basis of simple but plausible assumptions, Muller was able to calculate (numerically) how current density varies with radial distance, and the total voltage V_T with current. Figure 6.28 shows this schematically, and implies a switching action. Low currents are associated with high values of V_T, and this represents the pre-switching (OFF) state of the system. Once filament formation is initiated, V_T drops, and even at constant V_T, the current is increased by making the filament thinner. This is the ON-state, represented by the insert in Fig. 6.28.

The same arguments have been applied to the interpretation of threshold switching in multicomponent amorphous alloys ((Adler *et al.* 1978). The results confirm that the filament radius is always of the order of $(Dt_t)^{1/2}$ where t_t is the transit time of carriers through the high field region. The arguments which lead to filament formation on *thermal* grounds are actually quite similar and, in practice, reinforce the same

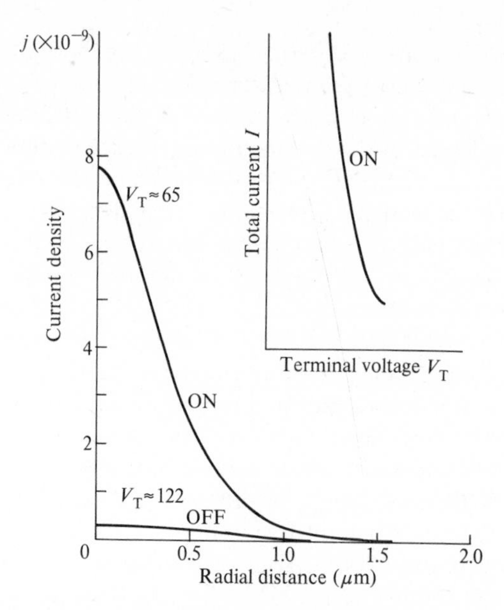

FIG. 6.28. Filament formation during avalanche breakdown; switching. Relationships between current density, radial distance, and voltage. High V_T associated with the OFF state, low V_T with the ON state. Insert: voltage–current relationship after filament formation; microplasmas in Si. Schematically after Muller (1968).

tendency. Smith and Henisch (1973) and Henisch *et al.* (1972) have described a method which may be used to distinguish between thermal and non-thermal switching processes. Filaments are often discussed under the heading of 'microplasmas', and the fact that localized defects can nucleate microplasmas has been well established for p–n junctions (e.g. see Haitz *et al.* 1963). Multiplication factors as high as 10^6 have been measured.

The fact that *any* non-equilibrium situation that involves an excess of carriers can support luminescence in the presence of a direct gap and/or of suitable recombination centres has already been described in Section 1.2.3. Light emission as a result of avalanche formation is merely a special case. That it happens at all is convincing evidence for the fact that some carriers are lost from the electron–hole plasma. Correspondingly, the carrier build-up must proceed somewhat more slowly than it does in models based on zero-less assumptions envisage, but the loss is actually small. Allen (1973) has presented simple estimates of quantum and power efficiency, which have been found to be in good agreement with observations of light emission from contacts with ZnSe:Mn. Because a very high field is required, which exists only over a fraction of the Schottky barrier thickness, quantum efficiencies are never high. In InZnSe, for instance, they are of the order of 10^{-3}, corresponding to a luminous efficiency up to

150 millilumens per watt at 10 mA and 15 V. For *injection* luminescence the situation is considerably more favorable (see below).

Avalanche-sustained light-emission occurs in many materials (in the visible or infrared parts of the spectrum), and one of the best opportunities for observing it is in ZnSe, an n-type zincblende phosphor with a direct band gap of 2.67 eV. Fischer (1961, 1965) has described the preparation of specimens, and the technique of luminescence measurements. A Ga–In alloy (brushed on without heating) can be used as a metal electrode, with a diffusion potential of about 0.4 eV. Yellow-green electroluminescence is observed when reverse voltages (4–5 volts or higher) are applied. Contacts of greater barrier height can also be made, using Cu, Ag, or Au, but are more highly resistive, and give lower emission levels. Only one in every 10^4 electrons passing actually produces a photon, which means that the process is much less efficient than injection luminescence (for which the ratio can be greater than 1 in 10). Fischer (1965) reports that the efficiency is voltage-independent, and suggests that it is governed by two balancing tendencies: (a) extra carrier generation by the high barrier field, and (b) increased rate of minority-carrier injection by the high barrier field.

Injection luminescence observed at the anode is characterized by a spectrum identical to that of photoluminescence, with a maximum at 6250 Å. In contrast, the spectrum associated with avalanche-sustained emission at the cathode has a peak which is shifted towards the shorter wavelengths; 5900 Å (Fig. 6.29). The conclusion is that some of the recombination events involve carrier pairs which are far from the band edges. Avalanche-sustained emission also has a long-wavelength tail, and

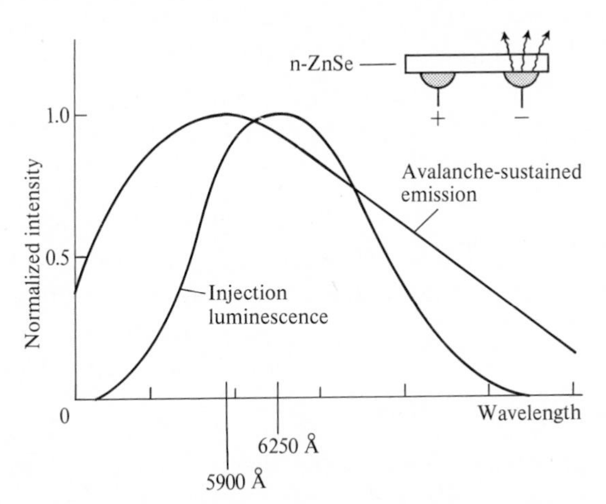

FIG. 6.29. Avalanche supported light emission from ZnSe. Comparison of spectra. Schematically after Fischer (1965).

this has been tentatively ascribed to the interaction of the high barrier field with the luminescence centres (here, cation vacancies), a matter which calls for further investigation. Another aspect which is not yet properly understood (but is very relevant to the interpretation of emission spectra) concerns the situation at the edge of the barrier. The barrier field tends to separate electron–hole pairs, and electrons which have not undergone recombination within the barrier are swept into the adjoining bulk material. They do not encounter impact-generated holes there, but their space charge nevertheless attracts an influx of additional holes from the bulk. Recombination will then follow, making use of centres which are *not* situated in the high barrier field. However, at this stage we do not know how many carriers recombine where.

Recombination radiation from avalanche-generated electron–hole pairs has also been observed at point contacts on such ordinarily insulating crystals as NaI, KI, and RbI, with evaporated gold electrodes. Because the band gap in these materials is wide, only a few electrons gain sufficient energy to sustain the multiplication process (see Paracchini 1971). Accordingly, the observed emission intensities are very low. Much the same applies to avalanche-supported luminescence in organic materials (e.g. see Hsu 1972).

Allen (1973) has pointed out that there is, in principle, yet another mechanism of light emission, namely that corresponding to electron transitions within a band. The probability of such transitions being radiative is small, and the chances of observation are, accordingly, greatest when other sources of luminescence are absent. The corresponding spectra yield interesting information on band structure subtleties.

Further reading

On experimental methods of exploring breakdown microplasmas: Zaitsevskij *et al.* 1980.

On avalanche breakdown in Schottky barriers on highly doped Si: Okuto and Crowell 1974.

On the theory of avalanche breakdown in point contact barriers: Nishizawa and Watanabe 1957.

On qualitative aspects of impact ionization: Curie 1952, Piper and Williams 1952.

6.7 Mechanical effects; deformation

6.7.1 Voltage–current characteristics under pressure; contacts on Si GaAs, Ge

Contacts may in practice be pressure-sensitive for a variety of reasons, but we are concerned here only with mechanisms which cause reversible changes. Even those have not been studied as extensively as one might wish. Results on silicon, using tungsten–ruthenium alloy probes with a 0.25 μm radius of a curvature, have been reported by Kramer and van Ruyven (1972a). Their experiments were all made on the (111)-surfaces of n-type silicon, with constant uniaxial contact thrusts of 35 g. After deformation of the semiconductor beyond the elastic limit, the actual contact radius was 2.6 μm. The barrier height itself was evaluated from the temperature dependence of the zero-bias resistance which, in turn, was deemed to be controlled by a diode mechanism. The temperature dependence of the band gap E_g was negligible throughout, compared with the pressure-induced changes (see also Ohl 1952).

Surprisingly, p-type and n-type silicon gave distinctly different results. Contact on p-type Si were almost insensitive to pressure, and yielded a barrier height ϕ_{ps} of about 0.32 eV. Contacts on n-type material gave $\phi_{ns} = 0.81$ eV in the unstressed condition, but only $\phi_{ns} = 0.25$ eV under 35 g thrust. One expects $\phi_{ns} + \phi_{ps} = E_g$, assuming that the Fermi level is 'pinned' to a tightly bunched set of interface states. For unstressed n-type material, this relationship turns out to be correct, since $0.32 + 0.81 = 1.13$ eV (±0.05 eV under the conditions of the experiments). For stressed Si, the sum is only about 0.54 eV, that is to say much less than the normal band gap.

Because of the highly asymmetric behavior of n-type and p-type material, Kramer and van Ruyven ascribe the effects entirely to a change in the position of E_c, the bottom of the conduction band, i.e. to a change in the electron affinity.

For both n-type and p-type material, the Fermi level remained pinned at the same energy interval from the top of the valence band, which means that the '$E_g/3$ rule' (see Section 1.3.3) must be pressure dependent. The interpretation that E_c is responsible for all changes is also supported by the results of measurements at zero pressure and varying temperature. The temperature-dependence of the band gap is then the significant factor (apart from the important e/kT term), and in that case, too, only ϕ_{ns} was found to change (See Crowell, Sze, and Spitzer 1964), and Aspness and Handler 1966). Moreover, experiments on n-type GaAs under *hydrostatic* pressure (Guétin and Schréder 1971) lead to the same conclusion, though $\Delta\phi_{ns}$ is positive for this case. These confirmatory indications are important, because they mitigate against an alternative explanation of the effects in terms of the space-charge discontinuity discussed in Section 2.1.3. Because so much depends on the position of ionized centres nearest to the interface, a pressure-dependence might be expected to arise from localized deformation. Moreover, the effects described here are between 10 and 100 times greater than those reported for Si p–n *junctions*, and that fact is also difficult to explain in terms of electron-affinity changes. A satisfactory model remains to be found.

Mobile point-contacts are items of great experimental convenience, since they permit measurements (e.g. of the local potential, or else of the local voltage–current relationship) in very small areas. However, the fact that the associated contact pressures are usually high (equal to the yield pressure of the semiconductor or metal, whichever is the smaller) has its drawbacks. Thus, Mil'shtein (1975) has reported on the generation of dislocations by contact pressure and, indeed, on the formation of rectifying junctions between undeformed and deformed regions (see also Mil'shtein and Nikitenko 1974). To avoid all such complications, Henisch and Noble (1966) used thin mercury jets as electrodes for the point-by-point exploration of freshly cleaved germanium surfaces.

With the virtual disappearance of pressure contacts from modern technology, the problems associated with localized deformation have become unfashionable research topics, but all these matters are in need of further clarification.

6.7.2 *Voltage–current characteristics under pressure; contacts on potassium tantalate*

One of the very few non-transistor materials ever systematically investigated from this point of view is potassium tantalate ($KTaO_3$), a material with a band gap of 3.5 eV, n-type when Ca-doped. Rideout and Crowell (1967) used gold contacts of 0.5 mm diameter for their tests, but applied pressure (with a stainless steel stylus) over only a fraction of the corresponding area. They reported a uniaxial pressure sensitivity of the barrier height amounting to 4×10^{-11} eV cm^2 dyn^{-1}, and correspondingly large changes of current (see also Wemple *et al.* 1967).

The voltage–current relationship of the system in the unstressed condition could be represented by

$$J = J_s\left\{1 - \exp\frac{(-V_T + JR_s)e}{\eta kT}\right\} \tag{6.7.1}$$

V_T being the total applied voltage (negative in the forward direction) and R_s a spreading resistance (11 ohms) in series with the barrier. For sufficiently high forward voltages, we have as usual:

$$\log|J| = \log J_s + \frac{(-V_T + JR_s)e}{\eta kT} \tag{6.7.2}$$

J being also negative. η is a correcting factor of the order of unity, as introduced in Section 2.2.5. For currents up to about 10^{-5} A, the shape of the voltage–current characteristic was little affected, η varying only between 1.10 (for zero thrust) and 1.05 (for a thrust of 30 g). On the other hand, J_s varied very sensitively; for thrusts increasing from 0–5 g, for instance, J_s increased by over two orders-of-magnitude, from 10^{-7} to

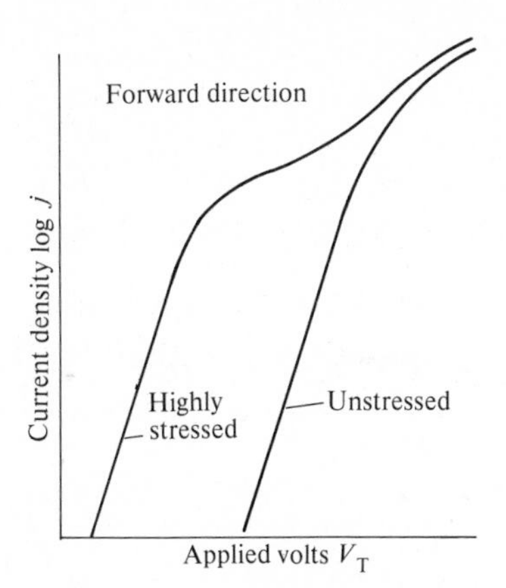

FIG. 6.30. Pressure dependence of contacts with potassium tantalate. From information reported by Rideout and Crowell (1967). Unstressed specimens presumably contained fewer dislocations than stressed ones.

2×10^{-5} A. At such low currents, the barrier resistance is still high enough to make the spreading resistance negligible. At higher currents the pressure sensitivity diminishes (Fig. 6.30) because the spreading resistance makes itself felt. A barrier height-lowering by an amount $\Delta\phi_{ns}$ was evaluated from the current changes as a function of pressure, the barrier height being (plausibly) regarded as the only significant factor. $\Delta\phi_{ns}$ varied between zero and 0.5 eV for pressures between zero and 2.3×10^{10} dynes cm^{-2}. In this case, as distinct from the Si case discussed above, nothing is known about the behaviour of p-type material or about band gap changes with deformation. Potassium tantalate appears to be an unusual material from this point of view, but is unlikely to be unique. Fonash (1974) has pointed to the potential usefulness of the pressure effects for the design of stress sensors.

REFERENCES

Mentioned in†

Abeles, B., Cody, G. D., Goldstein, Y., Tiedje, J., and Wronski, C. R. (1982). *Thin solid Films* **90,** 441. — 5.1 FR

Adams, A. C. and Pruniaux, B. R. (1973). *J. electrochem. Soc.* **120,** 408. — 2.1 FR

Adams, R. F. and Rosenberg, R. L. (1969). *Appl. Phys. Lett.* **15,** 414. — 5.1 FR

Adán, A. and Dobos. K. (1980). *Solid-state Electron.* **23,** 17. — 3.5.5

Adler, D., Henisch, H. K., and Mott, N. (1978). *Rev. mod. Phys.* **50,** 209. — 3.6.2, 6.6.2

Ali, M. P., Tove, P. A., and Ibrahim, M. (1979). *J. appl. Phys.* **50,** 7250. — 2.1 FR

Allen, J. W. (1973. *J. Lumin.* **7,** 228. — 6.6.1, 6.6.2

Allen, R. B. and Farnsworth, H. E. (1956). *J. appl. Phys.* **27,** 525. — 2.3 FR

Allyn, C. L., Gossard, A. C., and Wiegman, W. (1980). *Appl. Phys. Lett.* **36,** 373. — 2.1.1, 2.3.7

Ancker-Johnson, B. and Dick, C. L., Jr. (1969). *Appl. Phys. Lett.* **15,** 141. — 1.2.2, 3.3 FR, 2.1 FR

Ancker-Johnson, B., Robbins, W. P., and Chang, D. B. (1970). *Appl. Phys. Lett.* **16,** 377. — 3.3 FR

Anderson, C. A., Crowell, C. R., and Kao, J. W. (1975). *Solid-State Electron.* **18,** 705. — 5.1 FR

Anderson, C. L., Baron, R., and Crowell, C. R. (1976). *Rev. sci. Instrum.* **47,** 1366. — 4.1 FR

Anderson, W. A., Delahoy, A. E., and Milano, R. A. (1974). *J. appl. Phys.* **45,** 3913. — 5.2 FR

Anderson, W. A., Vernon, S. M., Mathe, P., and Lalevic, B. (1976). *Solid-State Electron.* **19,** 973. — 5.2 FR

Andrews, J. M. (1974). *J. Vac. Sci. Technol.* **11,** 972. — 1.3.3, 2.1.7, 6.5.2

Andrews, J. M. (1975). Electrochem. Soc. spring meeting, (Toronto) extended abstracts, **75–1,** No. 191. — 2.1.7

Andrews, J. M. and Koch, F. B. (1971). *Solid-State Electron.* **14,** 901. — 2.1 FR

Andrews, J. M. and Lepselter, M. P. (1970). *Solid-State Electron.* **13,** 1011. — 1.3.3, 2.1 FR, 2.2.6, 2.2.8, 6.4.1

Andrews, J. M. and Phillips, J. C. (1975). *Phys. Rev. Lett.* **35,** 56. — 1.3.3, 2.1.7

Ankerman, L. W. and Erler, J. W. (1957). *Optics Lett.* **1,** 178. — 6.2.1

† In showing where each reference is mentioned, 'FR' indicates the 'Further reading' lists at the end of most sections.

Mentioned in

Anonymous, (1982). *Phys. Bull.* **33,** 397. 6.3.2
Arnett, P. C. and DiMaria, D. J. (1975). *Appl. Phys. Lett.* **27,** 34. 3.5.5
Arnett, P. C. and DiMaria, D. J. (1976). *J. appl. Phys.*, **47,** 2092. 3.5.5
Arnold, E. and Poleshuk, M. (1975). *J. appl. Phys.* **46,** 3016. 3.5 FR
Ashok, S., Borrego, J. M., and Gutmann, R. J. (1978). *Electron. Lett.* **14,** 332. 3.5.5
Ashok, S., Borrego, J. M., and Gutmann, R. J. (1979). *Solid-State Electron.* **22,** 621. 3.5 FR
Ashok, S., Sharma, P. P., and Fonash, S. J. (1980). *I.E.E.E. Trans. Electron Dev.* **ED-27,** 725. 5.2 FR
Aspness, D. E. and Handler, P. (1966). *Surf. Sci.* **4,** 353. 6.7.1
Assimos, J. A. and Trivich, D. (1973). *J. appl. Phys.* **44,** 1687. 2.1 FR
Atalla, M. M. (1966). Proc. Munich Sympos. on Microelectronics, 123. 2.1.2
Aven, M. and Swank, R. K. (1969). in *Ohmic contacts in semiconductors* (ed. B. Schwartz). Electrochemical Society, New York. 6.5.1
Aydinli, A. and Mattauch, R. J. (1982). *Solid-State Electron.* **25,** 551. 2.1 FR
Baccarani, G. (1976). *J. appl. Phys.* **47,** 4122. 2.2.6, 2.3 FR
Baccarani, G., Calzolari, P. V., and Graffi, S. (1974). *J. appl. Phys.* **45,** 341. 3.5.2
Baccarani, G., Impronta, M., Ricco, B., and Perla, P. (1978). *Rev. Phys. appl.* **13,** 777. 6.4.1
Baccarani, G., Severi, M., and Soncini, G. (1973). *Appl. Phys. Lett.* **23,** 265. 4.1.4
Bachrach, R. Z. (1978). *J. Vac. Sci. Technol.* **15,** 1340. 6.5.2
Banbury, P. C. (1952). Ph.D. Thesis. University of Reading, U.K. 3.1.2
Banbury, P. C. (1953a). *Proc. Phys. Soc.* **B66,** 50. 3.5.1, 5.1.2
Banbury, P. C. (1953b). *Proc. Phys. Soc.* **B66,** 833. 3.2.1, 3.5.1
Banbury, P. C., Davis, E. A., and Green, G. W. (1962). *Proc. int. Conf. Phys. Semiconductors*, 813. Institute of Physics and The Physics Society, London. 2.1.7
Banbury, P. C. and Houghton, J. (1954). *Physica* **20,** 1050. 3.1.4, 3.5.1
Banbury, P. C. and Houghton, J. (1955). *Proc. Phys. Soc.* **B63,** 17. 3.1.4
Baraff, G. A. (1962). *Phys. Rev.* **128,** 2507. 6.6.1
Baraff, G. A. (1964). *Phys. Rev.* **133,** A26. 6.6.1
Barbe, D. F. (1971). *J. Vac. Sci. Technol.* **8,** 102. 3.6.2
Bardeen, J. (1947). *Phys. Rev.* **71,** 717. 1.3.3
Bardeen, J. (1950). *Bell Syst. Tech. J.* **29,** 469. 3.5.1, 5.1.2
Bardeen, J. and Brattain, W. H. (1948). *Phys. Rev.* **74,** 230. 1.2.1, 1.2.2
Bardeen, J. and Brattain, W. H. (1949). *Phys. Rev.* **75,** 1208. 3.2.1, 3.5.1
Barkhalov, B. Sh. and Vidadi, Yu.A. (1977). *Thin solid Films,* **40,** L5. 3.6.3
Barnes, P. A. and Williams, R. S. (1981). *Solid-State Electron.* **24,** 907. 6.5 FR
Barnett, A. M. and Rothwarf, A. (1980). *I.E.E.E. Trans. Electron. Dev.* **ED-27,** 615. 5.1 FR

Mentioned in

Baron, R. and Mayer, J. W. (1970). In *Semiconductors and semi-metals* (eds R. K. Willardson and A. C. Beer). Academic Press, New York. 5.1.6

Barret, C., Lebars, P., and Vapaille, A. (1975). *C.R. Acad. Sci.* (Paris), **280B,** 133. 1.3 FR

Barret, C. and Vapaille, A. (1976). *Solid-State Electron.* **19,** 73. 1.3 FR, 2.1 FR

Basterfield, J., Shannon, J. M., and Gill, A. (1975). *Solid-State Electron.,* **18,** 290. 2.1 FR

Becker, R. (1973). *Solid-State Electron.* **16,** 1241. 6.5 FR

Beguwala, M. and Crowell, C. R. (1974). *Solid-State Electron.* **17,** 203. 4.3 FR

Beichler, J., Fuhs, W., Mell, H., and Welsch, H. M. (1980). *J. noncryst. Solids* **35–36,** 587. 4.3.2

Beneking, H. (1968). *Z. angew. Phys.* **10,** 216. 3.5.1

Benenz, J. J., Scilla, G. J., Wrick, U. L., Eastman, L. F., and Morrison, G. H. (1976). *J. Vac. Sci. Technol.* **13,** 1152. 2.1.7

Bennet, A. I. and Hunter, L. P. (1951). *Phys. Rev.* **81,** 152. 6.2.1

Benzer, S. (1947). *Phys. Rev.* **71,** 141. 1.3.3, 6.4.1

Benzer, S. (1949). *J. appl. Phys.* **20,** 804. 1.3.3

Berger, H. H. (1972). *J. electrochem. Soc.* **119,** 507. 1.1.4

Bergh, A. A. and Dean, P. J. (1976). *Light-emitting diodes.* Oxford University Press, Oxford and New York. 1.2.3

Beringer, E. R. (1944). *Radiation Rep.,* No. 638. 4.2.2

Bernasconi, J., Klein, H. P., Knecht, B., and Strassler, S. (1976). *J. electron. Mat.* **5,** 473. 6.4.3

Berry, W. B. (1974). *Appl. Phys. Lett.* **25,** 195. 5.1.2, 5.2.1

Bethe, H. A. (1942). *M.I.T. Radiat. Lab. Rep.,* 43–12. 2.2.6

Betz, A. (1964). *Konforme Abbildung.* Springer, Berlin. 1.1.4

Bhargava, R. N. (1969). *Appl. Phys. Lett.* **14,** 193. 6.2.1

Bhattacharya, P. K. and Yeaman, M. D. (1981). *Solid-State Electron.* **24,** 297. 6.3.2

Bickford, L. R. and Kanazawa, K. K. (1976). *J. phys. Chem. Solids* **37,** 839. 2.1 FR

Billig, E. and Landsberg, P. T. (1950). *Proc. Phys. Soc.* **A63,** 101. 2.2.2, 2.3.1

Blakemore, J. S. (1982). *Solid-State Electron.,* **25,** 1067. 6.5.1

Blom, G. M. and Woodall, J. M. (1970). *Appl. Phys. Lett.* **17,** 373. 1.2 FR

Blötekjaer, K. (1966). *Ericsson Technics* **22,** 125. 2.2.7

Board, K. and Darwish, M. (1982). *Solid-State Electron.* **25,** 529. 6.4.1

Bonch-Bruevich, V. L. (1979). *J. noncryst. Solids* **32,** 105. 3.6.2

Boroffka, H. (1961). In *Electrical conductivity in organic solids* (ed. H. Kallmann and M. Silver). Interscience, New York. 3.6.3

Borrego, J. M. and Gutmann, R. J. (1976). *Appl. Phys. Lett.* **28,** 280. 3.5 FR

Borrego, J. M., Gutmann, R. J., and Ashok, S. (1977). *Appl. Phys. Lett.* **30,** 169. 1.3 FR, 2.2 FR

Bose, D. N., Sen Gupta, V., and Acharya, H. N. (1979). *Phys. stat. solidi* **(a)54,** K155. 3.5 FR

Mentioned in

Bosman, G. and Zijlstra, R. J. T. (1982). *Solid-State Electron.* **25,** 273. 1.1.2

Bosselar, C. A. (1973). *Solid-State Electron.*, **16,** 648. 6.3.1

Braun, I. and Henisch, H. K. (1966a). *Solid-State Electron.* **9,** 981. 3.1.4, 3.5.1

Braun, I. and Henisch, H. K. (1966b). *Solid-State Electron.* **9,** 1111. 5.1.2

Braun, S. and Grimmeis, H. G. (1973). *J. appl. Phys.* **44,** 2789. 2.1.8

Brillson, L. J. (1977). *Phys. Rev. Lett.* **38,** 245. 2.1.7

Brillson, L. J. (1978a). *Phys. Rev. Lett.* **40,** 260. 2.1.7

Brillson. L. J. (1978b). *J. Vac. Sci. Technol.* **14,** 885. 2.1.7

Brillson, L. J. (1978c). *J. Vac. Sci. Technol.* **15,** 1378. 2.1.7

Brillson, L. J. (1979). *J. Vac. Sci. Technol.* **16,** 1137. 2.1 FR

Brillson, L. J. (1982). *Thin solid Films* **89,** L1. 1.3.3, 2.1.7

Brillson, L. J. (1983). *J. Phys. Chem. Solids* **44,** 703. 2.1 FR

Brillson, L. J., Bauer, R. S., Bachrach, R. Z., and Hansson, G. (1980). *Appl. Phys. Lett.* **36,** 326. 1.3.3

Brillson, L. J., Brucker, C. F., Margaritondo, G., Slowik, J., and Stoffel, N. G. (1980). *J. Phys. Soc. Jap.* **49,** 1089. 2.1 FR

Brillson, L. J., Brucker, C. F., Stoffel, N. G., Katnani, A. D., and Margaritondo, G. (1981). *Phys. Rev. Lett.* **46,** 838. 2.1 FR

Brillson, L. J., Margaritondo, G., and Stoffel, N. G. (1980). *Phys. Rev. Lett.* **44,** 667. 1.3.3, 2.1.7

Brousseau, M., Barrau, J., Brabant, J. C., and van Tuyen, N. (1970). *Solid-State Electron.* **13,** 906. 3.3.2

Brown, G. W. and Lindsay, B. W. (1976). *Solid-State Electron.* **19,** 991. 2.1 FR

Bücher, H. K., Burkey, B. C., Lubberts, G., and Wolf, E. L. (1973). *Appl. Phys. Lett.* **23,** 617. 5.1.2

Buckley, W. D. and Moss, S. C. (1972). *Solid-State Electron.* **15,** 1331. 2.1 FR

Burgess, R. E. (1950). *Proc. Phys. Soc.* **B63,** 1036. 4.1.1

Burgess, R. E. (1953). *Proc. Phys. Soc.* **B66,** 430. 2.2.7

Burgess, R. E. (1955). *Proc. Phys. Soc.* **B68,** 908. 2.2.7, 6.2.1, 6.2.2

Burton, J. A., Hull, G. W., Morin, F. G., and Severiens, J. C. (1953). *J. phys. Chem.* **57,** 853. 4.2.1

Buturla, E. M. and Cottrell, P. E. (1980). *Solid-State Electron.* **23,** 331. 2.2 FR

Buxo, J., Esteve, D., and Sarrabayrouse, G. (1976). *Phys. stat. Solidi* **37,** K105. 3.5.5

Callarotti, R. C. (1981). Personal communication. 4.1.2, 4.3.1, 6.3.2

Card, H. C. (1974). *Proc. Manchester Conf. metal–semiconductor contacts*, 129 Inst. Physics, London. 6.6.1

Card, H. C. (1975a). *Solid State Commun.* **16,** 87. 2.1 FR

Card, H. C. (1975b). *Solid-State Electron.* **18,** 881. 2.3 FR

Card, H. C. (1976a). *J. appl. Phys.* **47,** 4964. 5.1.4

Card, H. C. (1976b). *I.E.E.E. Trans. Electron Dev.* **ED-23,** 538. 1.3.2, 6.5 FR

Card, H. C. (1977). *Solid-State Electron.* **20,** 971. 3.5.5, 5.2.2

Mentioned in

Card, H. C. and Hwang, W. (1980). *I.E.E.E. Trans. Electron. Dev.* **ED-27,** 700. 6.4 FR

Card, H. C. and Rhoderick, E. H. (1971a). *J. Phys. D. appl. Phys.* **4,** 1589. 3.5.5 4.1.6 5.2.2

Card, H. C. and Rhoderick, E. H. (1971b). *J. Phys. D. appl. Phys.* **4,** 1602. 3.5.5 5.2.2

Card, H. C. and Rhoderick, E. H. (1973). *Solid-State Electron.* **16,** 365. 3.5.5

Card, H. C. and Singer, K. E. (1975). *Thin solid Films* **28,** 265. 1.3 FR

Card, H. C. and Smith, B. L. (1971). *J. appl. Phys.* **42,** 5863. 1.2 FR

Card, H. C. and Yang, E. S. (1976). *Appl. Phys. Lett.* **29,** 51. 5.2.2

Card, H. C. and Yang, E. S. (1977). *I.E.E.E. Trans. Electron. Dev.* **ED-24,** 397. 6.4.1 6.4.4

Card, H. C., Yang, E. S., and Panayotatos, P. (1977). *Appl. Phys. Lett.* **30,** 643. 5.1.3

Carlson, D. E. (1977). *I.E.E.E. Trans. Electron. Dev.* **ED-24,** 449. 5.1.1 5.2 FR

Carlson, D. E. and Wronski, C. R. (1976). *Appl. Phys. Lett.* **28,** 671. 3.6.2

Carroll, J. E. (1970). *Hot electron microwave generation*, p. 171, Arnold, London. 6.6.1

Champness, C. H., Griffiths, C. H., and Sang, H. (1968). *Appl. Phys. Lett.* **12,** 314. 2.1 FR

Chandra, A. and Eastman, L. F. (1980). *Solid-State Electron.* **23,** 599. 6.4.2

Chang, C. Y. and Sze, S. M. (1970). *Solid-State Electron.* **13,** 727. 2.1.6 2.3.1 2.3.4 2.3.6 6.6.1

Charlson, E. J. and Lien, J. C. (1975). *J. appl. Phys.* **46,** 3982. 5.2 FR

Cheng, Y. C. (1977). *Prog. Surf. Sci.* **8,** 181. 2.3 FR

Cho, A. Y. and Casey, H. C., Jr. (1974). *J. appl. Phys.* **45,** 1258. 2.1 FR

Choo, S. C., Leong, M. S., and Kuan, K. L. (1976). *Solid-State Electron.* **19,** 561. 1.1 FR

Choo, S. C., Leong, M. S., Hong, H. L., Li, L., and Tan, L. S. (1977). *Solid-State Electron.* **20,** 839. 1.1 FR

Choo, S. C., Leong, M. S., Hong, H. L., Li, L., and Tan, L. S. (1978). *Solid-State Electron.* **21,** 769. 1.1 FR

Choo, S. C., Leong, M. S., and Tan. L. S. (1981). *Solid-State Electron.* **24,** 562. 1.1 FR

Chye, P. W., Lindav, I., Pianetto, P., Garner, C. M., Yu, C. Y., and Spicer, W. E. (1978). *Phys. Rev.* **B18,** 5545. 1.3.3 2.1.7

Chynoweth, A. G. (1965). In *Semiconductors and semi-metals*, (eds R. K. Willardson and A. C. Beer), Vol. 4, p. 263. Academic Press, New York. 6.6.1

Chynoweth, A. G. and McKay, K. G. (1957). *Phys. Rev.* **108,** 29. 6.6.1

Clark, G. L. and Roach, P. G. (1941). *Trans. electrochem. Soc.* **79,** 355. 1.3.4

Mentioned in

Clarke, D. R. (1978). *J. appl. Phys.* **49,** 2407. 6.4.3

Clarke, R. A., Green, M. A. and Shewchun, J. (1974). *J. appl. Phys.* **45,** 1442. 3.1.4

Coe, D. J. and Rhoderick, E. H. (1976). *J. Phys. D. appl. Phys.* **9,** 965. 2.1 FR

Coen, R. W. and Muller, R. S. (1980). *Solid-State Electron.* **25,** 35. 3.5 FR

Coleman, D. J., Jr. (1975). *J. appl. Phys.* **46,** 5309. 4.1.2

Conley, J. W., Duke, C. B., Mahan, G. D., and Tiemann, J. J. (1966). *Phys. Rev.* **150,** 466. 2.3.1, 2.3.3

Conley, J. W. and Mahan, G. D. (1967). *Phys. Rev.* **161,** 681. 2.3.1

Cook, R. K., Kasold, J. P., and Jones, K. A. (1980). *Solid-State Electron.* **23,** 391. 4.3 FR

Cornet, A., Siffert, P., Coche, A., and Triboulet, R. (1970). *Appl. Phys. Lett.* **17,** 432. 5.1.1

Courtens, E. and Chernow, F. (1966). *Appl. Phys. Lett.* **8,** 3. 2.1 FR, 6.3 FR

Cowley, A. M. (1970). *Solid-State Electron.* **12,** 403. 2.1 FR

Crowell, C. R. (1965). *Solid-State Electron.* **8,** 395. 2.1.7, 2.2.6

Crowell, C. R. (1969a). *Solid-State Electron.* **12,** 55. 2.2.6

Crowell, C. R. (1969b). In *Ohmic contacts to semiconductors* (ed. B. Schwartz). Electrochemical Society, New York. 2.3.3

Crowell, C. R. (1974). *J. Vac. Sci. Technol.* **11,** 951. 2.1.5

Crowell, C. R. (1977). *Solid-State Electron.* **20,** 171. 2.3.3

Crowell, C. R. and Beguwala, M. (1971). *Solid-State Electron.* **14,** 1144. 2.2.5

Crowell, C. R. and Nakano, K. (1972). *Solid-State Electron.* **15,** 605. 4.3.2

Crowell, C. R. and Rideout, V. L. (1969a). *Solid-State Electron.* **12,** 89. 2.1.6

Crowell, C. R. and Rideout, V. L. (1969b). *Appl. Phys. Lett.* **14,** 85. 2.3.1, 6.5.2

Crowell, C. R. and Roberts, G. I. (1969), *J. appl. Phys.* **40,** 3726. 4.1.6

Crowell, C. R. and Sze, S. M. (1966a). *Appl. Phys. Lett.* **9,** 242. 6.6.1

Crowell, C. R. and Sze, S. M. (1966b). *Solid-State Electron.* **9,** 1035. 2.3.2, 2.3.4, 2.3.5, 3.5.3

Crowell, C. R. and Sze, S. M. (1966c). *J. appl. Phys.* **37,** 2683. 2.1.5, 2.1.6, 2.3.4

Crowell, C. R., Sze, S. M., and Spitzer, W. G. (1964). *Appl. Phys. Lett.* **4,** 91. 6.7.1

Curie, D. (1952). *J. Phys. Radium* **13,** 317. 6.6 FR

Cuthrell, R. E. and Tipping, D. W. (1973). *J. appl. Phys.* **49,** 3277. 1.1 FR

Dacey, G. C. (1953). *Phys. Rev.* **90,** 759. 3.1 FR

Dalal, V. L. (1980). *I.E.E.E. Trans. Electron. Dev.* **ED-27,** 662. 5.2 FR

Darley, H. M. and Christopher, J. E. (1974). *Phys. Rev.* **B9,** 3447. 5.1.1

Davidov, B. (1939). *J. Phys. U.S.S.R.* **1,** 167. 2.1.2

Davis, G., Jr. and Grannemann, W. W. (1963). *J. appl. Phys.* **34,** 228. 2.1 FR

Davis, H. T. (1962). *Introduction to nonlinear differential and integral*

Mentioned in

equations, p. 58. Dover Publishing Co., New York. 6.6.1

Davison, S. G. and Levine, J. D. (1970). *Solid State Physics*, Vol. 25. Academic Press, New York. 1.3.2

Daw, A. N., Datta, A. K., and Ash, M. C. (1982). *Solid-State Electron.* **25,** 1205. 5.2 FR

Day, H. M., Gleason, K. R., and Macpherson, A. C. (1970). *Solid-State Electron.* **13,** 1111. 4.1.1

Dean, R. H. (1968). *Appl. Phys. Lett.* **13,** 164. 3.2.1

de Boer, J. H. and van Geel, W. C. (1935). *Physica* **2,** 309, 321. 1.3.4

Delacote, G. M., Fillard, J. P., and Marco, F. J. (1964). *Solid State Commun.* **2,** 373. 3.6.3

Demoulin, E. and Van der Wiele, F. (1974). *Solid-State Electron.* **17,** 825. 3.5.4

Deneuville, A. (1974). *J. appl. Phys.* **45,** 3079. 4.1.4, 4.3.1

Deneuville, A. and Chakraverty, B. K. (1972). *Phys. Rev. Lett.* **28,** 1258. 2.1 FR

Descalu, D., Brezeanu, Gh., Dan, P. A., and Dima, C. (1981). *Solid-State Electron.* **24,** 897. 1.1.4

Dewald, J. F. (1960). *Bell Syst. Tech. J.* **39,** 615. 4.1.2

Diligenti, A., Pellegrini, B., Salardi, G., and Bagnoli, P. E. (1980). *Solid-State Electron.* **23,** 799. 4.3 FR

Dilworth, C. C. (1948). *Proc. Phys. Soc.* **60,** 315. 1.3.4

DiMaria, D. J. (1974). *J. appl. Phys.* **45,** 5454. 3.5.5

DiMaria, D. J. and Arnett, P. C. (1975). *Appl. Phys. Lett.* **26,** 711. 3.5.5

DiStefano, T. H. (1971). *Appl. Phys. Lett.* **19,** 280. 2.3.6

Djurić, Z. and Smiljanić M. (1975). *Solid-State Electron.* **18,** 817. 3.5 FR

Djurić, Z., Smiljanić, M., and Tjapkin, D. (1975). *Solid-State Electron.* **18,** 827. 4.1 FR

Dmitriev, A. P., Stefanovich, A. E., and Tsendin, L. D. (1978). *Phys. stat. Solidi* (**a**)**46,** 45. 3.2 FR

Döhler, G. H. and Brodsky, M. H. (1974). *Amer. Inst. Phys. Conf. Proc.*, No. 20, 351. Yorktown Heights, New York. 6.4.2

Douglas, E. C. and D'Aiello, R. V. (1980). *I.E.E.E. Trans. Electron. Dev.* **ED-27,** 792. 5.1 FR

Dousmanis, G. C. and Duncan, R. C., Jr. (1958). *J. appl. Phys.* **29,** 1627. 3.1.3

Dubey, P. K. and Paranjape, V. V. (1977). *J. appl. Phys.* **48,** 324. 5.1.2

Dudek, I. and Kassing, R. (1979). *Solid-State Electron.* **22,** 361. 3.3.2

Duke, C. B. (1969). *Solid State Phys.* **Suppl. 10.** Academic Press, New York. 3.5.5

Duncan, W., Lamb, J., Mcintosh, K. G., and Smellie, A. R. (1973). *Appl. Phys. Lett.* **23,** 330. 6.4.2

Dunlap, H. L. and Marsh, O. J. (1969). *Appl. Phys. Lett.* **15,** 311. 2.1 FR

Edwards, D. A. and Jones, P. L. (1978). *Solid-State Electron.* **21,** 1163. 6.3 FR

Edwards, W. D., Hartman, W. A., and Torrens, A. B. (1972). *Solid-State Electron.* **15,** 387. 6.5 FR

Eger, D., Many, A., and Goldstein, Y. (1976). In *Electronic properties of quasi-two-dimensional systems* (eds J. J. Quinn and P. J. Stiles). North-Holland Publishing Co., Amsterdam. 2.1.2, 2.1.6, 6.1.1

Mentioned in

Ehrenberg, W. (1950). *Proc. Phys. Soc.* **A63,** 75. 6.5.1

Einzinger, R. (1975). *Ber. Deut. Keram. Ges.* **52,** 244. 6.4.3

El-Gabaly, M. (1980). *Solid-State Electron.* **23,** 9. 4.2 FR

Elsharkawi, A. R. and Kao, K. C. (1976). *Solid-State Electron.* **19,** 939. 6.5 FR

Emtage, P. R. (1977). *J. appl. Phys.* **48,** 4372. 6.4.1 6.4.3

Engemann, J. (1981). *Solid-State Electron.* **24,** 467. 4.1 FR

England, J. B. A. and Hammer, V. W. (1971). *Nucl. Instr. Meth.* **96,** 81. 6.4.2

Fabre, E. (1976). *Appl. Phys. Lett.* **29,** 607. 5.2 FR

Fahrenbruch, A. L. and Bube, R. H. (1983). *Fundamentals of solar cells.* Academic Press, New York. 5.1 FR

Fan, H. Y. (1948). *Phys. Rev.* **74,** 1505. 6.3.2

Fan, H. Y. and Becker, M. (1951). *Semiconducting materials* (ed. H. K. Henisch), p. 132. Butterworth Scientific Publications, London. 1.2.3

Fang, Y. K., Chang, C. Y., and Su, Y. K. (1979). *Solid-State Electron.* **22,** 933. 1.1.4

Fedorov, M. I. and Benderskii, V. A. (1971a). *Sov. Phys.-Semiconductors* **4,** 1198. 3.6.3

Fedorov, M. I. and Benderskii, V. A. (1971b). *Sov. Phys.-Semiconductors* **4,** 1720. 3.6.3

Feltl, H. (1978). *Solid-State Electron.* **21,** 1227. 4.1 FR

Fertig, D. J. and Robinson, G. Y. (1976). *Solid-State Electron.* **19,** 407. 2.1 FR

Feuchtwang, T. E., Leipold, W. C., and Martino, R. C. (1977). *Surf. Sci.* **64,** 109. 2.1.6

Feuerstein, A. and Kalbitzer, S. (1973). *Appl. Phys. Lett.* **22,** 19. 6.5 FR

Fischer, A. G. (1961). *Solid-State Electron.* **2,** 232. 6.6.2

Fischer, A. G. (1965). In *Radiative recombination in semiconductors,* p. 259. Academic Press, New York. 6.6.2

Florida, C. D., Holt, F. R., and Stephen, J. H. (1954). *Nature* **173,** 397. 3.4.2

Fonash, S. J. (1972). *Solid-State Electron.* **15,** 783. 2.2 FR

Fonash, S. J. (1974). *J. appl. Phys.* **45,** 496. 6.7.2

Fonash, S. J. (1975). *J. appl. Phys.* **46,** 1286. 5.2.2

Fonash, S. J. (1976). *J. appl. Phys.* **47,** 3527. 5.2 FR

Fonash, S. J. (1980). Photovoltaic devices, CRC critical reviews. *Solid State Mat. Sci.* **9,** 107. 5.1 FR

Fonash, S. J. (1981). *Solar cell device physics.* Academic Press, New York. 5.1 FR

Fonash, S. J. and Ashok, S. (1981). *Solid-State Electron.* **24,** 1075. 5.2.1 5.2.2

Fortin, E. and Masson, D. (1982). *Solid-State Electron.* **25,** 281. 5.1 FR

Fortin, E. and Sears, W. M. (1982). *J. phys. Chem. Solids,* **43,** 881. 5.1 FR

Franceschetti, R. D. and Macdonald, J. R. (1979). *J. appl. Phys.* **50,** 297. 2.2.9

Frankl, D. R. (1961). *Solid-State Electron.* **2,** 71. 6.1.1

Frankl, D. R. (1967). *Electrical properties of semiconductor surfaces.* Pergamon Press, New York. 1.3.2 2.1.2 3.1.1 6.1.1

Freeman, E. C. and Slowik, J. H. (1981). *Appl. Phys. Lett.* **39,** 96. 3.2 FR

Frenkel, J. (1930). *Phys. Rev.* **36,** 1604. 2.3.3

Mentioned in

Fritzsche, H. (1974). In *Amorphous and liquid semiconductors* (ed. J. Tauc). Plenum Press, New York. 3.6.2

Fuchs, H. and Heime, K. (1966). In *Microminiaturization* p. 335. R. Oldenbourg, Munich/Vienna. 2.1 FR 6.3 FR

Gale, R. W., Glew, R. W., and Bryant, F. J. (1975). *Solid-State Electron.* **18,** 839. 6.4.2

George, E. V. and Bekefi, G. (1969). *Appl. Phys. Lett.* **15,** 33. 1.1.1

Geppert, D. V. (1963). *J. appl. Phys.* **34,** 490. 6.3.2

Gerzon, P. H., Barnes, J. W., Waite, D. W., and Northrop, D. C. (1975). *Solid-State Electron.* **18,** 343. 4.2 FR

Ghosh, A. and Feng, T. (1973). *J. appl. Phys.* **44,** 2781. 2.1 FR 6.3 FR

Ghosh, A. K., Morel, D. L., Feng, T., Shaw, R. F., and Rowe, C. A. (1974). *J. appl. Phys.* **45,** 230. 3.6.3 6.3 FR

Gibson, A. F. (1953). *Proc. Phys. Soc.* **B66,** 588. 1.2.3

Gill, W. D. and Bube, R. H. (1970). *J. appl. Phys.* **41,** 3731. 5.1.2

Glover, G. H. (1973). *Solid-State Electron.* **16,** 973. 4.1 FR

Goldberg, C. (1969). *J. appl. Phys.* **40,** 4612. 2.2.7

Goldsmid, H. J. (1968). (Ed.) *Problems in solid state physics.* Academic Press, New York. 2.1.2

Gooch, C. H. (1973). *Injection electroluminescent devices.* John Wiley, New York. 1.2.3

Goodman, A. M. (1963). *J. appl. Phys.* **34,** 329. 4.1.2

Goodman, A. M. and Perkins, D. M. (1964). *J. appl. Phys.* **35,** 3351. 2.1.2 4.1.2

Goodman, B., Lawson, A. W., and Schiff, L. I. (1947). *Phys. Rev.* **71,** 191. 4.2.1

Goodman, N. B., Fritzsche, H., and Ozaki, H. (1979). In *Amorphous and liquid semiconductors*, p. 599. North-Holland Publishing Company, Amsterdam. 6.1.2

Goodwin, E. J. (1939). *Proc. Camb. phil. Soc.* **35,** 205, 221, 232. 1.3.2

Gopen, H. J. and Yu, A. Y. C. (1971). *Solid-State Electron.* **14,** 515. 6.5 FR

Gossick, B. R. (1956). *J. appl. Phys.* **27,** 905. 3.5.1

Gossick, B. R. (1960). *J. appl. Phys.* **31,** 29. 3.5 FR

Gossick, B. R. (1963). *Solid-State Electron.* **6,** 445. 3.5.1 3.5.4

Gossick, B. R. (1969a). *Surf. Sci.* **18,** 181. 1.1.2 6.5 FR

Gossick, B. R. (1969b). *Phys. stat. solidi* **35,** 997. 2.1 FR

Gossick, B. R. (1970). *Surf. Sci.* **21,** 123. 6.5 FR

Gossick, B. R. (1971). *Surf. Sci.*, **28,** 469. 6.5 FR

Gradshteyn, I. S. and Rjyzhik, I. W. (1965) *Table of integrals, series and products*, integral 2.315. Academic Press, New York. 3.1.2

Greebe, C. A. A. J. and Van der Maesen, F. (1963). *Philips Res. Rep.* **18,** 65. 3.3 FR

Green, M. A. (1976). *Solid-State Electron.* **19,** 421. 3.1.1 4.1.6

Green, M. A. (1977). *Solid-State Electron.* **20,** 265. 5.1.3

	Mentioned in
Green, M. A. (1980). *Appl. Phys. Lett.* **33,** 178.	5.2.1
Green, M. A. and Godfrey, R. B. (1976). *Appl. Phys. Lett.* **29,** 610.	5.2 FR
Green, M. A. and Shewchun, J. (1973). *Solid-State Electron.* **16,** 1141.	3.1.4 3.2.4 3.5.4 4.1.6
Green, M. A. and Shewchun, J. (1974). *Solid-State Electron.* **17,** 349.	3.5.5
Greene, R. F., Frankl, D. R., and Zemel, J. N. (1960). *Phys. Rev.* **118,** 2538.	6.1.1
Grimmeiss, H. G. (1974). In *Metal–semiconductor contacts,* Conf. Rep. p. 181. Institute of Physics, London.	2.1.8 3.6.2 4.3.1
Grinolds, M. and Robinson, G. Y. (1977). *J. Vac. Sci. Technol.* **14,** 75.	2.1 FR
Grinolds, M. and Robinson, G. Y. (1980). *Solid-State Electron.* **23,** 973.	1.3 FR 2.1 FR
Grosvenor, C. R. M. (1981). *Solid-State Electron.* **24,** 792.	6.5 FR
Grove, A. S. (1967). *Physics and technology of semiconductor devices.* John Wiley, New York.	3.1.4
Grove, A. S., Snow, E. H., Deal, B. E., and Sah, C. T. (1964). *J. appl. Phys.* **35,** 2458.	4.3 FR
Grover, N. B., Goldstein, Y., and Many, A. (1961). *J. appl. Phys.* **32,** 2538.	6.1.1
Gu, J., Kitahara, T., Kawakami, K., and Sakaguchi, T. (1975). *J. appl. Phys.* **41,** 1184.	6.5.1
Guétin, P. and Schréder, G. (1971). *Solid State Commun.* **9,** 591.	6.7.1
Guggenheim, E. A. (1953). *Proc. Phys. Soc.* **A66,** 121.	2.1.8
Guha, S., Arora, B. M., and Salvi, V. P. (1977). *Solid-State Electron.* **20,** 431.	2.1.7
Gundjian, A. A. and Habert, B. S. (1970). *Solid-State Electron.* **13,** 1507.	6.2.1
Gundlach, K. H. and Kadlec, J. (1972). *Appl. Phys. Lett.* **20,** 445.	3.5.5
Gunn, J. B. (1954). *Proc. Phys. Soc.* **B67,** 575.	3.1.4
Gunn, J. B. (1957). In *Progress in semiconductors* (ed. A. F. Gibson), Vol. 2. Heywood, London.	6.6.2
Gupta, H. M. and van Overstraeten, R. J. (1976). *J. appl. Phys.* **47,** 1003.	6.3 FR
Gutai, L. and Mojzes, I. (1975). *Appl. Phys. Lett.* **26,** 325.	2.2.5
Gutknecht, P. and Strutt, M. J. O. (1972). *Appl. Phys. Lett.* **21,** 405.	2.3.8
Gutkowicz-Krusin, D. (1981), *J. appl. Phys.* **52,** 5370.	5.1.2
Gutmann, F. and Lyons, L. E. (1967). (Eds) *Organic semiconductors.* John Wiley, New York.	3.6.3
Haitz, R. H., Goetzberger, A., Scarlett, R. M., and Shockley, W. (1963). *J. appl. Phys.* **34,** 1581.	6.6.2
Hall, P. M. (1967). *Thin solid Films* **1,** 277.	1.1.4
Hall, R. N. (1981). *Solid-State Electron.* **24,** 595.	5.1 FR
Hamilton, B. (1974). *Metal–semiconductor contacts,* Conf. Rep., p. 218. Institute of Physics, London.	3.6.2
Harnik, E., Goldstein, Y., Grover, N. B., and Many, A. (1960). *J. phys. Chem. Solids* **14,** 193.	6.1.1

Mentioned in

Harris, A. J., Walker, R. S., and Sneddon, R. (1980). *J. appl. Phys.* **51,** 4287. 3.6.2

Harrison, W. A. (1961). *Phys. Rev.* **123,** 85. 3.5.5, 5.2.3

Hartmann, A. (1936). *Physik. Zeit.* **37,** 862. 1.3.4

Harutunian, V. M. and Buniatian, V. V. (1977). *Solid-State Electron.* **20,** 491. 3.5.2

Haynes, J. R. and Nilsson, N. G. (1965). In *Radiative recombination in semiconductors*, Proc. 1964 Symp., p. 21. Dunod, Paris. 1.2.3

Heasell, E. L. (1979). *Solid-State Electron.* **22,** 89. 6.5 FR

Heasell, E. L. (1981). *Solid-State Electron.* **24,** 889. 5.1.2

Heiblum, M., Nathan, M. I., and Chang, C. A. (1982). *Solid-State Electron.* **25,** 185. 6.5.2

Heijne, L. (1962). *J. phys. Chem. Solids* **22,** 207. 2.1 FR

Heime, K. (1970a). *Solid-State Electron.* **13,** 710. 4.1 FR

Heime, K. (1970b). *Solid-State Electron.* **13,** 1505. 2.1 FR

Heime, K. (1972). *Z. angew. Phys.* **32,** 374. 4.1 FR

Heine, V. (1965). *Phys. Rev.* **138,** A1689. 1.3.3

Henisch, H. K. (1957). *Rectifying semiconductor contacts.* Clarendon Press, Oxford. 1.3.4, 2.2.1, 6.7.1

Henisch, H. K. and Ewels, J. (1950). *Proc. Phys. Soc.* **B63,** 861. 2.3.7

Henisch, H. K., Manifacier, J.-C., Callarotti, R. C., and Schmidt, P. E. (1980). *J. appl. Phys.* **51,** 3790. 6.3.2

Henisch, H. K., Manifacier, J.-C., Moreau, Y., and Park, J. W. (1982). *Solar Energy Mat.* **8,** 91. 3.5.3

Henisch, H. K. and Marathe, B. R. (1960). *Proc. Phys. Soc.* **76,** 782. 1.2.1, 1.2.3

Henisch, H. K. and Noble, W. P., Jr. (1966). *Surf. Sci.* **4,** 486. 2.1.7, 6.7.1

Henisch, H. K., Pryor, R. W., and Vendura, G. J. (1972). *J. non-cryst. Solids*, **8–10,** 415. 6.2.2

Henisch, H. K., Rahimi, S., Moreau, Y., and Szepessy, L. *Solid-State Electron.* (1984). (In press) 2.4.3

Henisch, H. K. and Roy, R. (1969). (Eds) *Proc. Int. Conf. Silicon Carbide.* Pergamon Press, Oxford. 6.4 FR

Henisch, H. K. and Tipple, P. M. (1954). *Proc. Phys. Soc.* **A67,** 651. 6.2.1

Hess, K., Neugroschel, A., Shiue, C. C., and Sah, C. T. (1975). *J. appl. Phys.* **46,** 1721. 6.1.1

Hess, K. and Sah, C. T. (1974). *J. appl. Phys.* **45,** 1254. 3.1 FR

Hesse, K. and Strack, H. (1972). *Solid-State Electron.* **15,** 767. 4.2.1

Hodges, D. T. and McColl, M. (1977). *Appl. Phys. Lett.* **30,** 5. 4.2 FR

Hoefflinger, B. (1966). *I.E.E.E. Trans. Electron. Dev.* **ED-13,** 151. 6.6.2

Hoffmann, A. (1950). *Z. angew. Phys.* **2,** 353. 4.1.1

Hoffmann, A., Rose, F., Walddötter, E., and Nitsche, E. (1950). *Z. Naturforsch.*, **5a,** 465. 2.3 FR

Hofstein, S. R. and Warfield, G. (1965). *Solid-State Electron.* **8,** 321. 4.2 FR

Hohnke, D. K. and Holoway, H. (1974). *Appl. Phys. Lett.* **24,** 633. 2.1 FR

Hökelek, E. and Robinson, G. Y. (1981). *Solid-State Electron.* **24,** 99. 1.3 FR

Holm, R. (1946). *Electric contacts.* Hugo Gerbers Förlag, Stockholm. 1.1.3

Mentioned in

Holm, R. (1951). *J. appl. Phys.* **22,** 569. 1.3.4

Holm, R. and Kirchstein, B. (1935). *Z. tech. Physik* **16,** 488. 1.3.4

Hosack, H. H. (1972). *Appl. Phys. Lett.* **21,** 256. 2.1 FR

Howorth, J. R., Harmer, A. L., Trawny, E. W. L., Holton, R., and Sheppard, C. J. R. (1973). *Appl. Phys. Lett.* **23,** 123. 2.1.5

Hsu, S. T. (1972). *Appl. Phys. Lett.* **20,** 20. 6.6.2

Huang, C. H. and Crowell, C. R. (1976). *J. Vac. Sci. Technol.* **13,** 876. 3.1 FR

Hunter, L. P. (1951). *Phys. Rev.* **81,** 151. 6.2.1

Ihm, J., Lovie, S. G., and Cohen, M. L. (1978). *Phys. Rev. Lett.* **40,** 1208. 2.1 FR

Iliadis, A. and Singer, K. E. (1983). *Solid-State Electron.* **26,** 7. 6.5 FR

Illegems, M. and Queisser, H. J. (1975). *Phys. Rev.* **B12,** 1443. 3.3.2

Inkson, J. C. (1971). *J. Phys.* **C4,** 591. 2.1.5

Itoh, M., Suzuki, S., Itoh, T., Yamamoto, Y., and Stephens, K. G. (1980). *Solid-State Electron.* **23,** 447. 6.5 FR

Jaffé, G. (1952). *Phys. Rev.* **85,** 354. 2.3.7

Jäger, H. (1969). *Solid-State Electron.* **12,** 85. 2.3.6

Jäger, H. and Kosak, W. (1969). *Solid-State Electron.* **12,** 511. 2.1 FR

Janney, R., Seibt, W., and Norde, H. (1976). *Solid-State Electron.* **19,** 645. 5.1 FR

Jaros, M. and Hartnagel, H. L. (1975). *Solid-State Electron.* **18,** 1029. 6.5 FR

Johnson, W. C. and Panousis, P. J. (1971). *I.E.E.E. Trans. Electron. Dev.* **ED-18,** 965. 4.1.2

Jonscher, A. K. (1968). In *Problems in solid state physics* (ed. H. J. Goldsmid), p. 392. Academic Press, New York; Pion Ltd., London. 2.1.2

Jonscher, A. K. (1983). *Dielectric relaxation in solids.* Chelsea Dielectrics Press, London. 4.1.1

Kahng, D. (1963). *Solid-State Electron.* **6,** 281. 2.1.5, 2.1.7, 2.3.8

Kahng, D., Brews, J. R., and Sundburg, W. J. (1977). *I.E.E.E. Trans. Electron Dev.* **ED-24,** 531. 3.5.5

Kajiyama, K., Mizushima, Y., and Sakata, S. (1973). *Appl. Phys. Lett.* **23,** 458. 2.1 FR

Kajiyama, K., Onuki, K., and Seki, Y. (1979). *Solid-State Electron.* **22,** 525. 6.3.1

Kallmann, H. and Pope, M. (1960a). *J. chem. Phys.* **32,** 300. 3.6.3

Kallmann, H. and Pope, M. (1960b). *Nature* **186,** 31. 3.6.3

Kaneda, T., Takanashi, H., Matsumoto, H., and Yamaoka, T. (1976). *J. appl. Phys.* **47,** 4960. 6.6.1

Kaplan, G. (1980). *Solid-State Electron.* **23,** 513. 4.2.1

Kaplan, G. (1981). *Solid-State Electron.* **24,** 88. 4.3.1

Kar, S. (1974). *Appl. Phys. Lett.* **25,** 587. 3.5.5

Kar, S. and Dahlke, W. E. (1971). *Appl. Phys. Lett.* **18,** 401. 3.5.5

Kar, S. and Dahlke, W. E. (1972). *Solid-State Electron.* **15,** 221. 4.1.1

Kellner, W., Enders, N., Ristow, D., and Kniekamp, H. (1980). *Solid-State Electron.* **23,** 9. 4.2 FR

Kennedy, D. P., Murly, P. C., and Kleinfelder, W. (1968). *IBM J. Res. Dev.* **12,** 399. 4.1.2

Kennedy, D. P. and O'Brien, R. R. (1969). *IBM J. Res. Dev.* **13,** 212. 4.1.2

Kepler, R. G. (1960). *Phys. Rev.* **119,** 1226. 3.6.3

Mentioned in

Kessler, A. (1981). *J. Phys.* **C14,** 4357. 6.3 FR

Kiess, H. and Rose, A. (1973). *Phys. Rev. Lett.* **31,** 153. 3.3.2 3.4.1

Kingston, R. H. (1957). (Ed.) *Semiconductor surface physics.* University of Pennsylvania Press, Philadelphia. 6.1.1

Kingston, R. H. and Neustadter, S. F. (1955). *J. appl. Phys.* **26,** 718. 6.1.1

Kircher, C. J. (1971). *Solid-State Electron.* **14,** 507. 1.3.3 2.1 FR

Klarmann, H. (1939). *Wiss. Veröff. Siemens Werke* **18,** 198. 1.3.4

Klimpke, C. M. H. and Landsberg, P. T. (1981). *Solid-State Electron.* **24,** 401. 5.2.2

Kober, H. (1957). *Dictionary of conformal representations.* Dover Publishing Co., New York. 1.1.4

Kolk, J. and Heasell, E. L. (1980a). *Solid-State Electron.* **23,** 223. 2.4 FR 6.3.1

Kolk, J. and Heasell, E. L. (1980b). *Solid-State Electron.* **23,** 229. 6.3.1

Korol, A. N., Kitsai, M. Ye., Strikha, V. I., and Sheka, D. I. (1975). *Solid-State Electron.* **18,** 375. 4.1 FR

Korwin-Pawlowski, M. L. and Heasell, E. L. (1975). Solid-*State Electron.* **18,** 849. 2.1 FR 2.2.5

Kottwitz, A., Leimer, F. and Stötzel, E. (1977). In *Amorphous semiconductors* (ed. I. K. Somogyi). Akadénirai Kiodó, Budapest. 3.6 FR

Kramer, P. and Van Ruyven, L. J. (1972a), *Appl. Phys. Lett.* **20,** 420. 6.7.1

Kramer, P. and Van Ruyven, L. J. (1972b). *Solid-State Electron.* **15,** 757. 1.1 FR

Kroemer, H. and Chien, W.-Y. (1981). *Solid-State Electron.* **24,** 655. 4.1.2

Kröger, F. A., Diemer, G., and Klasens, H. A. (1956). *Phys. Rev.* **103,** 279. 6.5.2

Krömer, H. (1953). *Z. Physik* **134,** 435. 2.2.7

Kumar, D. V. and Sharma, S. K. (1981). *Solid-State Electron.* **24,** 91. 5.1.3

Kumar, R., Jindal, S. and Bhattacharya, A. B. (1975). *Solid-State Electron.* **18,** 999. 6.6.1

Kuphal, E. (1981). *Solid-State Electron.* **24,** 69. 6.5 FR

Kurtin, S., McGill, T. C., and Mead, C. A. (1969). *Phys. Rev. Lett.* **22,** 1433. 1.3.3 2.1.6

Kurtin, S. and Mead, C. A. (1968). *J. phys. Chem. Solids* **29,** 1865. 2.1.7

Kurtz, S. K. and Warter, P. J., Jr. (1966). *Bull. Am. Phys. Soc.* **11,** 34. 2.1.1

Kuznicki, Z. T. (1976). *Solid-State Electron.* **19,** 894. 6.4 FR

Laflere, W. H., van Meirhaeghe, R. L., and Cardon, F. (1962). *Solid-State Electron.* **19,** 759. 3.5.5

Lampert, M. A. (1956). *Phys. Rev.* **103,** 1648. 6.3.1

Lampert, M. A. and Mark, P. (1970). *Current injection in solids.* Academic Press, New York. 1.2.4 3.3.2 6.3.1

Landauer, R. and Swanson, J. A. (1953). *Phys. Rev.* **91,** 555. 3.2.1

Landsberg, P. T. (1949). *Nature* **164,** 967. 2.2.8

Landsberg, P. T. (1951a). *Proc. R. Soc.* **A206,** 463. 2.2.4 2.2.5 2.2.8

Mentioned in

Landsberg, P. T. (1951b). *Proc. R. Soc.* **A206,** 477. — 2.3.7

Landsberg, P. T. (1952). *Proc. R. Soc.* **A213,** 226. — 2.2.2, 2.2.5

Landsberg, P. T. (1954). *Univ. Pennsylvania Tech. Report No.* 10 (N6-onr-24914). — 2.2.7

Landsberg, P. T. (1955). *Proc. Phys. Soc.* **B68,** 366. — 2.2.7

Landsberg, P. T. (1975). *Solid-State Electron.* **18,** 1043. — 5.1.3

Landsberg, P. T. and Hope, S. A. (1977). *Solid-State Electron.* **20,** 421. — 2.2 FR

Landsberg, P. T. and Klimpke, C. M. (1977). *Proc. R. Soc.* **A354,** 101. — 5.2 FR

Landsberg, P. T. and Klimpke, C. M. (1980). *Solid-State Electron.* **23,** 1139. — 5.2.2

Lang, D. V. (1974). *J. appl. Phys.* **45,** 3023. — 4.1 FR

Lavagna, M., Pique, J. P., and Marfaing, Y. (1977). *Solid-State Electron.* **20,** 235. — 5.1.2

Learn, A. J., Scott-Monck, J. A., and Spriggs, R. S. (1966). *Appl. Phys. Lett.* **8,** 144. — 2.1 FR, 6.3 FR

Leblanc, O. (1960). *J. chem. Phys.* **33,** 626. — 3.6.3

Lee, K. and Nussbaum, A. (1980). *Solid-State Electron.* **23,** 655. — 3.5 FR

Lee, S. (1969). *Solid-State Electron.* **12,** 299. — 5.1.1

Lee, S. and Henisch, H. K. (1968). *Solid-State Electron.* **11,** 301. — 4.1 FR, 5.1.1

Lee, S. and Henisch, H. K. (1969). *J. phys. Chem. Solids* **30,** 1286. — 4.3 FR, 5.1.1

Lee, S. and Pearson, G. L. (1980). *I.E.E.E. Trans. Electron. Dev.,* **ED-27,** 844. — 5.1.3, 5.1 FR, 6.4.2

Lehovec, K. (1949). *J. appl. Phys.* **20,** 123. — 4.1.1

Lehovec, K. (1951). *J. appl. Phys.* **22,** 934. — 2.3 FR

Lehovec, K. (1952). *Proc. I.R.E.* **40,** 1407. — 1.2.3

Lei, T. F., Lee, C. L., and Chang, C. Y. (1978). *Solid-State Electron.* **21,** 385. — 2.1 FR

Lei, T. F., Lee, C. L., and Chang, C. Y. (1979). *Solid-State Electron.* **22,** 1035. — 2.1 FR

Lemke, H. (1970). *Phys. stat. solidi* **(a)42,** 885. — 3.3 FR

Lempicki, A. and Wood, C. (1954). *Proc. Phys. Soc.* **A67,** 328. — 6.2.1

Lepek, A., Halperin, A., and Levinson, J. (1976). *J. Lumin.* **12/13,** 897. — 1.2 FR

Lepselter, M. P. and Andrews, J. M. (1969). In *Ohmic contacts to semiconductors* (ed. B. Schwartz), p. 159. Electrochemical Society, New York. — 1.3.3, 6.5.2

Lepselter, M. P. and Sze, S. M. (1968). *Bell Syst. Tech. J.* **48,** 195. — 2.3.8

Levine, D. (1975). *Crit. Rev. Solid State Sci.* **5,** 597. — 6.4.3

Levine, J. D. (1971). *J. appl. Phys.* **42,** 3991. — 2.3.3, 2.3.5

Levinson, J., Halperin, A., and Bar, V. (1973). *J. Lumin.* **6,** 1. — 1.2 FR

Levinson, L. M. and Philipp, H. R. (1974). *Appl. Phys. Lett.* **24,** 75. — 6.4.3

Levinson, L. M. and Philipp, H. R. (1975). *J. appl. Phys.* **46,** 1332. — 6.4.3

Levinson, L. M. and Philipp, H. R. (1976). *J. appl. Phys.* **47,** 1117. — 6.4.3

Lewicki, G. W. and Mead, C. A. (1966). *Appl. Phys. Lett.* **8,** 98. — 6.3.2

Li, S. S. (1978). *Solid-State Electron.* **21,** 435. — 5.1.3

Mentioned in

Lillington, D. R. and Townsend, W. G. (1976). *Appl. Phys. Lett.* **28,** 97. 5.2.2

Lim, G. C. and Leaver, K. D. (1980). *Solid-State Electron.* **23,** 935. 2.3.6

Lindau, I., Shye, P. W., Garner, M., Pianetta, P., Su, C. Y., and Spicer, W. E. (1978). *J. Vac. Sci. Technol.* **15(4),** 1332. 1.3.3 2.1.7

Livingstone, A. W., Turvey, K., and Allen, J. W. (1973). *Solid-State Electron.* **16,** 351. 1.2 FR

Logothetis, E. M., Holloway, H., Varga, A. J., and Wilkes, E. (1971). *Appl. Phys. Lett.* **19,** 318. 5.1 FR

Losee, D. L. (1972). *Appl. Phys. Lett.* **21,** 54. 4.3 FR

Lossew, O. W. (1940). *C. R. Acad. Sci. U.S.S.R.* **29,** 363. 6.4.1

Lou, L. F. (1979). *J. appl. Phys.* **50,** 555. 6.4.2 6.4.3

Lou, L. F. (1980). *Appl. Phys. Lett.* **36,** 570. 6.4.1

Low, G. G. E. (1955). *Proc. Phys. Soc.* **B68,** 310. 1.2.1

Lubberts, G. and Burkey, B. C. (1975). *Solid-State Electron.* **18,** 805. 2.1.1 2.3.7 4.1.2

Ludeke, R. and Rose, K. (1983). (Eds) *Interfaces and Contacts.* North-Holland, New York. 2.2 FR

Ludman, J. E. and Nowak, W. B. (1976). *Solid-State Electron.* **19,** 759. 3.5 FR

Lue, J. T. (1980). *Solid-State Electron.* **23,** 263. 5.2.1

Lue, J. T. (1982). *Solid-State Electron.* **25,** 869. 5.2.2

Lue, J. T. and Hong, Y. D. (1978). *Solid-State Electron.* **21,** 1213. 5.1.3

Lundstrom, M. S. Schwartz, R. J. and Gray, J. L. (1981). Solid-*State Electron.* **24,** 195. 6.5.2

McColl, M., Hodges, D. T., and Garber, W. A. (1977). *I.E.E.E. Trans. Microwave Theory Tech.* **MIT-25,** 463. 4.2 FR

McColl, M. and Millea, M. F. (1976). *J. Electron. Mat.* **5,** 191. 2.2.5 4.1.2 6.5.2

McColl, M., Millea, M. F., and Mead, C. A. (1971). *Solid-State Electron.* **14,** 677. 6.5.2

McColl, M., Millea, M. F., and Silver, A. H. (1973). *Appl. Phys. Lett.* **23,** 263. 4.2.2

McColl, M., Pedersen, R. J., Bottjer, M. F., Millea, M. F., Silver, A. H., and Vernon, F. L., Jr. (1976). *Appl. Phys. Lett.* **28,** 159. 4.2.2

Macdonald, J. R. (1954). *J. chem. Phys.* **22,** 1317. 2.1.1

Macdonald, J. R. (1958). *J. chem. Phys.* **24,** 1346. 2.1.1 6.1.1

Macdonald, J. R. (1959). *J. chem. Phys.* **30,** 806. 2.1.1

Macdonald, J. R. (1962). *Solid-State Electron.* **5,** 11. 2.2.9 2.4.2 4.1.3

McDougall, J. and Stoner, E. C. (1938). *Phil. Trans. R. Soc.* **237,** 67. 6.5.1

McGill, J., Wilson, J. I. B., and Kinmond, S. (1979). *J. appl. Phys.* **50,** 548. 5.2.2

McKelvey, J. P., Longini, R. L., and Brody, T. P. (1961). *Phys. Rev.* **123,** 51. 1.1.2

McNutt, M. J. and Sah, C. T. (1974). *Solid-State Electron.* **17,** 377. 4.1.6

Macpherson, A. C. and Day, H. M. (1972). *Solid-State Electron.* **15,** 409. 4.2 FR

Mentioned in

McQuat, R. F. and Pulfrey, D. L. (1976). *J. appl. Phys.* **47,** 2113. — 5.1.3

Madan, A., LeComber, P. G., and Spear, W. E. (1976). *J. non-cryst. Films* **20,** 239. — 6.1.2

Mahan, G. D., Levinson, L. M., and Philipp, H. R. (1979). *J. appl. Phys.* **50,** 2799. — 6.4.1, 6.4.3

Malhotra, A. K. and Neudek, G. W. (1975). *J. appl. Phys.* **46,** 2690. — 6.1.2

Mandelkorn, J. and Lamneck, J. H., Jr. (1973). *J. appl. Phys.*, **44,** 4785. — 5.1.2

Manifacier, J.-C. (1979). Personal communication. — 3.1.3

Manifacier, J.-C. and Fillard, J. P. (1976). *Solid-State Electron.* **19,** 289. — 2.2.5, 3.2.1

Manifacier, J.-C. and Henisch, H. K. (1978a). *Phys. Rev.* **B17,** 2640. — 3.2.1, 3.2.2, 3.2.3, 3.3.2

Manifacier, J.-C. and Henisch, H. K. (1978b). *Phys. Rev.* **B17,** 2648. — 3.2.1, 3.2.4, 3.3.2

Manifacier, J.-C. and Henisch, H. K. (1978c). *Nature* **272,** 521. — 3.2.3

Manifacier, J.-C. and Henisch, H. K. (1979). *Solid-State Electron.* **22,** 279. — 1.2.2, 3.2.2, 3.4.1

Manifacier, J.-C. and Henisch, H. K. (1980a). *J. phys. Chem. Solids* **41,** 1285. — 3.3.2, 3.6.1

Manifacier, J.-C. and Henisch, H. K. (1980b). *J. non-cryst. Solids* **35–36,** 117. — 3.2.3

Manifacier, J.-C. and Henisch, H. K. (1981). *J. appl. Phys.* **52,** 5195. — 3.3.2, 3.5.2

Manifacier, J.-C., Henisch, H. K., and Gasiot, J. (1979). *Phys. Rev. Lett.* **43,** 708. — 3.3.2

Manifacier, J.-C., Jimenez, J., Fillard, J. P., and Henisch, H. K. (1980). *Solid-State Electron.* **23,** 197. — 3.5.2

Manifacier, J.-C., Moreau, Y., and Henisch, H. K. (1983). Part I. *Solid-State Electron.*, **26,** 795. — 3.5.3

Manifacier, J.-C., Moreau, Y., Henisch, H. K., and Rieder, G. (1984). *J. appl. Phys.* (In press). — 5.1.6

Many, A. (1954). *Proc. Phys. Soc.* **B67,** 9. — 3.1.4, 3.2.1

Many, A. (1955). *Physica* **20,** 905. — 3.1.4

Many, A., Goldstein, Y., and Grover, N. B. (1965). *Semiconductor surfaces.* Interscience/Wiley, New York. — 1.3.2, 2.1.2, 3.1.1, 6.1.1

Many, A., Grover, N. B., and Harnik, E. (1960). In *Semiconductor surfaces* (ed. T. N. Zemel). Pergamon Press, New York. — 6.1.1

Mar, H. A., Simmons, J. G., and Taylor, G. W. (1977). *Solid-State Electron.* **30,** 241. — 3.5 FR

Mentioned in

Marfaing, Y., Loscaray, J., and Triboulet, R. (1974). *Metal–semiconductor contacts* (Conf. Proc.). Institute of Physics, London. 2.1.8 4.3.1

Marshak, A. H. and van Vliet, K. M. (1978). *Solid-State Electron.* **21,** 417. 6.4.2

Matsuoka, M. (1971). *Jap. J. appl. Phys.* **10,** 736. 6.4.3

Maue, A. W. (1935). *Z. Phys.* **94,** 717. 1.3.2

Meacham, L. A. and Michaels, S. E. (1950). *Phys. Rev.* **78,** 175. 3.4.2

Mead, C. A. (1965). *Appl. Phys. Lett.* **6,** 103. 1.3.3 2.1.7

Mead, C. A. (1966a). *Solid-State Electron.* **9,** 1023. 1.3.3 2.1.7

Mead, C. A. (1966b). *Solid-State Electron.* **9,** 1128. 1.3.3 2.1.7

Mead, C. A. (1969). In *Ohmic contacts to semiconductors* (ed. B. Schwartz), p. 3. Electrochemical Society, New York. 6.5.2

Mead, C. A., Snow, E. H., and Deal, B. E. (1966). *Appl. Phys. Lett.* **9,** 53. 1.3.1

Mead, C. A. and Spitzer, W. G. (1963a). *Phys. Rev. Lett.* **10,** 471. 1.3.3

Mead, C. A. and Spitzer, W. G. (1963b). *Phys. Rev. Lett.* **11,** 358. 1.3.3

Mead, C. A. and Spitzer, W. G. (1964). *Phys. Rev.* **134(3A),** A713. 1.3.3

Mehbod, M., Thijs, W., and Bruywseraede, Y. (1975). *Phys. stat. solidi* **(a)32,** 203. 2.2 FR

Mehely, M. A. (1968). *Int. J. Electron.* **24,** 41. 6.4.2

Mehta, R. R. and Sharma, B. S. (1973). *J. appl. Phys.* **44,** 325. 5.1.2

Meirsschaut, S. (1976). *Solid-State Electron.* **7,** 633. 2.2 FR

Meyer, W. and Neldel, H. (1937). *Z. techn. Physik,* **12,** 588. 2.3.8

Meyerhof, W. E. (1947). *Phys. Rev.* **71,** 727. 1.3.3

Migliorato, P., Margaritondo, G., and Perfetti, P. (1976). *J. appl. Phys.* **47,** 656. 3.3.2

Mil'Shtein, S. Kh. (1975). *J. appl. Phys.* **46,** 3894. 6.7.1

Mil'Shtein, S. Kh. and Nikitenko, V. (1974). *Sov. Phys. solid State* **16,** 346. 6.7.1

Millea, M. F., McColl, M., and Mead, C. A. (1969). *Phys. Rev.* **177,** 1164. 4.1.2 4.1.3 6.5.2

Milnes, A. G. and Feucht, D. L. (1972). *Heterojunctions and metal–semiconductor junctions.* Academic Press, New York/London. 6.3.1 6.4.2

Minucci, J. A., Kirkpatrick, A. R., and Mathei, K. W. (1980). *I.E.E.E. Trans. Electron. Dev.* **ED-27,** 802. 5.1 FR

Mitchell, E. W. J. and Sillars, R. W. (1949). *Proc. Phys. Soc.* **B62,** 509. 6.4.1

Miyauchi, T., Sonomura, H., and Yamamoto, N. (1969). *Jap. J. appl. Phys.* **8,** 711. 1.2 FR

Montojo, M. J. and Sanchez, C. (1974). *Solid-State Electron.* **35,** 1437. 3.3 FR

Montrimas, E. (1976). In *Amorphous semiconductors* (ed. A. Kósa Somogyi). Akadémiai Kiado, Budapest. 3.6.2

Moreau, Y. (1983a). Personal communication I. 5.1.5

Moreau, Y. (1983b). Personal communication II. 2.4.3

Moreau, Y. and Manifacier, J.-C. (1980). Personal communication. 2.4.1

Mentioned in

Moreau, Y., Manifacier, J.-C., and Henisch, H. K. (1981). *Solid-State Electron.* **24,** 883. 5.1.1

Moreau, Y., Manifacier, J.-C., and Henisch, H. K. (1982). *Solid-State Electron.* **25,** 137. 2.1.2 2.4.2

Moreau, Y., Manifacier, J.-C., and Henisch, H. K. (1984). Part II. *Solid-State Electron* **27,** 255. 3.5.2

Morel, D. L., and Moustakas, T. D. (1981). *Appl. Phys. Lett.* **39,** 612. 5.1.2

Mott, N. F. (1939). *Proc. R. Soc.* **A171,** 27, 144. 2.1.4 2.2.3

Mott, N. F. and Davies, E. A. (1979). *Electronic processes in non-crystalline materials.* Clarendon Press, Oxford. 3.6.1

Mott, N. F. and Gurney, R. W. (1940). *Electronic processes in ionic crystals.* Oxford University Press, Oxford. 1.2.4 6.3.1

Mott, N. P. and Sneddon, I. N. (1948). *Wave mechanics and its application,* p. 23 *et seq.* Oxford University Press. 2.1.6

Moustakas, T. D., Friedman, R., and Weinberger, B. R. (1982). *Appl. Phys. Lett.* **40,** 587. 5.1.2

Moustakas, T. D., Wronski, C. R., and Tiedje, T. (1981). *Appl. Phys. Lett.* **39,** 721. 5.1 FR

Muller, M. W. (1968). *Appl. Phys. Lett.* **12,** 218. 6.6.2

Muraka, S. P. (1974). *Solid-State Electron.* **17,** 985. 2.1 FR

Murphy, E. L. and Good, R. H., Jr. (1956). *Phys. Rev.* **102,** 1464. 2.1.6 2.3.4

Murrmann, H. and Widmann, D. (1969). *Solid-State Electron.* **12,** 879. 1.1.4

Müser, H. A. (1957). *Z. Phys.* **148,** 380. 5.1.2

Myszkowski, A., Sansores, L. E., and Tangueña-Martinez, J. (1981). *J. appl. Phys.* **52,** 4288. 5.2.3

Nagata, S. and Agata, K. (1951). *J. Phys. Soc. Jap.* **6,** 523. 1.3.4

Nakamura, K., Olowolafe, J. O., Lau, S. S., Nicolet, M.-A., and Mayer, J. W. (1976). *J. appl. Phys.* **47,** 1278. 2.1 FR

Namordi, M. R. and Thompson, H. W., Jr. (1975). *Solid-State Electron.* **18,** 499. 5.1 FR

Nasledov, D. N. and Yashukova, I. M. (1960). *Sov. Phys. solid State* **1,** 1087. 6.2.1

Nathan, M. I. and Heiblum, M. (1982). *Solid-State Electron.* **25,** 1063. 6.5 FR

Neamen, D. A. and Grannemann, W. W. (1971). *Solid-State Electron.* **14,** 1319. 2.1 FR

Neudeck, G. W. and Malhotra, A. K. (1975a). *J. appl. Phys.* **46,** 239. 6.1.2

Neudeck, G. W. and Malhotra, A. K. (1975b). *J. appl. Phys.* **46,** 2262. 6.1.2

Neugebauer, C. A. and Burgess, J. F. (1976). *J. appl. Phys.* **47,** 3182. 3.5.5

Neville, R. C. and Hoeneisen, B. (1975). *J. appl. Phys.* **46,** 350. 4.1.2

Neville, R. C. and Mead, C. A. (1970). *J. appl. Phys.* **41,** 3795. 6.4.3

Newcomb, W. A. (1982). *Amer. J. Phys.* **50,** 601. 6.3.2

Ng, K. K. and Card, H. C. (1981). *J. phys. Chem. Solids* **42,** 719. 5.1.1

Nishida, M. (1979). *I.E.E.E. Trans. Electron. Dev.* **ED-26,** 1081. 4.1.2

Nishizawa, J. and Watanabe, Y. (1957). *Sci. Rep. R.I.T.U. (Tokoku)* **B9,** 39. 6.6 FR

Noble, W. P., Jr., Braun, I., and Henisch, H. K. (1967). *Solid-State Electron.* **10,** 45. 2.1.7 5.1.2

Mentioned in

Noble, W. P., Jr. and Henisch, H. K. (1967). *J. appl. Phys.* **38,** 2472. 2.1.7
Norde, H. and Tove, P. A. (1977). *Vacuum* **27,** 201. 6.4.2
Norman, C. E. and Thomas, R. E. (1980). *I.E.E.E. Trans. Electron. Dev.* **ED-27,** 731. 5.1 FR
Numata, T. (1959). *J. phys. Soc. Jap.* **14,** 902. 6.2.1
Nwachuku, A. and Kuhn, M. (1968). *Appl. Phys. Lett.* **12,** 163. 3.6 FR
O'Connor, J. R. and Smiltens, J. (1960). (Eds) *Silicon carbide* (Proc. Boston Conf. 1959). Pergamon Press, New York/Oxford. 6.4 FR
Ohdomari, I., Tu, K. N., D'Heurle, F. M., Kuan, T. S., and Petersson, S. (1978). *Appl. Phys. Lett.* **33,** 1028. 2.1 FR
Ohl, R. S. (1952). *Bell Syst. Tech. J.* **31,** 104. 6.7.1
Okumura, K. (1974). *J. appl. Phys.* **45,** 5317. 5.2 FR
Okuto, Y. and Crowell, C. R. (1974). *J. Jap. Soc. appl. Phys.* **43,** 390. 6.6 FR
Oldekop, W. (1952). *Z. Physik* **134,** 66. 4.1.1
Oldham, W. G. and Milnes, A. G. (1964). *Solid-State Electron.* **7,** 153. 6.4.2
Olsen, L. C. (1977). *Solid-State Electron.* **20,** 741. 5.2.1, 5.2.2, 5.2.3
O'Reilly, T. J. and DeLucia, J. (1975). *Solid-State Electron.* **18,** 965. 3.3 FR, 6.3.1
Ottaviani, G. and Tu, K. N. (1980). *Phys. Rev. Lett.* **44,** 284. 2.1 FR
Ovshinsky, S. R. (1968). *Phys. Rev. Lett.* **21,** 1450. 3.6.2
Padovani, F. A. (1966). *J. appl. Phys.* **37,** 921. 5.1.3
Padovani, F. A. (1969). *Solid-State Electron.* **12,** 135. 2.2 FR
Padovani, F. A. (1971). In *Semiconductors and semi-metals* (ed. R. K. Willardson and A. C. Beer). Academic Press, New York. 2.3 FR
Padovani, F. A. and Stratton, R. (1966a). *Phys. Rev. Lett.* **16,** 1202. 2.3.4, 6.5.2
Padovani, F. A. and Stratton, R. (1966b). *Solid-State Electron.* **9,** 965. 2.1.6, 2.3.1, 2.3.4
Panayotatos, P. and Card, H. C. (1980). *Solid-State Electron.* **23,** 41. 5.1.4
Pande, P. K. (1980). *I.E.E.E. Trans. Electron. Dev.* **ED-27,** 631. 5.2 FR
Pankove, J. I. (1971). *Optical processes in semiconductors.* Dover Publications, New York. 1.2.3
Paracchini, C. (1971). *Phys. Rev.* **B4,** 2342. 6.6.2
Park, J. W. (1982). Personal communication. 2.4.1
Parker, G. H., McGill, T. C., Mead, C. A., and Hoffman, D. (1968). *Solid-State Electron.* **11,** 201. 1.3.3, 2.2.8
Parker, G. H. and Mead, C. A. (1969). *Appl. Phys. Lett.* **14,** 21. 2.1 FR
Parker, P. C., Sirrine, R. C., and Lemieux, P. (1976). *Solid-State Electron.* **19,** 493. 2.1 FR
Parrott, J. E. (1971). *Solid-State Electron.* **14,** 885. 3.2.2
Parrott, J. E. (1974). *Solid-State Electron.* **17,** 707. 3.2.2
Paul, W., Lewis, A. J., Connell, G. A. N., and Moustakas, T. D. (1976). *Solid-State Electron.* **20,** 969. 3.6.2
Pauwells, H. and De Vos, A. (1981). *Solid-State Electron.* **24,** 835. 5.1.3
Pearton, S. J. (1982). *Solid-State Electron.* **25,** 499. 4.3 FR
Pell, E. M. (1953). *Phys. Rev.* **90,** 278. 3.4.2
Pellegrini, B. (1974). *Solid-State Electron.* **17,** 217. 1.3.3

	Mentioned in
Pellegrini, B. (1975a). *Solid-State Electron.* **18,** 417.	2.1 FR 2.3.8
Pellegrini, B. (1975b). *Solid-State Electron.* **18,** 887.	6.4.2
Pellegrini, B. and Salardi, G. (1975). *Phys. Rev.* **18,** 791.	6.5.2
Pellegrini, B. and Salardi, G. (1978). *Solid-State Electron.* **21,** 465.	4.1.2
Perel', V. I. and Efros, A. L. (1968). *Sov. Phys. Semiconductors* **1,** 1403.	4.2.1
Persky, G. (1972). *Solid-State Electron.* **15,** 1345.	2.2.6
Pfann, W. G. and Garrett, C. G. B. (1959). *Proc. I.R.E.* **47,** 2011.	6.1.1
Pfeifer, J. (1976). *Solid-State Electron.* **19,** 927.	6.5 FR
Pfotzer, G. (1949). *Z. Naturforsch.* **4a,** 691.	4.1 FR
Philipp, H. R. and Levinson, L. M. (1975), *J. appl. Phys.* **46,** 3206.	6.4.3 6.4.4
Philipp, H. R. and Levinson, L. M. (1977). *J. appl. Phys.* **48,** 1621.	6.4.3
Pickett, W. E. and Cohen, M. L. (1978). *Solid-State Commum.* **25,** 225.	6.4.2
Pickett, W. E., Louie, S. G., and Cohen, M. L. (1977). *Phys. Rev. Lett.* **39,** 109.	6.4.2
Pike, G. E. and Sweet, J. N. (1975). *J. appl. Phys.* **46,** 4904.	2.3.8
Piotrowska, A., Guivarch, A., and Pelous, G. (1983). *Solid-State Electron.* **26,** 179.	6.5.1
Piper, W. W. and Williams, F. E. (1952). *Phys. Rev.* **87,** 151.	6.6 FR
Ponpon, J. P. and Siffert, P. (1976). *J. Appl. Phys.* **47,** 3248.	5.2.2
Ponpon, J. P. and Siffert, P. (1978). *J. appl. Phys.* **49,** 6004.	2.1 FR
Popescu, C. (1981). *Rev. Roum. Phys.* **26,** 979.	3.3 FR 3.5 FR
Popescu, C. and Henisch, H. K. (1975). *Phys. Rev.* **B11,** 1563.	3.2.2 3.2.3 3.3.1 3.3.2
Popescu, C. and Henisch, H. K. (1976a). *J. phys. Chem. Solids* **37,** 47.	3.2.3 5.1.7
Popescu, C. and Henisch, H. K. (1976b). *Phys. Rev.* **B14,** 517.	3.2.1 3.2.4 3.3.2
Popović, R. S. (1978). *Solid-State Electron.* **21,** 1133.	6.5.1
Popović, Z. D. (1979). *Appl. Phys. Lett.* **34,** 694.	4.1.2
Pounds, R. S., Saifi, M. A., and Hahn, W. C., Jr. (1974). *Solid-State Electron.* **17,** 245.	6.5 FR
Pruniaux, B. R. (1971). *J. appl. Phys.* **42,** 3575.	6.5.2
Pryor, R. (1981). *J. appl. Phys.* **52,** 3702.	3.5.5
Pugh, D. (1964). *Phys. Rev. Lett.* **12,** 390.	1.3.3
Pulfrey, D. L. (1976). *I.E.E.E. Trans. Electron. Dev.* **ED-23,** 587.	5.2.2
Pulfrey, D. L. (1978). *Solid-State Electron.* **21,** 519.	5.1.3
Pulfrey, D. L. and McQuat, R. F. (1974). *Appl. Phys. Lett.* **24,** 167.	5.1.3
Queisser, H. J., Casey, H. C., and Van Roosbroeck, W. (1971). *Phys. Rev. Lett.* **26,** 551.	3.3.1
Quinn, J. J. and Stiles, P. J. (1976). (Eds) *Electronic properties of quasi-two-dimensional systems.* North-Holland Publishing Co., Amsterdam.	2.1.6 6.1.1
Rahimi, S. and Henisch, H. K. (1981). *Appl. Phys. Lett.* **38,** 896.	1.3.2
Rahimi, S., Manifacier, J.-C., and Henisch, H. K. (1981). *J. appl. Phys.* **52,** 6723.	1.2.2

Mentioned in

Rajkanan, K. and Shewchun, J. (1979). *Solid-State Electron.* **22,** 193. 5.1.3

Rajkanan, K., Singh, R., and Shewchun, J. (1980). *I.E.E.E. Trans. Electron. Dev.* **ED-27,** 250. 5.2.1

Ramachandran, T. B., and Santosuosso, R. P. (1966). *Solid-State Electron.* **9,** 733. 6.5 FR

Redfield, D. (1980). *I.E.E.E. Trans. Electron. Dev.* **ED-27,** 766. 5.1 FR

Reed, C. E. and Scott, C. G. (1975). *Solid-State Electron.* **18,** 29. 5.1.4

Reith, T. M. (1976). *Appl. Phys. Lett.* **28,** 192. 2.1 FR

Reith, T. M. and Schick, J. D. (1974). *Appl. Phys. Lett.* **25,** 524. 2.1 FR

Restelli, G. and Rota, A. (1968). In *Semiconductor detectors* (ed. G. Restelli and A. Rota). North-Holland Publishing Co., Amsterdam. 5.1 FR

Rhoderick, E. H. (1975). *J. appl. Phys.* **46,** 2809. 2.1 FR, 2.3 FR

Rhoderick, E. H. (1978). *Metal-semiconductor contacts.* Clarendon Press, Oxford. 2.1.2, 2.2.5, 4.1.6

Rhodes, P. (1950). *Proc. R. Soc.* **A204,** 396. 6.5.1

Riben, A. R. and Feucht, D. L. (1966). *Int. J. Electron.* **20,** 583. 6.4.2

Rideout, V. L. (1974). *Solid-State Electron.* **17,** 1107. 2.1 FR

Rideout, V. L. (1975). *Solid-State Electron.* **18,** 541. 6.5 FR

Rideout, V. L. (1978). *Thin solid Films* **48,** 261. 1.1.3, 2.2 FR, 2.3.1, 2.3.3

Rideout, V. L. and Crowell, C. R. (1967). *Appl. Phys. Lett.* **10,** 329. 6.7.2

Rideout, V. L. and Crowell, C. R. (1970). *Solid-State Electron.* **13,** 993. 2.1.5, 2.1.6, 2.2.8, 2.3.1, 2.3.3, 4.1.2

Ridley, B. K. (1963). *Proc. Phys. Soc.* **82,** 954. 6.6.2

Rieder, G. (1979). Personal communication. 6.4.1

Rieder, G., Henisch, H. K., Rahimi, S., and Manifacier, J.-C. (1980). *Phys. Rev.* **B21,** 723. 3.2.2, 3.4.2

Rieder, G., Manifacier, J.-C., and Henisch, H. K. (1982). *Solid-State Electron.* **25,** 133. 3.2.4, 3.4.2

Rijanow, S. (1934). *Z. Phys.* **89,** 806. 1.3.2

Robbins, W. P. and Ancker-Johnson, B. (1971). *J. appl. Phys.* **42,** 5776. 3.2.1

Roberts, G. L. and Crowell, C. R. (1970). *J. appl. Phys.* **41,** 1767. 4.3.1, 4.3.2

Roberts, G. L. and Crowell, C. R. (1973). *Solid-State Electron.* **16,** 29. 4.3.1

Robinson, G. Y. (1974). *Appl. Phys. Lett.* **25,** 158. 2.1 FR

Robinson, G. Y. (1975). *Solid-State Electron.* **18,** 331. 2.1.7

Robinson, R. J. and Kun, *Z. K.* (1969). *Appl. Phys. Lett.* **15,** 371. 1.2 FR

Rolland, M., Nougier, J. P., Gasquet, D. and Alabedra, R. (1977). *Solid-State Electron.* **20,** 323. 6.5 FR

Rose, A. (1956). *Helv. Phys. Acta* **29,** 199. 3.6.1, 6.1.1

Mentioned in

Rose, A. (1960). *J. appl. Phys.* **31,** 1640. 5.1.2
Rose, D. J. (1957). *Phys. Rev.* **105,** 413. 6.6.2
Rose, F. (1951). *Ann. Phys.* **9,** 97. 4.1.1
Rose, F. and Schmidt, H. (1947). *Z. Naturforsch.* **2a,** 226. 2.3.7
Rose, F. and Spenke, E. (1949). *Z. Phys.* **126,** 632. 2.2.5
Rosenthal, A. and Sapar, A. (1974). *J. appl. Phys.* **45,** 2787. 6.3.1
Rosenzweig, W., Hackett, W. H., Jr., and Jayson, J. S. (1969). *J. appl. Phys.* **40,** 4477. 1.2 FR
Rothwarf, A. and Pereyra, I. (1981). *Solid-State Electron.* **24,** 1067. 5.2.2
Roy, S. B. and Daw, A. N. (1980). *Solid-State Electron.* **23,** 949. 2.3 FR
Ruppel, W. and Würfel, P. (1980). *I.E.E.E. Trans. Electron. Dev.* **ED-27,** 877. 5.1 FR
Rusu, A., Bulucea, C., and Postolache, C. (1977). *Solid-State Electron.* **20,** 499. 1.1 FR
Ryder, C. J. (1953). *Phys. Rev.* **90,** 766. 2.2.7
Ryder, E. J. and Shockley, W. (1951). *Phys. Rev.* **81,** 139. 2.2.7
Ryvkin, S. M., Tarkhin, D. V., Zinchik, Yu. S., and Sanin, K. V. (1979). *Sov. Tech. Phys. Lett.* **5(9),** 455. 3.2.1
Sah, C. T., Rosier, L. L., and Forbes, L. (1969). *Appl. Phys. Lett.* **15,** 316. 5.1.1
Salardi, G. and Pellegrini, B. (1979). *Solid-State Electron.* **22,** 435. 4.3 FR
Saltich, J. L. and Clark, L. E. (1970). *Solid-State Electron.* **13,** 857. 2.2 FR, 6.6.1
Sanchez, D., Carchano, M., and But, A. (1974). *J. appl. Phys.* **45,** 1233. 3.5 FR
Sarnot, S. L. and Dubey, P. K. (1972). *Solid-State Electron.* **15,** 745. 3.5 FR
Sarrabayrouse, G., Buxo, J., and Esteve, D. (1977). *J. Physique* **38,** 1443. 2.1.6
Sato, T. and Kaneko, H. (1950). *Tech. Rep. Tokyo Univ.* **15,** 1. 1.3.4
Sawyer, D. E. (1968). *App. Phys. Lett.* **13,** 11. 5.1 FR
Saxena, A. N. (1971). *Appl. Phys. Lett.* **19,** 71. 1.3.2, 2.3.8, 6.5 FR
Schade, H., Nuese, C. J., and Gannon, J. J. (1971). *J. appl. Phys.* **42,** 5072. 2.1 FR
Scharfetter, D. L. (1965). *Solid-State Electron.* **8,** 299. 3.1.1, 3.1.4
Schiavone, L. M. and Pritchard, A. A. (1975). *J. appl. Phys.* **46,** 452. 6.5.1
Schibli, E. and Milnes, A. G. (1968). *Solid-State Electron.* **11,** 323. 4.2.1
Schmidlin, F. W., Roberts, G. G., and Lakatos, A. I. (1968). *Appl. Phys. Lett.* **13,** 353. 2.1 FR
Schmidt, A. (1941). *Z. Phys.* **117,** 754. 1.3.1, 1.3.3, 2.3.7
Scholten, P. C. (1966). *Solid-State Electron.* **9,** 1142. 6.5 FR
Schoolar, R. B. (1970). *Appl. Phys. Lett.* **16,** 446. 5.1 FR
Schottky, W. (1938). *Naturwiss.* **26,** 843. 1.1.2, 2.2.2
Schottky, W. (1942). *Z. Physik.* **118,** 539. 2.1.1, 4.1.2
Schottky, W. (1952). *Z. Physik.* **132,** 261. 4.2.1, 4.2.2
Schottky, W. and Deutschmann, W. (1929). *Phys. Z.* **30,** 839. 4.1.1

Mentioned in

Schrieffer, J. R. (1955). *Phys. Rev.* **97,** 641. 3.1.1, 6.1.1

Schuldt, S. B. (1978). *Solid-State Electron.* **21,** 715. 1.1.4

Schultz, W. (1954). *Z. Physik* **138,** 598. 2.3.2

Schultz, W. (1971). *Solid-State Electron.* **14,** 227. 4.2.1

Schwartz, B. (1969). (Ed.) *Ohmic contacts to semiconductors.* Electrochemical Society New York. 6.5.1

Schwarz, R. F. and Walsh, J. F. (1953). *Proc. I.R.E.* **41,** 1715. 4.1.6

Schwob, H. P. and Williams, D. F. (1974). *J. appl. Phys.* **45,** 2638. 6.3 FR

Sebestyén, T. (1976). In *Amorphous semiconductors* (ed. I. Kósa Somogyi). Akadémai Kiadó, Budapest. 6.5.2

Sebestyén, T. (1982). *Solid-State Electron.* **25,** 543. 6.5 FR

Seefeld, von H., Cheung, N. W., Mäewpää, M., and Nicolet, M.-A. (1980). *I.E.E.E. Trans. Electron. Dev.* **ED-27,** 873. 5.2 FR

Seib, D. H. (1971). *Appl. Phys. Lett.* **18,** 422. 5.1.1

Seiwatz, R. and Green, M. (1958). *J. appl. Phys.* **29,** 1034. 3.1.1

Senechal, R. R. and Basinsky, J. (1968). *J. appl. Phys.* **39,** 4581. 4.1.2, 4.2.1

Shannon, J. M. (1974). *Appl. Phys. Lett.* **25,** 75. 2.1 FR

Shannon, J. M. (1976). *Solid-State Electron.* **19,** 537. 2.1 FR, 2.3.7, 6.5.1

Shannon, J. M. (1977a). *Appl. Phys. Lett.* **31,** 707. 2.3.8

Shannon, J. M. (1977b). *Solid-State Electron.* **20,** 869. 2.3.4

Shannon, J. M. (1979). *Appl. Phys. Lett.* **35,** 63. 2.3 FR

Sharma, B. L. and Purohit, R. K. (1974). *Semiconductor heterojunctions.* Pergamon Press, Oxford/New York. 6.4.2

Shewchun, J., Burk, D., and Spitzer, M. B. (1980). *I.E.E.E. Trans. Electron. Dev.* **ED-27,** 873. 5.2 FR

Shewchun, J., Dubow, J., Myszkowski, A., and Singh, R. (1978). *J. appl. Phys.* **49,** 855. 5.2.1, 5.2.3

Shewchun, J., Green, M. A., and King, J. D. (1974). *Solid-State Electron.* **17,** 563. 5.2.2

Shockley, W. (1939). *Phys. Rev.* **56,** 317. 1.3.2, 1.3.3

Shockley, W. (1950). *Electronics and holes in semiconductors.* Van Rostrand, New York. 3.5.0, 6.5.1, 6.6.1

Shockley, W. (1951). *Bell Syst. Tech. J.* **30,** 990. 2.2.7, 6.6.1

Shockley, W. and Queisser, H. J. (1961). *J. appl. Phys.* **32,** 510. 5.1.3

Shockley, W. and Read, W. J. (1952). *Phys. Rev.* **87,** 835. 3.2.4, 6.4.4

Shur, M. (1981). *I.E.E.E. Trans. Electron. Dev.* **ED-28,** 1120. 2.2.7

Shur, M. and Eastman, L. F. (1981). *Solid-State Electron.* **25,** 11. 2.2.7

Sillars, R. W. (1955). *Proc. Phys. Soc.* **B68,** 881. 1.1.3

Silver, M. (1962). In *Organic semiconductors* (ed. J. J. Brophy and J. W. Buttrey). Macmillan Co., New York. 3.6.3

Silver, M., Cohen, L., and Adler, D. (1982). *Appl. Phys. Lett.* **40,** 261. 3.6.2

Silver, M., Giles, N. C., Snow, E., Shaw, M. P., Canella, V., and Adler,

Mentioned in

D. (1982). *Appl. Phys. Lett.* **41,** 935. 3.6 FR
Silver, M. and Moore, W. (1960). *J. chem. Phys.* **33,** 1671. 3.6.3
Silver, M., Schoenherr, G., and Baessler, H. (1982). *Phys. Rev. Lett.* **48,** 352. 3.6 FR
Simmons, J. G. (1963). *J. appl. Phys.* **34,** 1793. 1.3.4, 2.1 FR, 6.3.2, 6.5.2
Simmons, J. E. and Taylor, E. W. (1983). *Solid-State Electron.* **26,** 705. 2.2 FR
Simpson. J. H. and Armstrong, H. L. (1953). *J. appl. Phys.* **24,** 25. 3.5.1
Singh, R., Rajkanan, K., Brodie, D. E., and Morgan, J. H. (1980). *I.E.E.E. Trans. Electron. Dev.* **ED-27,** 656. 5.2 FR
Sinha, A. K. and Poate, J. M. (1973). *Appl. Phys. Lett.* **23,** 666. 2.1 FR
Sinha, A. K., Smith, T. E., Read, M. H., and Poate, J. M. (1976). *Solid-State Electron.* **19,** 489. 2.1 FR
Sinkkonen, J. (1983). *Solid-State Electron.* **26,** 1111. 2.2 FR
Skinner, S. M. (1955a). *J. appl. Phys.* **26,** 498. 2.4.3, 6.3.2
Skinner, S. M. (1955b). *J. appl. Phys.* **26,** 509. 2.4.3, 6.3.2
Sleger, K. and Cristou, A. (1978). *Solid-State Electron.* **21,** 677. 2.1 FR
Slotboom, J. W. (1977). *Solid-State Electron.* **20,** 167. 3.1 FR
Slowik, J. H. (1981a). *Appl. Phys. Lett.* **38,** 269. 3.2 FR
Slowik, J. H. (1981b). *J. Vac. Sci. Technol* **19,** 411. 3.3 FR
Smith, B. L. (1972). *Appl. Phys. Lett.* **21,** 350. 2.1.8, 3.6.2
Smith, B. L. (1974). In *Metal–semiconductor contacts* (Conf. Proc). Institute of Physics, London. 2.1.8, 3.6.2
Smith, B. L. and Rhoderick, E. H. (1971). *Solid-State Electron.* **14,** 71. 2.1 FR
Smith, R. A. (1979). *Semiconductors* (2nd edn). Cambridge University Press, Cambridge 6.5.1
Smith, W. and Henisch, H. K. (1973). *Phys. stat. Solidi* **(a)17,** K81. 6.6.2
Snell, A. J., Mackenzie, K. D., Lecomber, P. G., and Spear, W. E. (1979). *Phil. Mag.* **B40,** 1. 3.6.2, 4.3.1
Socha, J. B. and Eastman, L. F. (1982). *I.E.E.E. Electron. Dev. Lett.* **EDL-3,** 27. 2.2.7
Sommerfeld, A. and Bethe, H. (1933). *Handbuch der Physik*, Vol. 24/2. Springer Verlag, Berlin. 1.3.4
Soukup, R. J. (1975). *J. appl. Phys.* **46,** 463. 6.3.2
Spear, W. E., LeComber, P. G., and Snell, A. J. (1978). *Phil. Mag.* **38,** 303. 3.6.1, 4.3.1, 4.3.2
Spenke, E. (1941). *Wiss. Veröff. Siemeus-Werk.* **20,** 40. 4.1.3
Spenke, E. (1949). *Z. Phys.* **126,** 67. 2.2.5
Spenke, E. (1950). *Z. Phys.* **128,** 586. 4.1.3
Spenke, E. (1955). *Elektronische Halbleiter.* Springer Verlag, Berlin. 2.1.1, 2.2.1, 2.2.2, 3.5.1

Mentioned in

Spicer, W. E., Eglash, S., Lindau, I., Su, C. Y., and Skeath, P. (1982). *Thin solid Films* **89,** 447. 2.1.7

Spicer, W. E., Lindau, I., Skeath, P., Su, C. Y., and Chye, P. (1980). *Phys. Rev. Lett.* **44,** 420. 2.1.7, 2.1 FR

Spitzer, W. G. and Mead, C. A. (1963). *J. appl. Phys.* **34,** 3061. 1.3.3

Spitzer, W. G. and Mead, C. A. (1964). *Phys. Rev.* **133,** A872. 1.3.3

Spitzer, W. G., Trumbore, F. A., and Logan, R. C. (1961). *J. appl. Phys.* **32,** 1822. 1.2.3

Srivastava, A. K., Arora, B. M., and Guha, S. (1981). *Solid-State Electron.* **24,** 185. 2.2 FR

Srivastava, G. P., Bhatnagar, P. K., and Dhariwal, S. R. (1979). *Solid-State Electron.* **22,** 581. 5.2.2

Stafeev, V. I. (1958). *Sov. Phys.: Tech. Phys.* **3,** 1502. 3.2 FR

Statz, H. (1950). *Z. Naturforsch.* **5a,** 534. 1.3.2

Statz, H., deMars, G. A., Davis, L., Jr., and Adams, A., Jr. (1956). *Phys. Rev.* **101,** 1272. 6.1.1

Stirn, R. J. and Yeh, Y. C. M. (1975). *Appl. Phys. Lett.* **27,** 95. 5.2.1

Stoica, T. and Popescu, C. (1978). *Phys. Rev.* **B17,** 3972. 3.3.2

Stoisiek, M., Wolf, D., and Queisser, H. J. (1971). *Appl. Phys. Lett.* **19,** 228. 3.3.2

Stokoe, T. Y. and Parrott, J. E. (1974). *Solid-State Electron.* **17,** 477. 2.2.7

Stratton, R. (1956a). *Proc. Phys. Soc.* **B69,** 491. 1.1.3, 6.4.1

Stratton, R. (1956b). *Proc. Phys. Soc.* **B69,** 513. 6.4.1

Stratton, R. (1962a). *J. phys. Chem. Solids* **23,** 1177. 3.5.5

Stratton, R. (1962b). *Phys. Rev.* **125,** 67. 2.2 FR

Stratton, R. (1964). *Phys. Rev.* **135,** 794. 2.3.4

Stratton, R. (1969). In *Tunneling phenomena in solids* (ed. E. Burstein and S. Lundquist). Plenum Press, New York. 2.1.6

Stratton, R. and Padovani, F. A. (1967). *Solid-State Electron.* **10,** 813. 2.1.6

Studer, B. (1980). *Solid-State Electron.* **23,** 1181. 2.3.7

Suits, G. H. (1957). *J. appl. Phys.* **28,** 454. 6.3.2

Sullivan, L. and Card, H. C. (1974). *J. Phys. D., Appl. Phys.,* **7,** 1531. 6.3.2

Sutherland, J. C. and Hauser, J. R. (1977). *I.E.E.E. Trans. Electron. Dev.,* **ED-24,** 363. 6.4.2

Swanson, J. A. (1954). *J. appl. Phys.,* **25,** 314. 3.1.4, 3.2.1, 3.5.1

Sze, S. M. (1969). *Physics of semiconductor devices.* Wiley-Interscience, New York. 2.2.6, 3.1.4, 5.1.3, 6.6.1

Sze, S. M., Coleman, D. J., Jr., and Loya, A. (1971). *Solid-State Electron.* **14,** 1209. 2.2.9, 3.5.2, 4.1.5

Sze, S. M., Crowell, C. R. and Kahng, D. (1964). *J. appl. Phys.* **35,** 2534. 2.1.5, 2.3.3

Sze, S. M. and Gibbons, G. (1966). *Appl. Phys. Lett.* **8,** 111. 6.6.1

Mentioned in

Sze, S. M., Lepselter, M. P., and Macdonald, R. W. (1969). *Solid-State Electron.* **12,** 107. 6.1.1

Szydlo, N. and Poirier, R. (1973). *J. appl. Phys.* **44,** 1386. 2.1 FR

Szydlo, N. and Poirier, R. (1980). *J. appl. Phys.* **51,** 3310. 2.1 FR 4.1.2

Tamm, I. (1932). *Phys. Z. Sowietunion,* **2,** 128. 1.3.2

Tantraporn, W. (1970). *J. appl. Phys.* **41,** 4669. 2.2.9 6.5 FR

Tarr, G. N. and Pulfrey, D. L. (1980). *I.E.E.E. Trans. Electron. Dev.* **ED-27,** 771. 5.1.2

Tauc, J. (1974). (Ed.) *Amorphous and liquid semiconductors.* Plenum Press, New York/London. 3.6.1

Taylor, P. D. and Morgan, D. V. (1976). *Solid-State Electron.* **19,** 935. 3.5 FR

Taylor, W. E., Odell, N. H. and Fan, H. Y. (1952). *Phys. Rev.* **88,** 867. 1.3.2 6.4.1

Temple, V. A. K., Green, M. A., and Shewchun, J. (1974). *J. appl. Phys.* **45,** 4934. 5.2.1

Temple, V. A. K. and Shewchun, J. (1974). *Solid-State Electron.* **17,** 417. 3.5.5 5.2.1

Terman, L. M. (1962). *Solid-State Electron.* **5,** 285. 6.1.1

Terry, L. E. and Saltich, J. (1976). *Appl. Phys. Lett.* **28,** 229. 2.1 FR

Thiel, F. A., Bacon, D. D., Buehler, E. and Bachmann, K. J. (1977). *J. Electrochem. Soc.* **124,** 317. 2.1 FR

Thomchick, J. and Buonocristiani, A. M. (1980). *J. appl. Phys.* **51,** 6265. 1.1.2

Thomchick, J. and Buonocristiani, A. M. (1981). *J. appl. Phys.* **52,** 7296. 1.1.2

Thredgold, R. H. (1966). *Space charge controlled conduction in solids.* Elsevier Publishing Co., Amsterdam. 1.2.4

Tidje, T. (1982). *Appl. Phys. Lett.* **40,** 627. 5.1 FR

Ting, C.-Y. and Chen, C. Y. (1971). *Solid-State Electron.* **14,** 433. 1.1.4

Tipple, P. M. and Henisch, H. K. (1953). *Proc. Phys. Soc.* **B66,** 841. 6.2.1

Torrey, H. C. and Whitmer, C. A. (1948). *Crystal rectifiers.* McGraw-Hill, New York. 2.2.6 4.2.1

Tove, P. A., Ali, M. P., Ibrahim, M. M., and Norde, H. (1979). *Phys. stat. solidi* **51,** 491. 6.4.2

Tove, P. A., Susila, G., and Hyder, S. A. (1974). *Solid-State Electron.* **17,** 411. 1.1.4

Triboulet, R. and Rodot, H. (1968). *C. R. Acad. Sci.* Paris **266AB,** 498. 6.5 FR

Tsao, K. Y. and Leenov, D. (1976). *Solid-State Electron.* **19,** 27. 4.2.1

Turner, M. J. and Rhoderick, E. H. (1968). *Solid-State Electron.* **11,** 291. 2.1.7

Uebbing, J. J. and Bell, R. L. (1967). *Appl. Phys. Lett.* **11,** 357. 2.1 FR

Uemura, Y. (1976). In *Electronic properties of quasi-two-dimensional systems* (ed. J. J. Quinn and P. J. Stiles). North-Holland Publishing Co., Amsterdam. 6.1.1

Valois, A. J. and Robinson, G. Y. (1982). *Solid-State Electron.* **25,** 1089. 6.5 FR

van der Ziel, A. (1970). *Noise: Sources, characterization, measurement.* Prentice-Hall, Englewood Cliffs, New Jersey. 1.1.2

van der Ziel, A. (1976). *J. appl. Phys.* **47,** 2059. 5.1 FR

Mentioned in

van Gurp, G. J. (1975). *J. appl. Phys.* **46,** 4308. 2.1 FR

van Gurp, G. J. and Langereis, C. (1975). *J. appl. Phys.* **46,** 4301. 2.1 FR

van Meirhaege, R. L., Laflere, W. H. and Cardon, F. (1982). *Solid-State Electron.* **25,** 1089. 5.2 FR

van Vliet, K. M., Friedmann, A., Zijlstra, R. J. J., Gisolf, A., and van der Ziel, A. (1975a). *J. appl. Phys.* **46,** 1804. 1.1.2

van Vliet, K. M., Friedmann, A., Zijlstra, R. J. J., Gisolf, A., and van der Ziel, A. (1975b). *J. appl. Phys.* **46,** 1814. 1.1.2

Vanadamme, J. K. J. and Brugman, J. C. (1980). *J. appl. Phys.* **51,** 4240. 6.4.1, 6.4.3

van Roosbroeck, W. (1950). *Bell. Syst. Tech. J.* **29,** 560. 5.1.1

van Roosbroeck, W. (1953). *Phys. Rev.* **91,** 282. 3.2.1

van Roosbroeck, W. (1961). *Phys. Rev.* **123,** 474. 3.2.4

van Roosbroeck, W. and Casey, H. C. (1970). *Proc.* 10*th. int. Conf. Phys. Semiconductors,* p. 832. U.S. Atomic Energy Commission, Springfield, Virginia. 3.3.1

van Roosbroeck, W. and Casey, H. C. (1972). *Phys. Rev.* **B5,** 2154. 3.3.1

Varma, R. R., McKinley, A., Williams, R. H., and Higgenbotham, I. G. (1977). *J. Phys.* **D10,** L171. 2.1.7

Vasilieff, G., Martinot, H., and Rey, G. (1977). *Solid-State Electron.* **20,** 35. 1.2 FR

Vasudev, P. K., Mattes, B. L., Petras, E., and Bube, R. H. (1976). *Solid-State Electron.* **19,** 537. 4.1.2

Verwey, J. F. (1972). *Appl. Phys. Lett.* **21,** 417. 3.5.5, 6.6.1

Verwey, J. F. (1973). *J. appl. Phys.* **44,** 2681. 6.3.1

Viktorovitch, P. and Kamarinos, G. (1976). *Solid-State Electron.* **19,** 1041. 2.3 FR

Vilms, J. and Wandinger, L. (1969). In *Ohmic contacts on semiconductors* (ed. B. Schwartz). Electrochemical Society, New York. 2.3.4

Vincent, G., Bois, D., and Pinard, P. (1975). *J. appl. Phys.* **46,** 5173. 4.3 FR

Visvanathan, S. (1960). *Phys. Rev.* **120,** 376. 1.2.3

Wada, O., Majerfeld, A., and Robson, P. N. (1982). *Solid-State Electron.* **25,** 381. 2.1 FR

Wagner, C. (1980). *Z. phys. Chem.* **B11,** 139. 2.2.2

Wagner, C. (1931). *Phys. Z.* **32,** 641. 2.2.2

Waite, M. S. and Vecht, A. (1971). *Appl. Phys. Lett.* **19,** 471. 1.2 FR

Waldrop, J. R. and Grant, R. W. (1979). *Appl. Phys. Lett.* **34,** 630. 2.1 FR

Walpole, J. N. and Nill, K. W. (1971). *J. appl. Phys.* **42,** 5609. 2.1 FR, 3.1.1, 4.1.6

Walser, R. M. and Bené, R. W. (1976). *Appl. Phys. Lett.* **28,** 624. 2.1 FR

Warner, R. M. (1979). *Solid-State Electron.* **22,** 215. 3.3 FR

Wemple, S. H. (1969). In *Ohmic contacts in semiconductors* (ed. B. Schwartz). Electrochemical Society, New York. 2.1.1

Wemple, S. H., Kahng, D., and Braun, H. J. (1967). *J. appl. Phys.* **38,** 353. 6.7.2

Wey, M. Y. and Fritzsche, H. (1972). *J. non-cryst. Solids,* **8–10,** 336. 3.6.2, 5.2 FR

Wilkinson, J. M., Wilcock, J. D. and Brinson, M. E. (1977). *Solid-State Electron.* **20,** 45. 2.2.5

Williams, R. (1974). *J. appl. Phys.* **45,** 1239. 6.3 FR

Mentioned in

Williams, R. H., Varma, R. R., and McKinley, A. (1977). *J. Phys.* **C10,** 45. 2.1.7

Wilson, A. H. (1932). *Proc. R. Soc.* **A136,** 487. 2.3.1, 2.3.3

Wilson, J. I. B. and Allen, J. W. (1975). *Solid-State Electron.* **18,** 759. 2.3.8

Wilson, J. I. B. and McGill, J. (1978). *I.E.E.E. Solid State & Electronic Devices Special Issue,* 57. 5.1.3, 5.2 FR

Wintle, H. J. (1975). *Solid-State Electron.* **18,** 1039. 6.3 FR

Wittmer, M., Lüthy, W., Studer, R., and Melchior, H. (1981). *Solid-State Electron.* **24,** 141. 2.1 FR

Wittmer, M., Pretorius, R., Mayer, J. W., and Nicolet, M.-A. (1977). *Solid-State Electron.* **20,** 433. 2.1.7

Wolf, M. (1980). *I.E.E.E. Trans. Electron. Dev.* **ED-27,** 751. 5.1.3

Wong, J. (1976). *J. appl. Phys.,* **47,** 4971. 6.4.3

Wong, J., Rao, P., and Koch, E. F. (1975). *J. appl. Phys.* **46,** 1827. 6.4.3

Wortmann, A. K. and Kohn, E. E. (1975). *Solid-State Electron.* **18,** 1095. 4.2 FR

Wronski, C. R. (1977). *I.E.E.E. Trans. Electron. Dev.* **ED-24,** 351. 3.6.2, 4.3.1, 5.2 FR

Wronski, C. R., Abeles, B., Daniel, R. E., and Arie, Y. (1974). *J. appl. Phys.* **45,** 295. 2.3.6

Wronski, C. R., Carlson, D. E., and Daniel, R. E. (1976). *Appl. Phys. Lett.* **29,** 602. 3.6.2

Wu, C. M. and Yang, E. S. (1979). *Solid-State Electron.* **22,** 241. 6.4.2

Wu, C. M., Yang, E. S., Hwang, W., and Card, H. C. (1980). *I.E.E.E. Trans. Electron. Dev.* **ED-27,** 687. 5.1 FR

Wu, C.-Y. (1980). *J. appl. Phys.* **51,** 3786. 3.5.5

Wu, C.-Y. (1981). *Solid-State Electron.* **24,** 857. 2.3 FR

Wu, S. Y. and Campbell, R. B. (1974). *Solid-State Electron.* **17,** 683. 2.1 FR, 2.3.8

Wyeth, N. C. (1977). *Solid-State Electron.* **20,** 629. 5.1.3

Yamaguchi, E., Nishioka, T., and Ohmachi, Y. (1981). *Solid-State Electron.* **24,** 263. 6.5 FR

Yamaguchi, J. and Katayama, S. (1950). *J. Phys. Soc. Jap.* **5,** 386. 1.3.4

Yamaguchi, J., Nagata, S., and Matsuo, Y. (1951). *J. Phys. Soc. Jap.* **6,** 521. 1.3.4

Yaron, G. and Frohman-Bentchkowsky, D. (1980). *Solid-State Electron.* **23,** 433. 4.1.2

Young, C. E. (1961). *J. appl. Phys.* **32,** 329. 6.1.1

Yu, A. Y. C. (1970). *I.E.E.E. Spectrum* **7,** 83. 6.6.1

Yu, A. Y. C. and Snow, E. H. (1968). *J. appl. Phys.* **39,** 3008. 6.6.1

Yu, A. Y. C. and Snow, E. H. (1969). *Solid-State Electron.* **12,** 155. 3.1.4

Zaitsevskij, I. L., Konakova, R. V., Shakhovtsov, V. I., and Tkhorik, Yu. A. (1980). *Solid-State Electron.* **23,** 401. 6.6 FR

Zakharov, A. K. and Neizvestnyi, I. G. (1975). *Phys. stat. Solidi* **(a)30,** 419. 4.1 FR

Zallen, R. (1983). *The physics of amorphous solids.* Wiley-Interscience, New York. 3.6 FR

Zemel, J. N. (1961). (Ed.) *Semiconductor surfaces.* Pergamon Press, New York/Oxford. 6.1.1

Mentioned in

Zettler, R. A. and Cowley, A. M. (1969). *I.E.E.E. Trans. Electron. Dev.* **16,** 58. 6.6.1

Ziegler, K., Klausmann, E., and Kar, S. (1975). *Solid-State Electron.* **18,** 189. 4.1 FR

Zuidberg, B. F. J. and Dymanus, A. (1976). *Appl. Phys. Lett.* **29,** 643. 4.2 FR

INDEX

absorption constant
  optical 23, 241, 276
absorption length 261
acceptors 128
accumulation (of majority carriers) layers 42, 50, 54, 203, 265, 283 ff., 320
accumulation (of minority carriers) 19, 23, 165
accumulation ratio 23
activation energy, donor 223
adatoms 69
admittance studies 229
adsorption 30, 66, 68
affinity
  electron 29, 309, 337 ff.
aluminium contacts 68, 130 ff., 212 ff., 257 ff., 269, 299
aluminium gallium arsenide 312
aluminium nitride 299, 322
aluminium oxide 35, 124, 295
ambipolar Debye length 154, 167
ambipolar diffusion constant 169
ambipolar diffusion length 179, 207
amorphous semiconductors 49, 206 ff., 241, 288, 333; *see also* silicon, amorphous
amplification, current 193, 204, 248, 330 ff., 334, 336
anisotropy 91, 329
anthracene 213, 301
antimony
  contacts 212
  implantation 127
antimony oxide 314
area contacts 12, 122
argon sputtering 66
artificial barriers 39 ff.; *see also* oxide layers
avalanche breakdown 193, 326, 329, 332 ff., 336
avalanche buildup time 332, 334

back-to-back systems 100 ff.
  capacitance of 226
ballistic models 11, 97
band bending 69
band structure (band gap) 150, 213, 253, 257, 258, 264, 271, 277, 309, 325, 327, 329, 332, 336 ff., 338 ff.
  near surface 285 ff.
barium titanate 268
barrier capacitance 121, 129, 143, 315
  effect of traps on 233 ff.
  high frequency reduction of 232
barrier contours (profile) 44, 74, 106, 146, 152, 204, 252
barrier height 9, 34 ff., 37 ff., 47, 50, 64 ff., 72, 74, 118 ff., 131, 158, 160, 212, 222, 224, 225, 228, 250, 265, 268, 285, 289 ff., 318 ff., 325, 335, 337, 339
  average (mean) 122, 202 ff.; *see also* barrier height, effective
  effective 56, 58, 127, 128, 149, 153, 307, 323
  non-constant 120 ff., 218
barrier lowering 28, 39, 98, 105, 122, 132
barrier resistance 55, 68, 217, 324, 339
barrier thickness (width) 30, 47 ff., 52, 54, 71, 74, 77, 186, 194, 208, 212, 219, 223, 224, 237, 246, 251 ff., 260, 300, 304, 323 ff., 334
barrier voltage 9, 75, 186
barriers
  interacting 104, 143
  negative 19; *see also* accumulation layers
  static 43 ff., 52
  thin 323 ff., 332
  visual demonstration of 44
bismuth oxide 312, 314
blocking direction 6
Boltzmann contour (profile) 128, 151, 157, 189, 195
Boltzmann statistics 69, 70, 78, 80, 81, 90, 94, 189, 194, 197 ff., 201, 223, 271, 274, 294, 320 ff.
boron 254
boron arsenide 322
boron nitride 34
boundary conditions (varied contexts) 78 ff., 106, 133, 167, 199, 223, 241, 248
boundary conditions
  emission controlled 108 ff., 156, 199
boundary disequilibrium 197, 199; *see also* recombination in interface states
breakdown
  dielectric 104, 283, 301
  hard and soft 327 ff.
  thermal 290
breakdown avalanche 326, 329, 332
breakdown processes 124, 304, 329, 332; *see also* breakdown, thermal; breakdown dielectric

bulk conductivity
  non-linear 210
bulk resistance 76, 89, 130, 138, 141, 150, 197, 213, 215, 255, 318

cadmium fluoride 301
cadmium selenide 35, 73, 173, 252
cadmium sulphide 35, 73, 124, 252, 258, 301, 312, 325
cadmium telluride 35, 73, 238, 244, 268, 322, 325
camel diode 127, 134
capacitance
  barrier 121, 215 ff., 244, 284
  differential (dynamic) 218 ff., 224 ff., 305
  effect of interface occupancy on 224 ff.
  measurement by bridge methods 229
  space charge 217
  static 218
  system 224
capacitance creep 129, 247
capacitance hysteresis 237
capacitance measurements 132, 254
capacitance relationships 86, 121, 215 ff., 244, 284, 312
carrier concentration contours (profiles) 82, 146, 162, 168 ff., 172 ff., 182, 183, 187, 261 ff., 274, 293
carrier generation 254, 265, 335 ff.
ceramics 301
cermet 124
chalcogenide alloys 212, 213, 279
chromium carbonate 314
chromium contacts 148
cleaved surfaces 148
co-sputtering 124
cobalt oxide 314
Cole–Cole plots 215 ff.
collection region 255
collection velocity 259 ff.
collector contacts 184, 193, 232, 252
collision time 286
collodion layers 247
composite barriers 143 ff.; *see also* oxide layers
  photoresponse of 269
conductance
  differential 245
conduction
  lateral 281 ff.
conductivity
  effective 21 ff., 161 ff., 165, 176
  thermal 290
conductivity modulation 22
conformal mapping 14, 16
connectivity 207
contact characteristics
  dynamic 185
contact potential 27
contact radius 12, 49, 159, 160; *see also* curvature, contact
contact temperature 289 ff.
contacts
  blocking 42
  intergranular 302
  low resistance 10, 42, 105, 107, 319 ff.
contamination
  sodium 124
continuity relationships 168
copper 252
copper contacts 213, 214, 335
coupling
  thermal 290
creep
  capacitance 129, 247
  current 128, 134, 218
cupric iodide contacts 214
cuprous oxide 73, 220, 231, 268
cuprous sulphide contacts 252
current amplification 193, 204, 248, 326, 330 ff., 334.
current creep 128, 134, 218
current composition ratio 18 ff., 156 ff., 185, 192, 197, 199, 244, 269, 275
current composition ratio, resistance network simulation 203
current continuity 194, 200, 306 ff.
current saturation 85, 90, 93, 117, 121, 189, 248, 308
current transient 211
curvature
  contact 49, 290, 337 ff.; *see also* contact radius

Debye length 28, 32, 47 ff., 54, 100, 134, 136, 139 ff., 179, 221, 261, 296; *see also* screening length
Debye length
  ambipolar 154, 167
Debye temperature 96
decay
  photoconductive 267 ff.
deep level spectroscopy 233
deformation 336 ff.
degeneracy (incl. non-degeneracy) 42, 47, 53, 71, 78, 146, 148, 153, 195, 222, 257, 276, 308, 319 ff., 324
delocalization of defect states 213
Dember effect 242, 245, 249
density of states, effective 79
depletion layer 54
diamond 229
dielectric breakdown 104, 283, 301

dielectric constant 28, 39 ff., 47, 219, 247, 277; *see also* permittivity
  effective 315
dielectric relaxation time 21, 47, 168, 174, 180, 223, 230, 232, 267 ff.
dielectrics
  interface 269; *see also* oxide layers
diffusion 75, 148, 325, 330
  ionic 222, 272, 325
  lateral 275
Diffusion constant 47, 188, 330
  ambipolar 169
diffusion length 21, 47, 157, 160, 172, 176 ff., 186, 253, 254, 291
  ambipolar 179, 207
diffusion potential 32, 45, 76, 152 ff., 222, 225, 229, 265, 287, 296, 298 ff., 302, 311, 314, 324, 375
  effective 53, 221
  low assessment of 227, 228
diffusion-diode model 107 ff., 115 ff.
diode detectors 233
diode theory 11, 74, 89 ff., 102, 105, 111, 112, 120, 131, 307, 337
diodes
  equilibrium and non-equilibrium 272
  light emitting 203; *see also* light emission
dipolar correction 99
direct gap material 24, 324
discontinuity
  lattice 8
dislocations 338
disorder 206
donor activation energy 223
  donor clustering 322
donor concentration 30, 124, 128, 142
donor concentration contours (profiles) 49, 77, 80 ff., 96, 109, 119, 124 ff., 221, 229, 320
donor ionization 223, 227, 230
  incomplete 69 ff., 223
donor profiles
  saw-tooth shaped 128
donor seggregation 325
donors
  deep 69 ff.
  relaxation of 229
doping optimization 325
double injection 173, 180
drift velocity 45, 94, 96, 110
  effective 112, 113
  saturation of 80, 95, 96
drift-diffusion 10, 42 ff., 77, 92, 94, 96, 99, 100, 105, 111, 143 ff., 182, 187, 307

edge effects 17, 77, 99, 124, 132, 247, 332
effective fields 309
effective mass 91, 93, 97, 119, 270
efficiency 252 ff., 259, 268, 270, 277 ff., 334
$E_g/3$ rule 34, 35, 337
Einstein's relation 78, 94
elastic limit 337
electro-optical devices 241
electrode geometry 11 ff., 21
electrode resistance
  lateral 258
electroluminescence 24, 203 ff., 305, 335; *see also* light emission
electrolysis
  dry (incl. electromigration) 40, 68, 128, 129, 218
electrolyte contacts 213, 215, 326
electron affinity 29, 309 ff., 337 ff.
electron beam detectors 269
electron hole plasma 334
electron multiplication 326 ff., 330; *see also* current amplification
electron reflection 63, 107; *see also* tunnelling
electron temperature 96, 291; *see also* hot carriers
electronegativity 64
electrothermal effects 289 ff., 292
emission spectra 336
emission velocity 199
equilibrium
  thermodynamic 27, 32, 108; *see also* Boltzmann statistics
equilibrium constant (governing trap occupancy) 297
equivalent circuits 215 ff., 269, 290, 292
Esaki diode 114
etching 66
exciton 214
exclusion (of minority carriers) 19, 22, 156, 165, 180 ff.
exclusion ratio 23
exhaustion layer 223
extraction (of minority carriers) 19, 22, 132, 180 ff., 259
extraction ratio 23
extraction transient 185

Fermi integral 321
Fermi level pinning 31 ff., 68, 246, 337
Fermi–Dirac statistics 117, 271
ferroelectric materials 44
field
  effective 309
  maximum 163, 164, 170, 181, 182
field contours 163, 168, 170, 172
field curvature 164

field effect 281 ff., 285 ff.
field emission 14; *see also* thermionic field emission
field ionization 301, 327
field penetration 27 ff.
filament formation 333
fill factor 254 ff., 277
flat band condition 101, 102, 283 ff., 298, 320
floating potential 4, 245 ff., 262 ff.
fluorine 254
flux method 11
forming 128, 218
forward direction 6, 9, 20, 74
free carrier effects 52, 73, 74, 117, 132, 141, 146, 156, 194, 196, 224; *see also* space charge effects
frequency characteristics 205, 215 ff., 223, 229, 233 ff., 293 ff., 313

gain
  photoconductive 251 ff.; *see also* current amplification
gallium antimonide 35, 68
gallium arsenide 67, 68, 72, 73, 116, 118, 119, 124, 179, 193, 195, 211, 220, 222, 224, 231, 233, 246, 257, 258, 269, 270, 309 ff., 322, 324 ff., 331, 337
gallium–indium alloy contacts 335
gap
  direct 24, 324 ff.
  mobility 208, 288; *see also* amorphous semiconductors
generation of majority carriers 252, 335
generation of minority carriers 132, 264
geophysical analogy 46
germanium 33, 34, 37, 38, 68, 72, 89, 94, 149, 150, 152, 163, 166, 168, 173, 175, 183, 190, 191, 195, 196, 220, 231, 240, 244, 247, 249, 250, 258, 290 ff., 302, 307 ff., 310 ff., 314, 331, 338
germanium
  amorphous 288
  neutron bombarded 173
gold alloy contacts 324
gold contacts 66 ff., 116, 118, 124, 129 ff., 148, 180, 205, 211, 212, 222, 229, 231, 236, 238, 244, 246, 247, 252, 257, 258, 270, 273, 322, 324, 335 ff., 338
gradients
  temperature 291
grain boundaries 269, 312 ff., 319 ff.
grain size (distribution) 315, 318
guard ring methods 105

hafnium contacts 33, 132, 148
Hall effect 171
heat of formation 35, 64, 67
hemispherical contacts 76, 77, 159 ff., 187, 247 ff., 249
heterocontacts 308 ff.
  amorphous 312
heterojunctions 143, 252, 258, 295, 309
  graded 258, 309 ff., 326
high field effects 14, 45, 97, 301, 327; *see also* hot carriers
homojunctions; *see* junctions
hopping conduction 210, 272, 297, 322
hot carriers 80, 94 ff., 113, 161, 301, 327, 329 ff.
hot spots 55, 117, 123
hybrid models 105 ff.
hybrid regimes 176, 182

ideality 88, 89, 114 ff., 117, 205, 255, 278, 322, 338; *see also* non-ideality factor
illumination 145, 184, 210, 213, 237 ff., 241 ff., 259, 261, 267 ff., 273 ff., 277 ff., 318
  non-uniform 173
illumination in the injection region 267
image force
  multiple reflection effect 300
image forces 58 ff., 63, 90, 105, 107, 113, 115, 117, 119, 127, 131 ff., 149 ff., 153, 191, 214, 222, 224, 227, 246, 259, 313, 322, 326; *see also* Schottky effect
impact ionization 104, 213, 327 ff., 330, 336
impact ionization of impurities 332
impedance considerations 193, 215 ff., 223
indium 244
indium antimonide 34, 73, 166, 220, 322
indium arsenide 35, 73, 148, 228
indium phosphide 35, 41, 66, 280, 322, 325
indium–zinc selenide 334
inductive behaviour 173, 218, 228, 292
inertia
  thermal 291 ff.
inhomogeneity
  lateral 55, 56, 117
injection
  double 301
injection (of minority carriers) 10, 19, 23, 130, 146 ff., 156 ff., 161, 170, 213, 252, 272, 295, 307
injection diodes 205
injection luminescence 23 ff., 203; *see also* light emission
injection of majority carriers 180
injection pulse 173, 185
injection ratio 22, 156 ff., 166, 187, 188, 192 ff., 196, 199 ff., 232, 247, 249, 267, 268, 269, 272, 295 ff., 299, 301

insulator–semiconductor contacts 295
insulators 24, 25, 39, 40, 180, 202 ff., 269 ff., 273, 276 ff., 293, 295; *see also* oxide layers
interdiffusion 37
interface conductance
  measurement of 282
interface contamination 99
interface layers, optimum thickness of 274, 279
interface recombination 251; *see also* recombination in interface states
interface recombination velocity 259, 260
interface states 34 ff., 64, 120 ff., 133, 143, 218, 224, 247, 250, 270 ff., 276, 281, 287, 301 ff., 306, 310 ff., 319 ff.
  current controlled occupancy of 120, 133, 272 ff., 302
interface trapping 279
interface wave functions 72
interfaces 25, 41, 72, 204, 205, 265, 301, 337
interfacial bonding 67 ff.
interfacial layers 117, 133, 222, 227, 269, 272 ff.; *see also* oxide layers
ITO 270, 273, 276 ff.
intergranular capacitance 304
intergranular contacts (barriers) 302, 307, 312 ff.
intrinsic material 24, 52, 118, 144, 153 ff., 160, 180 ff., 194 ff., 205, 264, 291
inversion layers 146 ff., 156, 195, 197 ff., 205, 227 ff., 233, 269, 276, 283 ff.
ion implantation 326
ion-bombarded surfaces 133
ionic solids 35 ff.
ionicity 35, 64
ionization 128, 146, 301, 327, 329, 331
  incomplete 69 ff., 223
ionization energy
  lattice 327, 329
ionization rate 331

junctions (homo-) 14, 24, 186, 193, 195, 210, 242, 247, 257, 258, 269, 270, 330, 334, 337 ff.

Kelvin probe 68
kinetic gas laws 89; *see also* Boltzmann statistics

lattice distortion 322, 336 ff.; *see also* pressure effects
lead contacts 148, 232
lead oxide 73
lead selenide 73
lead sulphide 184, 268
lead telluride 73, 148, 228, 268
lifetime
  (diffusion length-) 21, 157, 160, 164, 166, 168, 172 ff., 178, 185, 195, 252, 261, 267, 277, 307, 319
  effective 165
  filament 186
lifetime regimes 173 ff.
lifetime semiconductors 19, 21 ff., 146, 150, 158, 161, 164, 166, 168 ff., 182, 186 ff., 195, 204, 265, 267
light detection 259
light emission 23 ff., 203, 333 ff., 336
light emission due to intraband transitions 336
light emitting diode 24, 203 ff.
linear superposition 250 ff., 254, 257, 260, 275
linearization 168 ff., 171, 176 ff., 182, 208
low resistance contacts 10, 33, 44, 105, 107, 119, 132, 212, 215, 216, 260, 266, 276, 291, 319 ff., 325
luminescence 23, 334; *see also* light emission, light emitting diode
luminescence centres 336
luminous efficiency 334 ff.

magnesium contacts 299
magnetoresistance 89
majority carrier depletion 175 ff., 180; *see also* Schottky barrier
majority carrier excess 175; *see also* accumulation
manganese oxide 314
Many bridge 156, 166
mass
  effective 91, 93, 97, 201, 204, 286, 324
matrix methods 138, 143
mean free path 74, 90, 93, 108, 111, 112, 127, 295, 301, 320, 327, 329
mechanical deformation 336 ff.; *see also* pressure effects
mercury contacts 250, 338
metal–semiconductor–metal systems 199 ff., 205, 240
Meyer–Neldel rule 133
microcrystalline systems 312 ff.
microplasmas 333 ff., 336
microwave diodes 233
microwave effects 299 ff.
minority carrier exhaustion 184, 185; *see also* extraction, exclusion
mixing 233, 291
mobility 18, 205, 207, 214, 261 ff.
  close to surface 285
  field dependent 167; *see also* hot carriers
  high field 210; *see also* hot carriers
  effective 210

mobility edge 206, 208, 210, 238; *see also* amorphous semiconductors
mobility gap 208, 288
mobility ratio 173, 181, 182, 261
molecular crystals 301
molybdenum contacts 212
molybdenum sulphide 268
Mott barrier 57 ff., 76, 81 ff., 106, 121, 327

near-insulators 295; *see also* relaxation semiconductors
negative barrier height 53; *see also* accumulation layers
negative resistance
  differential 180, 290, 329, 333
networks 16, 17, 258
neutrality considerations 19, 22, 31, 74, 146 ff., 150, 158, 162, 164, 173 ff., 180, 195, 208 ff., 227, 248, 271, 281, 306, 321
neutron bombardment 33, 308
Newton's law (of cooling) 290 ff.
nickel contacts 68, 119, 124, 272 ff.
noise
  electrical 11, 232
non-homogeneity
  bulk 210
  lateral 122, 218
non-ideality factor 88, 130, 144, 145, 191, 196, 250, 251, 255 ff.
non-linear bulk conductivity 210
non-linearity 8, 23, 210
non-parabolic bands 205

Ohm's law 4, 8 ff., 44
ohmic conduction 10, 23, 42, 161, 163, 178, 322, 324
organic semiconductors 73, 93, 213 ff., 280, 301, 336
oscillations 180
oscillatory solutions 175, 179
overvoltage 333
oxide layers 37 ff., 57, 145, 200 ff., 213, 215, 229, 237, 270, 272, 279 ff., 283, 331

p–n junctions; *see* junctions
palladium contacts 212
particle (flux) excitation 243
peak back current 290
peak back power 290
percolation analysis 315
permittivity 44, 143, 144, 271, 274, 281, 287, 312, 323; *see also* dielectric constant
  high frequency 58 ff., 116
  optical measurement of 61
phonon scattering 23
phosphorus 254
photo-emission 27, 68, 124, 204, 336
photocapacitance 213, 229, 238, 240, 246, 247
photoconductance
  negative 244
photoconductive decay 267, 268
photoconductive gain 251 ff.
photoconductivity (bulk) 210, 244, 246, 251, 255, 316, 318
photoconductivity transient 210
photogeneration 252, 269, 335
photoluminescence 23, 334 ff.; *see also* photo-emission
photosensitivity 312, 319
photovoltage 4, 247, 270 ff., 279 ff.; *see also* photovoltaic effect
photovoltaic effect 4, 212, 242 ff., 247, 249, 251, 254, 275
phthalocyanines 213
pinholes 275 ff.
plasma
  electron-hole 334
platinum contacts 104, 148, 212, 238
point contacts 10, 12 ff., 151, 160, 184, 233, 290 ff., 336
  mobile 338
  temperature of 290
Poisson's equation 8, 44, 96, 124, 126, 134, 140, 151, 153, 167, 172, 189, 209, 223, 235, 285, 294, 297
polarization 44, 217
polymers 301
  organic 280
potassium iodide 336
potassium tantalate 338 ff.
potential
  contact 27
  floating 4, 245 ff., 262 ff.
power dissipation 289 ff.
power output 254 ff., 273
pressure effects 313, 336 ff
probes
  potential 1, 2, 262 ff.
pseudo temperature 117, 121, 256

quantization of barrier states 286
quasi-Fermi levels 82, 146, 194, 196, 197, 263, 274, 283, 309
quasi-intrinsic material 205, 212
quasi-neutrality 162, 181 ff., 265, 267; *see also* neutrality considerations
quasi-ohmic behaviour 132, 193

radiation damage 213
Randschichtleitfähigkeit 83, 85

reach-through voltage 102
reactive properties 215 ff., 313; *see also* inductive behaviour, capacitance *topics*
recombination 24, 172, 177, 187, 195, 247, 254, 260, 265, 269, 274, 310, 327, 330, 335
  bimolecular 166
  radiative 203, 331, 334, 336; *see also* light emission, electroluminescence
  Shockley–Read 166, 172, 179
recombination centres 163, 324; *see also* recombination, Shockley–Read
recombination in interface states 202 ff., 217, 252 ff., 307, 319, 326
recombination rate 172
recombination resistance 217, 218
recombination with donors 230, 297
rectification
  direction of 114
  disappearance of 138
  enhancement of 126, 128
  maximum degree of 93
  parametric description of 113
rectification ratio 1, 6, 9, 10, 50, 89, 151, 185, 191, 212, 213, 291, 297, 299, 307, 311
  high frequency 231, 232
reflectivity, optical 241
relaxation regimes 173 ff.
relaxation semiconductors 21, 148, 168, 173 ff., 180, 214, 267, 268
reserve layer 223, 230; *see also* donor ionization, incomplete
resistance (electrical)
  spreading 13 ff., 21, 187, 275, 291, 338
  contact 1, 4, 25; *see also* low resistance contacts *and* barrier *topics*
  maximum 11
  negative 334
  tunnelling 217
resistance enhancement 162, 173, 182
reverse direction 6, 9, 20, 74
Richardson's constant 91, 105, 112, 118, 131
rubidium iodide 336
Runge–Kutta procedures 134

scattering by optical phonons 332
scattering centres 207; *see also* scattering length
scattering length (critical distance) 295, 329; *see also* mean free path
Schottky barrier 43 ff., 60, 84 ff., 106, 108, 117, 119, 133, 142, 143, 156, 205 ff., 220 ff., 224, 229, 246, 252, 259, 263, 268, 271, 274 ff., 284, 288, 306, 315, 322 ff., 326 ff., 334, 336
Schottky capacitance relationship 219 ff., 225
Schottky effect 58 ff., 226, 313; *see also* image force
Schrieffer curve 284
screening 32 ff., 64, 69, 149 ff., 153, 212, 214, 282, 288, 301, 305, 310
screening length 28, 47, 49, 54, 179, 206 ff.; *see also* Debye length
selenium 35, 40, 128, 220, 280
self-heating 99, 226, 281 ff., 301, 307 ff., 313, 332; *see also* thermal effects
self-oscillation 291
semi-insulators 268, 293 ff.
semiconductor(amorphous) – semiconductor(amorphous) contacts 312
semiconductor(amorphous) – semiconductor(crystalline) contacts 312, 325
semiconductor(crystalline) – semiconductor(crystalline) contacts 29, 38, 301
semiconductor–insulator–semiconductor systems 280
series resistance 89, 130, 141, 163, 173, 197, 215 ff., 247, 255 ff.
Shockley–Read recombination 166, 172, 179, 261, 270, 305, 319
shooting method 135, 138, 143
short circuit current 242 ff., 247, 272 ff.
short circuits
  local 117
shunt resistance
  effective 257
sign conventions 9, 76, 80, 259, 266
silicide contacts 67, 72, 73, 99, 127, 130 ff., 148
silicon 33, 34, 37, 41, 66, 94, 104, 118, 119, 121, 124, 127, 129 ff., 148 ff., 158, 161, 163, 175, 180, 220, 225, 228 ff., 246, 252, 257, 258, 261, 269 ff., 276 ff., 307, 312, 319, 325, 331, 336 ff.
  amorphous 209, 212 ff., 246, 253, 254, 269, 272, 279, 288
  intrinsic 212
silicon carbide 73, 130, 302, 314, 319
silicon dioxide 124, 204 ff., 225, 270 ff.
silicon nitride 204, 205, 331
silicon–germanium alloys 133
silver contacts 89, 168, 213, 214, 225, 335
small signal theory 164, 168, 175, 182, 207 ff., 248, 267
sodium contamination 124
sodium iodide 336
solar cells 210, 241, 254, 257 ff., 274 ff., 279

solar cells—*contd.*
  computations on 276 ff.
  optimization of 279
space charge continuity (incl. discontinuity) 44, 45, 54 ff., 322, 337
space charge effects 8, 9, 24, 42, 44, 46, 63, 69, 96, 117, 146, 167, 177, 180, 204, 210, 227, 233, 240, 268, 270, 293, 295 ff., 307, 310 ff., 318 ff., 330, 333, 336
spectral characteristics 246, 252, 268, 335 ff.
spectroscopy
  deep trap 211, 233
spin 69
spreading resistance
  electrical 13 ff., 21, 187, 275, 291, 338
  thermal 290
stainless steel contacts 212, 214, 338
strontium titanate 222
sulphur 211
sunlight 279
superconducting contacts 232
superlattice 54
superlinearity 313
surface contamination 49, 64, 66, 124, 260
surface (minority carrier) generation 184
surface recombination (velocity) 165, 186, 245, 246, 252 ff., 276; *see also* recombination in interface states
surface state distribution 30 ff., 35 ff., 50, 199, 282, 313; *see also* interface states
surface states 26, 30, 34 ff., 50, 53, 64, 68, 212, 224, 301 ff., 314; *see also* interface states
  bunched 224; *see also* interface states
  slow 283; *see also* interface states
surfaces
  cleaved 65, 66
  ultra-clean 29, 38, 65, 66
switching 212 ff., 311, 318, 333 ff.
switching alloys 212

Tamm states 30, 35
tellurium contacts 212
temperature
  electron 96, 291; *see also* hot carriers
  pseudo- 117, 121
thermal breakdown 290
thermal coupling (constant) 290
thermal effects 90, 218
thermal generation of carriers 180, 200
thermal inertia 291 ff.
thermal time constant 292 ff.
thermal treatment 72, 260
thermal turnover 281 ff., 290
thermal velocity 93, 259, 329
  effective 158
thermalization 90, 92, 96, 111
thermally stimulated currents 71, 210 ff.
thermionic emission 60, 74, 89 ff., 94, 107, 117, 118, 193, 200, 202, 217, 272, 274, 322 ff.
thermionic field emission 63, 107, 115, 143, 301, 319, 323
thermionic work function 26 ff., 37, 50, 53, 64, 212, 220, 246, 287, 301, 320, 322
thermoelectric measurement 289
thin systems 140 ff.
third phases 25, 38 ff.; *see also* oxide layers
threshold voltage 318
tin oxide 205
titanium 68
titanium dioxide 73, 220, 272 ff.
titanium nitride 280
transients 165, 185, 269
  extraction of 185
  photoconductive 210
transistors 19, 21, 146, 193, 248
  field effect 281, 285
transit time 330
transmission line models 16, 17, 291
transparency (to electrons) 29, 113, 118, 200; *see also* tunnelling
trap distribution (over energy) 206 ff., 285, 288
trap-free material 168 ff.
trapping in interface states 279; *see also* traps
trapping time 238 ff.
traps (trapping) 21, 25, 49, 105, 128, 129, 145, 164, 166, 171 ff., 176 ff., 180, 193, 205 ff., 218, 233 ff., 247, 260, 268, 274, 288, 293, 295, 297 ff., 301, 307, 309, 314
  deep 209, 213, 240
  field ionization of 316
  multiple level 240
tungsten contacts 116, 222, 290
tungsten–ruthenium alloy contacts 337
tunnel diodes 268
tunnelling 40, 58, 61 ff., 64, 72, 74, 80, 90, 99, 106, 118, 127, 132, 143, 149, 153, 191, 200, 204, 210, 217 ff., 226, 268, 272, 275 ff., 295, 303 ff., 312, 314 ff., 320 ff., 324, 332
  distance 119
  models 11, 107, 113 ff.
  resistance 217
turnover, thermal 281 ff., 290
two-carrier barrier models 186 ff., 192, 206
two-contact systems 5, 6, 40, 50, 142, 143, 192 ff.

varistors 312 ff.
velocity
  collection 259 ff.
  emission 199
  interface recombination 259 ff.
  surface recombination 245
  thermal 93, 158, 259, 329
voltage
  contact 1, 2
voltage–capacitance relationships 220, 224, 227 ff., 240
voltmeter measurements 46

WKB approximation 117 ff.
wave function overlap 34 ff.
whisker contacts; *see* point contacts
Wilson model 113 ff.

X-ray photoemission 68

yield point 13

Zener effect 104, 314, 318, 326 ff., 332 ff.
zero-recombination contour 173, 194
zinc oxide 35, 312, 314 ff.
zinc selenide 133, 322, 334 ff.
zinc sulphide 35, 240, 301, 322
zinc telluride 240